TRAFFIC PSYCHOLOGY TODAY

TRAFFIC PSYCHOLOGY TODAY

edited by

Pierre-Emmanuel Barjonet
European Association of Transport Psychologists

KLUWER ACADEMIC PUBLISHERS
Boston / Dordrecht / London

Distributors for North, Central and South America:
Kluwer Academic Publishers
101 Philip Drive
Assinippi Park
Norwell, Massachusetts 02061 USA
Telephone (781) 871-6600
Fax (781) 681-9045
E-Mail <kluwer@wkap.com>

Distributors for all other countries:
Kluwer Academic Publishers Group
Distribution Centre
Post Office Box 322
3300 AH Dordrecht, THE NETHERLANDS
Telephone 31 78 6392 392
Fax 31 78 6546 474
E-Mail <services@wkap.nl>

Electronic Services <http://www.wkap.nl>

Library of Congress Cataloging-in-Publication Data

A C.I.P. Catalogue record for this book is available
from the Library of Congress.

Printed on acid-free paper.

Printed in the United States of America

TABLE OF CONTENTS

PART THREE: CHANGING DRIVING COGNITIONS, ATTITUDES AND BEHAVIOR

PART FOUR: TRAFFIC PSYCHOLOGY IN PRACTICE: SOME SELECTED EXAMPLES

LIST OF CONTRIBUTORS

ÅBERG, Lars, Dalarna University, Sweden

BAÑULS, Rosa, University of Valencia, Traffic Research Institute, Spain

BARJONET, Pierre-Emmanuel, European Association of Transport Psychologists, France

BUKASA, Birgit, KfV, Intitute for Traffic Safety Research, Austria

CARBONELL, Enrique J., University of Valencia, Traffic Research Institute, Spain

CAVALLO, Viola E., INRETS, Laboratoire de Psychologie de la Conduite, Arcueil, France

COHEN, Amos S., Universität Zürich, Psychologisches Institut, Zürich, Switzerland

DE VRIEZE, Nicole, Psychologist, (IBSR) Belgian Road Safety Institute, Belgium

DIENES, Elizabeth, Hungarian Coordination Council for Work Psychology, Hungary

FULLER, Ray, Department of Psychology, Trinity College, Dublin, Ireland

HARMS, Lisbeth, Swedish National Road and Transport Research Institute (VTI), Sweden

HUGUENIN, Raphael D., Swiss Council for Accident Prevention, bfu, Berne, Switzerland

KROJ, Günter, BASt, Federal Highway Research Institute, Germany

MONTORO, Luis, University of Valencia, Traffic Research Institute, Spain

NILSSON, Lena, Swedish National Road and Transport Research Institute (VTI), Sweden

PETERS, Björn Swedish National Road and Transport Research Institute (VTI), Sweden

PETICA, Stefan, National Institute for Transport and Safety Research (INRETS), France

ROSKOVA, Eva, Comenius University, Slovakia, Bratislava

ROTHENGATTER, Talib, Center for Environmental and Traffic Psychology, University of Groningen, The Netherlands

RUMAR, Kare, Swedish Road and Transport Research Institute VTI, Linköping, Sweden

SCHANZ-LORENZ, Iris, Psychologist, Stuttgart, Germany

SCHNEIDER, Walter, KfV, Intitute for Traffic Safety Research, Austria

SIEGRIST, Stefan, Swiss Council for Accident Prevention, bfu, Berne, Switzerland

SUMMALA, Heikki, Department of Psychology, Traffic Research Unit, University of Helsinki, Finland

THEEUWES, Jan, Vrije Universiteit, Amsterdam, The Netherland

TORTOSA, Francisco, INTRAS, Institute for Traffic Safety Research, Spain

VALLET, Michel, INRETS, National Institute for Transport and Safety Research, France

FOREWORD

Pierre-Emmanuel Barjonet
Editor
European Association of Transport Psychologists
Chairman
Paris, April 2001

The moving of people and the carriage of goods are basic activities in our societies. More than ever, mobility is a need. The acceleration of mobility generates a traffic in constant progression in the sky and on the seas, the roads and the railways.

As we know, mobility and traffic generate economic value and social well-being; they also have harmful consequences on the health, the existence and the destiny of people as on the natural environment. Reducing harmful effects of traffic and preserving its advantages requires serious control and organization. A key element of regulation is the "human factor." Among social sciences, psychology is, essentially, the instrument of knowledge of the "human factor," and, more precisely, of man in action; this is why, since the beginning of 20th century, psychology was brought to study drivers abilities and mental structures. Today, this applied branch of psychology is arrived to maturity. It has provided and continues to provide scientific knowledge in the fields of perception, cognition, emotion, attitudes, learning related to drivers and also provide assistance to public administrations, companies and people with regard to drivers' education, communication, diagnosis and therapy.

However, psychologists do not replace public authorities which remain judges of the effective application of the results of psychology; yet we could imagine that with a little more psychology, traffic would be improved considerably, in particular on the side of road insecurity which level and human cost remain, in many country, unbearably high.

This work presents a photography of the state of traffic psychology at the whole beginning of the century. It is intended to researchers, students and for all those who want to really understand what is psychology of man or woman at the wheel or the commands of a haulage machine. The principal fields of traffic psychology such that they exist in the research institutes and universities, were covered. Their drafting was entrusted to academics and researchers of great experiment which expose in a clear way problems, methods, assumptions and results of the fields they know best.

The work is introduced and concluded by two major figures of our discipline: Y. Brown and H. Summala, who guarantee, to some extent, the excellence of the delivery. All the authors, often very busy with their activities of teaching and research, provided a particular effort for the achievement of this work; the Kluwer editions constantly helped and encouraged us; the readers and correctors faced the worst by correcting our texts; all are acknowledged for confidence that they granted to us.

Lastly, this work is published on the initiative of the European Association of Psychology Applied to Transport (EUROPSYT). One of the aim of EUROPSYT is to promote the psychology of transport; the publication of this book was one of its major objectives. The structure of the work, the choice of the authors, and the dynamics necessary to the achievement of work return to the executive committee and the board of directors of EUROPSYT. Many thanks for all.

PREFACE

Ivan Brown

Including the word "Today" in the title of a textbook risks built-in obsolescence, because today is tomorrow's yesterday. However, the risk seems worth taking with the present book, because the approaches to theory and practice of traffic psychology today are very different from those of a decade or two ago, yet they seem likely to remain of importance and interest for some considerable time to come. This book is also of importance because it represents the collected views and experiences of experts in the field, who have, until very recently, had no journal dedicated solely to their subject, with the result that potential readers of their publications have had to search a wide variety of journals specializing in traffic engineering, ergonomics, human factors, accident prevention, psychology, or medicine.

It is not difficult to understand why research publications in traffic and transport psychology have been so dispersed for so long. As motorized transport became more and more affordable, and accident rates increased, it seemed logical to attribute the resulting carnage to human error because the technological parts of transport systems were considered the best that could be devised at that time. Hence concern centered on individual differences in competence in vehicle control. Within public transport systems, psychophysics became the standard approach to the selection of individuals with apparently appropriate levels of perceptual-motor skill. Transport psychologists concentrated on the development of objective tests of vehicle control usage, of attentional distribution and of judgments of speed and distance. While sufficing for the selection and subsequent intensive training of professional drivers and pilots, such tests were clearly inappropriate for the assessment of skills among the rapidly increasing numbers of private motorists, whose competence continued to be tested subjectively by non-psychologist driving examiners and who often received no professional instruction at all prior to being awarded a license to drive.

Increasing road traffic density and associated casualty rates, plus the collection of official accident statistics, focused attention on individual differences in accident involvement. Statisticians became interested in the concept of "accident proneness," which attributed safety problems to a minority of individuals who were regularly over-involved in crashes. Road safety thus became simply a matter of identifying these individuals and preventing them from driving. Although psychologists were associated with this analytic approach to accident reduction, it is not clear that there was any attempt to "explain" accident proneness in psychological terms and the issue came to be regarded as a blind alley in safety research.

The development of ergonomics by psychologists during World War II transferred attention to the interaction between human operators and increasingly sophisticated technological systems. Designing such systems to match the performance limits of ordinary individuals was seen as a much more rational approach to the development of technology that could be used more efficiently than was the design of systems which paid little regard to human variability. This approach paid off for some twenty years or so, until "at-the-scene" studies revealed human error to be a direct contributory factor in the majority of road accidents, rather than simply contributing to problems at the interfaces with vehicles and roads. Clearly there was an urgent need to understand the nature and sources of these errors. At the same time, it was recognized that, although much remained to be done by way of improving secondary safety in vehicle accidents (injury prevention and reduction), further advances in road safety required a greater concentration on primary safety, the prevention and reduction of accidents by directing countermeasures at the undesirable attitudes, behavior and skills of certain road users, particularly new drivers.

It is not necessary to be a psychologist to recognize the many dichotomies among driver characteristics. Road users may readily be categorized as young or old, as skilled or unskilled, as experienced or inexperienced, as driving for commercial reasons or for leisure purposes, as familiar with their traffic environment or unfamiliar with it, and so on. It is not necessary to be a traffic expert to recognize that these individual differences are quite likely to be associated with differences in accident liability. Any layperson could also surmise that homogeneity in traffic behavior equates to safety, because behavioral variety is likely to provoke misunderstandings, conflicts and collisions. However, only the psychologist has the expertise to explore and understand these differences in accident liability validly and advises soundly on the design of accident countermeasures. Even that expertise has been limited by a lack of theory in traffic psychology. As Huguenin and Rumar point out, the cognitive theories, which have been adopted from mainstream psychology perhaps, fall short in terms of their emotional and social psychological aspects.

A persistent problem with the cognitive approach to transport safety research has been the development of methodology to examine the covert processes in driving. While overt behavior is obviously observable and thus potentially measurable, environmental inputs to the driver's information processing are usually obscure. On-road research suffers from the disadvantage that traffic conditions can seldom be repeated precisely from one trial to the next, while simulator studies are limited by restricted visual displays and the absence of injury risk to the "driver." Yet vision is the principal channel for the uptake of traffic and environmental information and perception is the essential component of drivers' cognitive interactions with their task demands. The importance of visual perception as an active and constructive process has been well demonstrated by Cavallo and Cohen's discussion of the value of drivers' perception of speed and distance and of "time-to-collision" in establishing personal safety margins on the road. These authors also emphasize the importance of "perceptual learning" among new drivers, which explains the current interest in hazard perception as an essential skill for safe driving. As Fuller points out, learners have little opportunity to practice this skill during the limited period they spend in professional training.

Relatively little attention is also given during driver training to the inculcation of appropriate attitudes towards the task and towards the behavior of fellow road users. Recent contributions to this field from theories of Reasoned Action and Planned Behavior are presented by Åberg and West.

Increasing traffic density throughout the world inevitably leads to an increasing number of conflicts between road users and frequently to an upsurge in emotional responding; often resulting in what the media have termed "road rage." In addition, anxiety will often be experienced by novice drivers and the elderly, as they contrast their perceived undeveloped or declining skills with those of the "average" motorist they see around them. Furthermore, drivers will occasionally (perhaps frequently) feel stressed in traffic because of personal or task-related reasons. All these potentially distracting emotional factors have adverse implications for road safety, in ways discussed by Bañuls and Montoro. A variety of other, perhaps more understandable factors influencing driver behavior, receive comment by Vallet.

Changing the attitudes and behavior of qualified and experienced drivers who break traffic regulations is as important as inculcating these characteristics among new drivers. Kroj and Dienes discuss certain professional approaches to driving improvement among offenders and consider some of the difficulties in this field. A similar psychological assessment of drivers in the German traffic system is later presented by Lorenz and also by Schneider and Bukasa. Siegrist and Roscova consider ways in which safety regulations and law enforcement influence traffic behavior. Research has categorized unwanted behavior into errors, mistakes, lapses and violations. The latter are outstandingly associated with road accidents and these authors give particular attention to the causes, types and prevention of such unsafe behavior.

De Vrieze considers the effectiveness of road safety campaigns; usually a difficult exercise, as their effects are often impossible to distinguish from the influences of other ongoing safety measures. However, the evaluation of such measures is usually considered cost-effective, given the high cost of road accidents. Carbonell pursues this issue of social interventions in traffic safety, arguing for a "cultural transformation" which "diminishes the tolerance of infractions and reckless driving behavior." Clearly this approach will never be 100% effective and there is room for an alternative method of mediating undesirable behavior by appropriate design of the road system, as is discussed by Theeuwes. Yet another alternative way of homogenizing the behavior of individual drivers and thus reducing conflicts and accidents is provided by research and development over the past decade in the field of Road Transport Telematics. In theory, such on-road and in-vehicle driver support systems could relieve individuals of many sub-tasks where their skills are found wanting. Nilsson, Harms and Peters suggest that the introduction of telematics systems "[...] may turn out to be the greatest and most comprehensive transformation of road transportation since the invention of the combustion engine!" Certainly the DRIVE program brought together European traffic psychologists and obliged them to work with representatives from traffic engineering and vehicle manufacturing, which perhaps broadened everyone's knowledge and understanding of certain practical interface issues between people, vehicles and roads. But it also focused attention on cognitive problems at these interfaces, because the psychologists concerned appreciated that it was in these areas that drivers particularly needed support. Nilsson *et al.* point out that task allocation between driver and technology is a crucial aspect of system efficiency, but the acceptability of these new systems at both the individual and societal levels is also crucial to their impact on road safety; an issue elaborated by Petica.

In summary, this book provides a fairly comprehensive overview of the more important issues confronting traffic psychologists today and it will be of interest to anyone considering research or practice in this difficult area of transport safety. Parts of it may even persuade reluctant vehicle designers and highway engineers that individual differences in attitudes, behavior and skills among drivers do not inevitably present insoluble problems that are best ignored! This is certainly not a "cook-book" aimed at presenting ready-made solutions to traffic problems, but it does provide a variety of recipes by which many current problems in traffic psychology might satisfactorily be resolved or ameliorated.

PART ONE
GENERAL OVERVIEW

Chapter 1

OBJECTIVES, TOPICS AND METHODS

Talib Rothengatter
Center for Environmental and Traffic Psychology, University of Groningen, The Netherlands

1. OBJECTIVES

Traffic and transport sciences in a general sense concem the analysis, explanation and prediction of all manifestations which are related to the mobility of people and goods (Michon, 1989b). The transport system has many components (rail, road and air infrastructure vehicles) in which many actors (haulage companies, public transport providers, infrastructure planners, transport consumers) can act within certain limits of freedom (economics, traffic law and transport regulations). Psychology can contribute to the interdisciplinary traffic and transport sciences with its specific models, metaphors and methodolologies which are specific to its disciplinary approach. It can determine and predict what effects the possibilities and constraints of the transport system will have on the decision making processes of the actors, and, reversely, it can determine and predict what demands the actors will pose on the transport system components. Also it can determine and predict the consequences of the decisions made by these actors. For example, it may determine the effects of transport mode characteristics (e.g., time required to travel from A to B) on transport mode choice of the individual transport users or, alternatively, may determine the effects of transport mode choice on safety and environmental pollution. The psychological approach is in that respect complementary to the other traffic and transport sciences, such as engineering, planning and economics, and it shares their objectives, that is, to optimize the transport system in the sense of fulfilling transport demand with minimum damage to the environment and human life.

People, unlike goods, have the capability to realize mobility independently; they are not only transport consumers, they are also road users. Traditionally, the various tasks involved in traffic participation are divided in three hierarchical levels, i.e., the strategic, the tactical and the operational levels (Michon, 1989a). The *strategic* level involves route choice and the realization of the selected route. The *tactical* level concerns manoeuvering, such as the decision to overtake. The *operational* level concerns vehicle control in terms of speed and course control. These levels are hierarchical which implies that decisions on a higher level determine the constraints for the decisions on a lower level (Summala, 1997).

Decisions at each of these levels can have direct consequences in terms of negative effects with regard to environmental pollution and safety. The strategic decision to drive through a residential area rather than follow the ring road around it directly affects the quality of life of the residents. The tactical decision to drive at high speeds and overtake frequently not only increases accident risk but also environmental pollution. The ability to control the vehicle's speed and course accurately likewise affects traffic safety. This is the domain of traffic psychology in its more limited definition. It concerns road user behavior and the psychological processes underlying that behavior (Rothengatter, 1997b). The task of traffic psychology is to understand, predict and provide measures to modify road user behavior at the levels identified with as general objective to minimize the harmful effects of traffic participation. Reduction of traffic participation, for example, by influencing travel mode choice or by reducing travel demand, is the domain of transport psychology in general, but not that of traffic psychology as defined above.

2. TOPICS

The terms "decision" and "ability" as used above are certainly not used to suggest that road user behavior is driven by rational considerations only. Undoubtly, much of the behavior displayed by road users can be regarded as adequate, that is input-determined, responses to the momentary road and traffic situations. Anyone who observes drivers, or cyclists, in dense city-traffic cannot but marvel at the expediency of the traffic participants observed. Yet, anyone who watches traffic long enough also cannot fail to notice the foolish, reckless manoeuvres that road user undertake, and this applies equally for drivers, cyclists and pedestrians. Obviously, if most road user behavior is rational, some is not, or at least, has a different rationale than efficiency and safety. Traffic psychology has dealt with this by using different models of road user behavior simultaneously (see Chapter X for a detailed discussion of the various models). Input-output models, such as the STI compensatory driver model (McRuer *et al.*, 1977), describe in detail how drivers maintain course and speed. At a high level, such models may also describe how drivers anticipate and deal with potentially hazardous situations (Fuller, 1984), but models such as Fuller's may also be regarded as motivational, notwithstanding the fact that their terminology is strictly behavioristic in nature.

At the other end the model spectrum, there are a plethora of models that assume that road user behavior is driven by internal – psychological – states. Taxonomic models in this category have been labeled "trait" models, and functional models as motivational or cognitive process models (Michon, 1985). Classic trait models focus on individual differences in accident involvement, or accident proneness; functional models focus on the way individuals deal with the driving task, and in particular, with the risk of traffic participation. Motivational modeling has been equivalent to risk modeling until attitude models were used in traffic psychology to explain risky behavior. These provide, in principle, the possibility to integrate motivational models and cognitive models, although few attempts have been made to incorporate these models (Van Winsum, 1996).

The various road user behavior models have to a large extent determined the topics that have been addressed in traffic psychology. The various appraoches can, for the purpose of this chapter, be summarized as "performance" approaches versus "motivational" approaches. In *performance* approaches the key question is *how* drivers, or other road users, perform specific tasks. The question then is what causes drivers to make errors, and what in the road or traffic environment evokes these errors. Although individual differences, in terms of age or experience, may be taken into account, the fact that different traffic participants are differently motivated is not. *Motivational* approaches, on the other hand, have little concern for driver error. Here the question is *why* drivers carry out certain task the way they do. It is, implicitly, assumed that safe driving, or cycling, or crossing is attainable for all road users. Hence, accident involvement needs to be explained in terms of personality, attitudinal or motivational factors. In reality, the distinction is of course not as clear cut, but it serves perfectly to outline the different topics that are currently at issue in traffic psychology (Rothengatter, 1997a).

3. METHODS

Most methods used in traffic psychology are not specifically *traffic* psychology methods but generally used in empirical social sciences. However, there are a number of techniques and indices, or parameters, that are specifically geared to the domain under study. In performance studies the research designs closely resemble those used in experimental psychology. Typically, the performance of groups of road users is compared under different conditions in a between-subjects design, or the performance of one group of road users is compared under different conditions in a within-subjects design. Which of the two designs is preferable depends largely on the question whether the subject after having been exposed to one condition will change his performance accordingly when exposed to the other condition(s). Such history effects can be counteracted by presenting the different conditions in random order to the different subjects, but if they are likely to be sizeable, between-subjects designs are preferred even though these deisgns generally require larger numbers of subjects due to the fact differences between the subjects tned to make it more difficult to demonstrate that the found effects are statistically significant. Since within-subjects designs have the same individual differences between conditions, they are inherently more powerful.

Experimental design implies that the conditions in which the performance is measured is under experimental control. Laboratory tasks representing an aspect of the driving task obvisouly allow the required level of control but have as a distinct disadvantage limited validity. Reactiont time tasks, for example, may have a certain face validity, but there is very little evidence that reaction times are indeed a valid measure of driving performance, or give any indication of how likely it is that a subject may get involved in a traffic accident. In view of this caveat, much research is carried out by simply putting subjects into a vehicle or onto a bicycle, and measuring their performance while they are negotiating, usually predetermined routes. If the experimenter does no more than systematically record the subjects' actions or the outcome of their actions, the role of the experimenter resembles closely that of the examinator during a driving proficiency test. In order to avoid the subjective element that is part of such procedures, experimental vehicles have been developed that allow the objective registration of driver performance. Typical parameters that can be measured include the speed of the vehicle, the lateral position, steering wheel position and heading, and many measures that are derived from these such as longitudinal and lateral acceleration and deceleration, and variability of speed, steering wheel position and lateral position or time-to-lane-crossing (TLC) (Godthelp *et al.*, 1984).

Even if the vehicle is fully instrumented and the subject's behavior is recorded in detail, the experimenter is no longer an experimenter in the strict sense as the traffic situation is not under control. Variation in the traffic situation due to momentary conditions create severe methodological problems. A driver may react very differently in crowded traffic conditions than on a virtually empty road. Situations in which the experimenter is interested may arise, but the test ride may just as easily be completed without once encountering a critical situation. In some experiments, this is solved by having one or several cars driving around the subject's car and performing manoeuvres that are under experimental control. However, this is an extremely cumbersome and costly procedure, which, moreover, does not exclude the possibility that momentary conditions affect the subject's behavior. To counteract the latter, it is necessary to use large numbers of subjects, which makes on-the-road studies time-consuming and costly.

The need to create experimental conditions that closely resemble everyday driving, and allow full control of the traffic situation has led to the development of driving simulators. While the early simulators were no more than simple tracking task devices or filmed or videotaped traffic scenes, the availability of powerful graphic computers now allow simulators in which the subject is presented with a simulated traffic and road environment which is fully controlled by the experimenter and the subject's action. Most research insitutes now have a graphic simulator at their disposal. These may vary widely in sophistication in terms of vehicles characteristics (a so-called moving base simulator simulates car movements), graphic display (techniques such as rendering allow a realistic presentation of the environment) and traffic (articficial intelligence can create other road users that interact with the subject). Much research is now carried out in such simulated environments. However, very few validation studies have been carried out that convincingly demonstrate that subject's reactions and behavior is similar to subject's reactions and behavior on the road. This is an issue that I very difficul to address as the simulators that are presently in use vary immensely in terms of vehicle characteristics and graphic representation. The rapid increase in sophistication of graphic simulators is likely to make them a very important research tool for the future, and it is likely that a certain standardization will arise, which makes validation of simulator studies achievable.

For some research topics, neither instrumented vehicles nor simulators are suitable as they involve subjects who are well aware that they are being observed even if they do not have a precise idea what aspect of their behavior is under study. This awareness may cause them to act more prudently or more normatively than they would otherwise have done. Even if that is not relevant, they may allocate more effort to perform optimally than they would otherwise have done. Knowing that they are observed, is in most cases likely to bring the best out of the subjects, not the worst. Although this effect can to some degree be counteracted by careful briefing of the subjects, it cannot be eradicated.

For this reason, there is a long tradition of unobtrusive observation of traffic behavior (Grayson *et al.*, 1987; Grayson *et al.*, 1988). Behavior observation in natural settings can produce detailed information about the way traffic participants deal with specific situations, such as intersections. Recording behavior on video and subsequent analysis allows careful timing of their behavior and has resulted in the development of measures such as gap acceptance and time-to-collision (TTC) for behavior at intersections and headway, following-distance and time-to-collision at road segments. It also allows registration of traffic law violations such as speeding or red light running. What it does not allow is linking these behavioral characteristics to the psychological processes underlying that behavior. It is impossible to tell from a single observation whether the road user observed failed to notice the red light or intentionally ignored it, or whether the short gap or TTC was the result of miscaculation or intent. It has been attempted to interview road users immediately after they were observed, but the results of such approaches have been disappointing. It seems fair to conclude that the method of behavioral observation can identify problems but not the cause of them.

Even behavioral observation can miss some behavior that is rarely displayed or difficult to observe or infer. Agression on the road or drunk-driving are two examples that are easily missed even through careful and prolonged observation.

Psychology, including traffic psychology, is essentially a *descriptive* science. It aims to describe and analyze processes with the goal to understand these processes and not necessarily to change them. Identifying the perceptual and cognitive processes that are related to performance deficits of older drivers is very relevant for the analysis of the problem of older driver involvement, but it does not necssarily tell what can be done to solve the problem. Traffic psychology has as yet not developed much in terms of *prescriptive* knowledge. There are a few exceptions. Attitude research has identified the motives of drivers to commit traffic law violations in detail, and has, on that basis, been able to develop effective measures to drivers (Parker *et al.*, 1996). Studies with respect to road safety education have resulted in child training programs that have effectively changed young children's road crossing behavior (Limbourg *et al.*, 1981; Preusser *et al.*, 1987). Driver improvement and rehabilitation is an important area of development in German-speaking countries (Kroj, 1997). But these approaches are few and far between. Traffic psychology has as yet not developed methods for designing programs to improve driver training, to reduce traffic violations. The systematic approach of problem analysis, developing objectives, developing methods to achieve the objective, and evaluating these methods in terms of achieving objectives and goals, as is, for example, prominent in educational research, is not on a regular basis applied to traffic psychology. This is only partly due to descriptive orientation of traffic psychology research. It is also due to the simple fact that sponsors of educational research may be aware of the necessity to carry out such costly studies, while those of traffic psychology research most are not. A similar situation exists with regard to road design and transport telematics. Traffic psychology research may very well be capable of identifying design problems that can lead to dangerous situations; it is not very well geared to specifying design problems that do not. This is exemplified in the issue of workload resulting from installing in-vehicle devices such as route guidance systems. There now are numerous studies that demonstrate that some systems increase workload, but it is at present impossible to state on the basis of these studies what level of workload would be acceptable (De Waard, 1996). Such methodological shortcomings severely limit the impact traffic psychology can have on the other areas of traffic science.

In reaction to the increasing sophistication in methods and statistical techniques, "a new, humanistic paradigm has emerged in some areas of psychological research, which could contribute to traffic psychology and traffic safety research, too. It concentrates more on human experience rather than explicit behavior, and is based on the concept that human beings exist within an experience of meaning and retain the possibility of acting within self-determining purpose" (Draskóczy, 1998). Some of the characteristics of this "new" approach are that it recognizes the value of qualitative design, focuses on the wholeness of the experience, concentrates on the meaning of experience rather than measurement and is based on first-person accounts in informal and formal conversation. Two aspects need to be considered in detail. The first is that of *individual experience*. In traffic psychology, there are many instances that suggest individual experience is either leading to erroneous conclusions or to no conclusions at all. One example is the issue of subjective versus objective road safety. Careful studies in the past (Miedema *et al.*, 1998) demonstrated a disparity between subjective and objective traffic safety, i.e., those environements which were considered dangerous were in fact not those which were objectively dangerous. Taking into account subjective experience only would have led to measures that would have increased subjective safety but would not have reduced accidents. Secondly, those involved in or witness to accidents appear to have very imprecise recollections of these accidents and tend to have biased attributions as to what caused the accident. Thirdly, estimations of accident risk appear very often confounded by factors such as traffic speed or intensity. While individual experience may be a basis to act to make people *feel* safe, it does not appear to be a reliable basis for objective safety improvements. There is, at present, in this respect no substitute for painstaking empiricism. The second issue is of a more philosophical nature and concerns the role of traffic psychology as a discipline within traffic and transport sciences, but, on a larger scale, also the role of science in addressing societal problems. This dilemma is perhaps best illustrated with the example of the phenomenon of drunk-driving. Obviously, a driver who is drunk has a higher risk of being involved in an accident. Thus, it is perfectly acceptable to recommend stringent BAC as a limit. However, these risks are predominant in young drivers, and the driver above 25 years of age appears to be relatively low even at levels far above the prevailing European Union limit (Simpson, 1994). Would that be an argument for a traffic psychologist to campaign for higher BAC limits? Or, for moral reasons, argue the opposite? Such dilemmas cannot be soived from a humanistic, action-oriented approach and it would be wrong to suggest otherwise.

Yet the humanistic approach is valuable where it fonctions to identify research issues which are considered societally relevant but which are not sufficiently taken into account in the mainstream empirical approach, often because such issues are very difficult to analyze using available technology or methodology. It also has a valuable contribution to make in the implementation stage of measures where these need to be adapted or modified to gain societal acceptance. In that sense humanistic and empirical approaches can be complimentary, or to put it differently, the humanistic approach can be considered part of the empirical cycle.

4. REFERENCES

DE WAARD, D. (1996), *The Measurement of Drivers' Mental Workload*, Thesis, University of Groningen.

DRASKÓCZY, M. (1998), "Traffic Safety and the new Research Paradigm in Human Sciences", in ROTHENGATTER, J. A., CARBONELL, E. (Eds.), Oxford: Pergamon.

FULLER, R. (1984), "A Conceptualisation of Driver Behaviour as Threat Avoidance", *Ergonomics*, 27, 1139-1155.

GODTHELP, J., MIGRAM, P., BLAAUW, G. J. (1984), "The Development of a Time-Related Measure to Describe Driving Strategy", *Human Factors*, 26, 257-268.

GRAYSON. G. B., HAKKERT, A. S. (1987), "Accident Analysis and Conflict Behaviour", in ROTHENGATTER, J. A., DE BRUIN, R. A. (Eds.), *Road Users and Traffic Safety*, Assen: Van Gorcum.

GRAYSON, G. B., MAYCOCK, G. (1988), "From Proneness to Liability", in ROTHENGATTER, J. A., DE BRUIN, R. A. (Eds.), *Road User Behaviour: Theory and Research*, Assen: Van Gorcum.

KROJ, G. (1997), "Verkehrspsychologie in Deutschland", in SCHLAG, B. (Ed.), *Fortschritte der Verkehrspsychoolige 1996*, Bonn: Deutscher Psychologenverlag.

LIMBOURG, M., GERBER, D. (1981), "A Parent Training Program for the Road Safety Education of Preschool Children", *Accident Analysis and Prevention*, 13, 255-267.

MCRUER, D. T., ALLEN, R. W., WEIR, D. H., KLEIN, R. H. (1977), "New Results in Driver Steering Control Models", *Human Factors*, 19, 381-397.

MICHON, J. A. (1985), "A Critical Review of Driver Behavior Models: What Do we Know, What Should we Do?", in EVANS, L., SCHWING, R. C. (Eds.), New-York: Plenum.

MICHON, J. A. (1989a), "Explanatory Pitfalls and Rule-Based Driver Models", *Accident Analysis and Prevention*, 21(4), 341-353.

MICHON, J. A. (1989b), "Sociale verkeerskunde: Een plaatsbepaling", in VAN KNIPPENBERG, C. W. F., ROTHENGATTER, J. A., MICHON, J. A. (Eds.), *Handboek Sociale Verkeerskunde*, Assen: Van Gorcum.

MIEDEMA, B., MENKEHORST, H., VAN DER MOLEN, H. H. (1998), "The Subjective Experience of Traffic Safety, Consequences for Design", in ROTHENGATTER, J. A., DE BRUIN, R. A. (Eds.), *Road User Behaviour: Theory and Research*, Assen: Van Gorcum.

PARKER, D., STRADLING, S. G., MANSTEAD, A. S. R. (1996), "Modifying Beliefs and Attitudes to Exceeding the Speed Limit: An Intervention Study Based on the Theory of Planned Behavior", *Journal of Applied Social Psychology*, 26 (1), 1-19.

PREUSSER, D. F., BLOMBERG, R. D. (1987), "Development and Validation of a Road Safety Public Education Process", in ROTHENGATTER, J. A., DE BRUIN, R. A. (Eds.), *Road Users and Traffic Safety*, Assen: Van Gorcum.

ROTHENGATTER, J. A. (1997a), "Errors and Violations as Factors in Accident Causation", in ROTHENGATTER, J. A., CARBONELL, E. (Eds.), *Traffic and Transport Psychology*, Amsterdam: Pergamon.

ROTHENGATTER, J. A. (1997b), "Psychological Aspects of Road User Behaviour", *Applied Psychology: An International Review*, 46 (3), 223-234.

SIMPSON, H. M. (1994), "BAC Levels and Risk of Collision: Implications for Public Policy", in VALLET, M., KHARDI, S. (Eds.), *Vigilance et transports: Aspects fondamentaux, dégradation et prévention*, Lyon: Presses Universitaires de Lyon.

SUMMALA, H. (1997), "Hierarchical Model of Behavioural Adaptation and Traffic Accidents", in ROTHENGATTER, J. A., CARBONELL, E. (Eds.), *Traffic and Transport Psychology*, Amsterdam: Pergamon.

VAN WINSUM, W. (1996), *From Adaptive Control to Adaptive Driver Behaviour*, University of Groningen.

Chapter 2

TRANSPORT PSYCHOLOGY IN EUROPE: A HISTORICAL APPROACH

Pierre-Emmanuel Barjonet,
European Association of Transport Psychologists, France
Francisco Tortosa,
INTRAS, Institute for Traffic Safety Research, Spain

1. INTRODUCTION

This chapter presents the stages of the development of transport psychology from the beginning of the century to the present days (see for a history of psychology applied to transport, on an European scale: Giese, 1921; Baumgarten, 1928; Dorsch, 1963; Hale, 1980; Jaeger and Stauble, 1981; Klebelsberg, 1982; Geuter, 1985; Métraux, 1985; Rüegsegger, 1986; Tortosa, Montoro and Carbonell, 1989; Echterhoff, 1991; Yela and Prieto, 1993; Dehue, 1995; Gundlach, 1993; 1996a, b, 1998).

This historical approach reveals the remarkable chronological similarities in the development of transport psychology in Europe based on the successful circulation of psychological theories during a period of great industrial expansion and rationalization of all the stages of production including the human element. Although the production based economy began turning fundamentally to the field of medicine, especially to the branch of physiology, in order to "solve the problems concerning the economy of energy of the workers, this sector progressively began to turn to the field of psychology in search of unified solutions" (Mallart, 1942), and finding them.

The sectors of the economy related to the transport of both people and goods were always aware of the need to develop the possibilities of the practical applications of psychological knowledge. The majority of advanced countries around the world soon began to regulate, through official institutions, and public and private agencies, access and continuation in jobs as motor vehicle drivers and operators in public transportation or mass transit by means of special requirements and selective and obligatory exams. Different professionals were called upon to design *aptitude exams* that would provide an "assessment of qualifications" for the performance of these professional activities.

The development of psychology applied (psychotechnics or psychotechnology or technopsychology) to transport in Europe, also has had its origins in practicalities linked to the driving of vehicles. This is an area of psychology which is pragmatic, rather than led by theoretical dynamics.

Although Germany was initially a leader in these developments, one must not forget how markedly widespread this phenomenon was during the first thirty years of the twentieth century. Loewental, Moede, Piorkowski, Stern, Sachs, Tramm and Rüpp in Germany; Bacquerisse, Lahy, Ioteyko, and Pieron, in France; Huysmans, Van Mayenburg, and Bigiel in the Netherlands; Haudemont, Christiaens, Ovide and Decroly in Belgium; Pistone, Ponzo, Bonaventura, and Gemelli in Italy; Suter, Fontègne, Bovet, and Claparede in Switzerland; Burt, Miles, Shellow, Farmer, Myers and Manning in Great Britain; Forster and Vana in Czechoslovakia; Studen, Lipszycowa, and Wojciechowski in Poland; Donajevski, Syrkin, Spielrein, Gastew, and Kolodnaja in Russia; Madariaga, Mira, Rodrigo, and Germain in Spain; Scott, Münsterberg, Parsons, Gerhardt, and Viteles in the United States... work, create and intervene, and... in addition, they offer their work to society as well as the scientific community.

On the theoretical side, the development of transport psychology relies on the development of psychology itself and remains faithful to its divisions between experimental (observation, experimentation, measurement) and comprehensive (clinical and therapeutic) sciences, and its extension into academic sectors.

2. EUROPEAN TRENDS IN PSYCHOLOGY

The ideal of a universal science without national boundaries of any description has, from time to time, obscured the reality of the consistent continental and national trends which have occurred in research and theorizing. However, it seems we are forced to admit the existence of these different *continental trends*. It is a well-known fact that the most prominent views printed in American psychological literature, are centered mainly upon their own traditions, paying much less attention to other national psychological traditions. As Hebb wrote some years ago, "to a great extent, American Psychology today is psychology" (Hebb, 1960). This fact cannot be denied, but it is also true that Hebb's assertion must be contrasted and limited. In the beginning, as Fraisse (1965) wrote "it is [in Germany] where Experimental Psychology was really founded, and their methods were diffused incredibly quickly to other countries. However, the beginnings of experimental psychology in England, France, Russia, and the United States, are branded with the intellectual tradition of every country, and institutions connected to the new science." This is coincidental with other sociological and historiographical analysis of the social origin of psychology, which showed how different national conditions favored the emergence and development of the discipline (see e.g., *Journal of the History of the Behavioural Sciences*, 1965, 1(1); Misiak and Sexton, 1966).

Since the 1930s, and especially since the end of World War II, North America's dominance, as well as what was for many its scientific colonialism, was evident. But, it is also true that European psychology progressively restored itself and recovered its own defining characteristics after suffering the terrible blow caused first by the wave of fascism and later by the Second World War (e.g., Sexton and Misiak, 1976; Gilgen and Gilgen, 1987; Perls, 1987). These events, aside from the material destruction of lives and institutions, provoked a massive exodus of intellectuals especially toward the North American continent (i.e., Fleming and Bailyn, 1969). Even the term "European psychology" itself could be misleading because important national differences existed during the period in which "psychology" was being formed and becoming institutionalized as a discipline within the different countries, and psychology applied to the transport sector has produced different results (and models of intervention) in each country. However, it must not be overlooked that the existence of a European psychological tradition is being increasingly acknowledged and that this recognition comes from a variety of sources (e.g., *European Review of Applied Psychology, European Psychologists, various European Societies*).

3. HISTORICAL DEVELOPMENT

Traffic psychology developed at the same time as psychology became established as an independent area of knowledge. The founding texts of scientific psychology, those written by Wundt, Ribot, Stumpf, Sechenov, Spencer, Galton... are from the 1870s and 1880s. The first International Congress of Psychology took place in Paris in 1889 and by the 1920s nearly all countries had psychological laboratories in their universities, there were many national societies of psychology, numerous journals, and international societies (the current IUPsyS and IAAP) (Mueller, 1960; Montoro *et al.* 1983; Carpintero, Peiró and Tortosa, 1988; Yela and Prieto, 1993; Tortosa,

1998; Osier and Wozniak, 1984). Because they broke with introspective and idealistic theory, experimental psychology and psychometrics formed the new bases of academic psychology and became part of a general scientific movement founded on factual analysis, positivism and empiricism (Fraisse, 1965; Krüger *et al.* 1987; Danziger, 1990).

After the Second World War came the rapid development of ergonomics which provided a new direction for the study of people in relation to machines. Then came social psychology with theories of communication and persuasion, and psycho-physiology with the development of electronic instrumentation. Psychology adopted instrumentation, information technology and became specialized. It adapted to the industrial world and advertising. At the same time, transport in Europe expanded considerably and led to a demand for training from employees of transport companies and road users. This demand also inspired a scientific analysis of accidents, hazards and a better knowledge about driving behavior. Thus, supply and demand matched.

Current psychology, which developed to a large extent according to the Wundtian model, has experienced dramatic expansion. Today this area of knowledge is one of the most productive and fruitful of the scientific-technological disciplines and is characterized by its extensive institutional development, the great importance of its technological applications, theoretical-conceptual pluralism, internationalism and still inadequately defined professional role in many areas.

Independently of other influences, and in spite of the pluralism and heterogeneity of this discipline, without a doubt, the core of current psychology is cognitive psychology, or more accurately the new cognitive psychology. The peak of cognitive psychology is usually presented as coinciding with the decline in the influence of behaviorism in its various forms. Also it is common to juxtapose these two perspectives in such a way that instead of referring to a subject that passively reacts to stimuli in the environment, one speaks of an active subject who processes and gives meaning to these stimuli. The latter organizes his behavior according to strategies and plans while the former emits elements in a "linear chain." Cognitive psychology's proliferation of internal concepts – strategies, plans, schemes – stands in sharp contrast to behaviorism's parsimonious reductionism, stimulus-response, explanations. However, while this is (partially) true within the American intellectual scene, it is not an accurate account of the developments in Europe.

Behaviorism's almost total dominance in the United States from the 1920s through the 1960s (Gardner, 1985), was not shared in the European countries where different schools of thought coexisted. These European perspectives can be grouped within the broader category of functionalism in Western Europe – as in the case of Piaget and the English factorialists – or within the quite different framework of Soviet psychology. European psychology maintained its interest in the study of higher level mental processes and in explanations based on human models rather than animal models, which was the case on the other side of the Atlantic. In Europe there was an important underlying continuity along with substantial modifications in the methodologies and technology employed. For this reason, the appearance of cognitive psychology can be better explained as a gradual evolution rather than in terms of revolutionary changes. The individual's immediate experience, lived by the subject and generated (more or less) actively (information processing) by the subject based on data extracted from the individual and from the environment, orients him toward one type of behavior or another and allows him to make decisions. This subject is epistemologically complex, (partially) conscious, a user of symbols, free, goal oriented, capable of making historic interpretations of situations and of acting according to norms and values. At the same time, however, this subject is unstable, contradictory, easily influenced (e.g., by suggestion), psycho-physiologically vulnerable and susceptible to disorders (which may be more or less transitory).

4. PSYCHOTECHNICS

In response to the demand for ways to evaluate the qualifications of prospective drivers, the then *new* psychology proposed a series of methods that constituted what, in Europe, was called Psychotechnics or Psychotechnology, and, in the U.S., was called Applied Psychology. Under either of these denominations, this psychology was concerned with the human factor in the workplace in all its dimensions, and tried to offer methods to evaluate (and in this case measure) the natural aptitudes of each person, whether physical, manual, technical or intellectual, in order to orient workers toward the most appropriate jobs and to select the most apt (or least likely to make mistakes) for a specific position. Thus, individual differences with regard to capabilities and temperaments, for example certain psycho-physiological factors (e.g., fatigue, pain, muscle functions, nervous reactions, control over emotions, etc.) and the ability to learn the tasks relating to the job, were topics of interest in the new and already existing laboratories.

It is a well known fact that scientific psychology was born as an autonomous discipline in laboratories in which concepts and instrumentation borrowed from other sciences (e.g., physiology, physics, astronomy) were initially employed. Traditionally, W. Wundt (1832-1920) is referred to as the "founder" of scientific psychology because, among other reasons, in Leipzig, this scientific instrumentation was used. The first methods which were employed to evaluate the level of aptitudes (and the lack of them) were very much linked to these instruments and to the statistical analyses used in order to obtain data from them. The first aptitude exams had their origin in experimental psychology of the senses, which employed measurement instruments, which explored different sensory capacities, and the higher-level mental functions that were associated with them (the differential-psychophysical approach), and the measurements of certain personal factors (differential-psychometric approach).

It was Hugo Münsterberg (1863-1915), the North American-German, who further defined this field of psychology, pointing out that "the different professions require different psychophysical capacities and even different classes and levels of the same capacity," and indicated who the professionals should be who work within it. "It is up to psychologists to determine the qualifications that a given profession requires and then to evaluate the individual who seeks to work in that profession in order to decide whether he truly possesses those qualities" (Münsterberg, 1912). It was Münsterberg who proposed the first experimental psychological study related to driving and the first aptitude exam (Hale, 1980; Caparrós, 1985). He formulated and expressed the problems and basic methodological questions of the period: (1) Reactive conception (combination of sensory-motor factors) of driving aptitude; (2) He considered that the selection of drivers had to be a professional selection and not one of personnel; (3) In order to create the professiograms (aptitudinal profile of a task), he proposed direct observation of the task and surveys be carried out by specialists; (4) He chose to consider and study the psychophysical structure just as it presents itself in the performance of the task, rather than analyze (and measure) the basic psychological functions in an isolated manner; (5) He chose to use an internal analogy referring to the mental situation rather than the external similitude of the apparatus and experimental conditions (experimental devices analogous to the real ones).

The first two cases received universal acceptance; this was not so for the other two. The strategy of observing genuine behavior did not necessarily mean that all analysis would be excluded; actually many observers looked for partial but integrated sets of aptitudes and functions. For example, although it is true that in the right handed performance of a task the different functions do not occur in an isolated manner, this does not mean that the performance of these functions (e.g., those pertaining to the senses) should not reach minimum (legally established) levels and therefore could be evaluated independently. Numerous laboratories chose to use miniature models that simulated the external situations that drivers had to respond to (e.g., Moede and Piorkowski, and Lahy, who adapted cinemagraphic methods to their driving *simulators*).

The real situation of driving a vehicle was simulated transferring the model of classic reaction time experiments that was used in the Experimental Psychology laboratories to the simulation. The candidate had to respond as quickly and precisely as possible, with responses that were explained beforehand, to a series of visual and acoustic signals. The speed and accuracy of the responses was measured. This procedure was imposed and consequently became the *paradigm* of Psychotechnics applied to transport (Gundlach, 1996b).

The field of psychology applied to driving became established during the years of the First World War, even though this paradigm was also utilized in the civilian realm (Gundlach, 1998). In fact, the different warring countries applied psychological knowledge to different ends, such as the selection of aviators, drivers, tank operators and engine drivers, as well as industrial jobs and the Army in general (selection and classification of personnel). The highly specialized and large scale psychotechnological selections that took place in the different armies, and the progress of psychology itself due to the development of important basic research and the honing of theories, instruments, methods, devices and tests, favored the spectacular development of applied psychology, which, without a doubt, benefited the increased use of psychometric and projective tests.

5. THE GOLDEN AGE: 1920-1950

The 1920s commenced with very favorable circumstances for psychotechnology. The enormous post-war demand directed toward economic reconstruction created a *golden age* of applied psychology (e.g., Baumgarten, 1928, 1957; Dorsch, 1963; Jaeger and Stauble, 1981; Gundlach, 1998). Of course there were great differences among countries, but, in all of them, psychological intervention was in growing demand.

Psychological services were provided in railroads and trolleys, employment offices, post offices, police stations, etc. Large companies (e.g., Krupp, Borsig, Zeiss, Osram, MAN, Siemens and AEG) established their own psychotechnics departments. These professionals were sought by universities, by public leaders and by private industry. Specialized professional journals were created and promoted (e.g., *Zeitschrift für Angewandte Psychologie; Praktische Psychologie; Zhurnal psikhologii; Pedologii y Psikhotekhniki; Industrielle Psychotechnik; Psychotechnische Zeitschrift; Revue de psychologie appliquée; Human Factors; Journal de psychologie théorique et pratique; Jurnal de psihotehnica; Archivio di psicologia, neurologia, psichiatria e psicoterapia*; *Revista de Organización Científica del Trabajo; Travail Humain*; *Psicotecnia*). The already broad spectrum of psychotechnics apparatus for job and aptitude testing grew (the manufacturers of apparatus in this field, such as E. Zimmerman – Germany – and Stoelting – U.S.A. – increased their line of products). The number of special sessions and conferences, as well as societies and professional groups proliferated.

These spectacular developments favored the appearance of a strong psychotechnology, that in many cases was a benefit rather than a detriment to the institutionalization of psychology in Central Europe, Scandanavia, the U.S.S.R. and its satellites, the French-speaking area, Great Britain, the Netherlands, Italy and Spain (e.g., Dehue, 1995; Rüegsegger, 1986; Saari, 1995; Maharaba; Tortosa *et al.*, 1993; Gundlach, 1996b; Petrovsky). This progress also fostered much participation in International Conferences on Psychology (Montoro *et al.*, 1992), and the establishment of the *Association Internationale de Psychotechnique* (since the XII Meeting – London 1955 – International Association of Applied Psychology – IAAP), that enabled professional psychologists from different countries to meet periodically at the International Cönferences on Psychotechnology that were held (Geneva, 1920; Barcelona, 1921; Milan, 1923; Paris, 1927; Utrecht, 1928; Barcelona, 1930; Moscow, 1931; Prague, 1934).

This Association emphasized the role of aptitude exams, advocating the existing point of view, which eventually evolved into the combined use of psychophysical examinations and psychometric evaluations, (paper and pencil) "tests." During the second Meeting, which took place Barcelona, the nucleus of topics that were discussed paved the way for the future. This Meeting concentrated on the relationship between psychology and the Public Administration, as well as the legal aspects that regulate professional activities, the instruments and materials that are employed in aptitude research, and methodology, calling for public administrations and the private sector to recognize and support these procedures.

The evaluation of aptitudes in the transport profession, which soon spread throughout all of industry, became a key step in the process of professionalization. Psychometrics were definitively adopted by transport companies and various public authorities in order to rationalize the selection of drivers according to the skills required by the job and, particularly, to establish a prognosis of their proneness to accidents and thereby eliminate those applicants whom the tests showed as being highly prone. This approach is still long-lived especially in public transports.

While certain tendencies where maintained – the analytical procedures of the exam still dominate – there were also important changes in orientation. In general, the measurement of "pragmatic aptitudes" – skills and abilities – (sensory-perceptive-attentional and psychoreactive aspects), became less important while more global aptitudes, such as personality (from organismic, typological, characteriologic and psychodynamic perspectives), and intelligence (especially from a factorial point of view) were emphasized. In general, the tests (psychometric and projective) gradually took the place of the more sophisticated apparatus for evaluating, which not all institutes and laboratories could afford. The early and generalized use of factorial analysis in psychology (Krüger *et al.*, 1987) facilitated the success of this perspective, and the factorial definitions of driving soon took the place of the professiographic ones.

When the applied psychologists met again after the Second World War, in order to organize themselves at the international level (Berne, 1949), following a 15 year parenthesis, as the academicians had done shortly before, the expansion of the areas that psychotechnology would encompass, as well as important changes in methodological orientation, were announced (Baumgarten, 1949).

6. THE ROLE PLAYED BY PUBLIC AUTHORITIES

This phase is marked by the growth of motorization, the increase in road accidents, the setting up of public safety policies, which included state research institutes, the rapid development of the mass media, the widespread use of ergonomics and social psychology. If the first two phases in the development of transport psychology are linked to the development of military psychology and transport companies and their need to select personnel with the aptitudes required for avoiding accidents, the third phase corresponds to the huge growth in the number of car owners in Europe at the beginning of the 1950s, particularly in Western Europe.

This motorization was accompanied by a constantly growing death rate from the end of World War II to the middle of the 1970s (U.N., 1986). Thus, in nearly all European countries, the number of road deaths doubled between 1955 and 1970 and statistics reveal 91.000 deaths in 1970. It is during this period that real public safety policies were put in place, particularly measures concerning drinking and driving, speed limits and the compulsory wearing of seat belts. It is also the period when research into the psychology of driving expanded in order to discover the human causes of accidents, the human causes of risks, and thus, be able to improve the training of drivers by learning theories, the observance of safety regulations, through social persuasion, the driving conditions by the analysis of the driving task and ergonomics and by promoting the rehabilitation of drivers with psychological expertise.

A notable effect of modernization was the development of state institutes of applied research, particularly research into road traffic; it was no longer left to the initiative of transport companies, University professors or psychological laboratories but was taken over by the authorities as a result of a growing need for action, for the preparation of safety measures or, quite simply, as the result of the need to understand road accidents. Thus, research is often spread between a state institute –BAST (Germany), TRL (UK), SWOV (the Netherlands), VTI (Sweden), BPA (Switzerland), KfV

(Austria), ONSER (France), etc. – and universities. In the Netherlands, for example, research is carried out at state institutes, SWOV and TNO, at the universities of Groningen, Delft, Leiden, etc. In Sweden, research takes place at VTI in Linköping and at the University of Uppsala. In Ireland, at the Trinity College. In Spain, until the very recent creation of INTRAS, research was done at the Universities of Valencia, Barcelona and Madrid, also in DGT. In Germany, there is BAST and university research in Mannheim or Wuppertal, for example, and then psychologists are employed as experts in the regional polytechnics. In France, research was almost entirely centralized at ONSER, then at INRETS and in car industry (Renault) and transport companies (SNCF). In Great Britain there was the TRL and the psychological research carried out in Cambridge or Manchester. There were also state institutes in Romania, Hungary, and Czechoslovakia where research was also carried out by the large nationalized transport companies. Thus, on a European level, research is extremely scattered both geographically and institutionally. In addition, it often depends on the academic orientations of the team and their own fields of interest. It depends also on the problems that each country has to (or wants to) take into consideration or from organizational concerns of transport companies. The determination belongs both to science, economy and politics.

Hence the difficulty of establishing a simple classification of the directions research work in this field has taken, directions, which also coincide with various movements in general psychology (Carpintero and Tortosa, 1991). Besides, this impression of dispersion is reinforced by the fact that there is no European Journal of Transport Psychology and that research in this field is to be found variously in road safety, ergonomics or general psychology publications.

The tradition of driving assessment and driving improvement which were the bases of transport psychology and originally confined to professional road or rail drivers, then extended to all road users, was still alive in Eastern Europe countries and in Germany. This aspect of transport psychology is both an area of research and an area of assessment. With the psychological expertise, a new field of psychology has found its place: psychotherapy.

Psychological modeling was extremely successful in the 1970-80s; it corresponded to the researchers natural wish for a broadening of horizons, for a synthesis of sparse and contradictory data. It was also due to their wish to update determinants – internal and external, social and mental, cognitive and motivational. This abstraction/integration was founded on existing and epistemological sound theoretical bases. Some of these models remain well known as the Threat Avoidance Model developed in Dublin by Ray Fuller; The Zero Risk Theory by Näätänen and Summala in Finland; Michon's work in Groningen and that of Van Der Molen and Bötticher (for a short presentation by the authors, see Rothengatter and De Bruin, 1988; Varii, 1988; for a general review Tortosa *et al.*, 1989, 1990, 1993, 1996).

One theory had a great influence on the world of road safety: the Risk Homeostasis Theory (for a complete view, see Wilde, 1995). As with some theories, this one was highly controversial for the reason that it considered that drivers were not only cognitive driving machines but also beings with motivations and emotions, that attitude to risk depended on factors which were impervious to cognition or the rational measures of the authorities and that the "accursed" side was stronger than the rational side when it came to driving a car. This theory was more inclusive than the others thus more controversial with regard to the narrow specialized and scattered world of traffic safety.

In many ways car driving was an ideal terrain for applied research in this new field in the 1950s. The explicit vocation of ergonomics is to adapt work and the machine to people and runs parallel to the aim of psychometrics, which is to select individuals according to their aptitudes. In an area where there is little selection, such as car driving, the ergonomic outlook seems perfectly adapted to the improvement of people/machine performance. By developing the concept of work analysis, psychological ergonomics showed how best to perform a task and thereby led to a more realistic design of the driving seat, road signs, road trajectories, etc. And above all, it considered that car driving was a task and therefore brought to driving psychology the concepts operating in industrial psychology and the analysis of work-related accidents, particularly the concepts of mental load, error, etc.

The following phase was the development of the concept of mental work, considering that work, human activity, is determined by internal, mental, intellectual components, not only by environmental factors and, moreover, that the study of human activity cannot just be an analysis of gestures and the outward signs of behavior. This approach restored the cognitive aspects of the working person by defining "thought," and the way it is processed by internal psychological functions like memory or mental representation, as an external information gathering by perception. This new approach was to lead to a whole trend in research, particularly in the area of visual perception which is in great demand in the road vehicle driving field as it concerns all the driving tasks: orientation, keeping of distances, obeying road signs, etc. This school of thought which assumes a necessary communication between person and machine led to a dialogue between the various designers of car interiors, road signs and all the different infrastructure and hence to a greater flexibility in their hitherto purely engineering approach which supposed that people would adapt to the machine.

Apart from the various ways this approach could be applied to driving and to improving infrastructure, apart from the scientific interest it created in the cognitive processes of people at work, it was also of extreme interest in that it modified the usual representations of fault and human error. It would, in a way, objectify human error which now appears to be the consequence of a cognitive malfunctioning and so relativises the ideological conception of voluntary error. It also influenced the debate on accident proneness. General psychology, visual perception, the study of errors, of mental load, of decision making, and of cognitive models are still one of psychologists' favorite research fields.

The study of fatigue at the workstation was one of the first objects of study of psychology applied to transport, particularly to professional drivers. This was the beginning of transport psychology, with or without the help of psychometrics. The Industrial Fatigue Research Board was set up in the UK in 1917, on the basis of physiological rather than psychological analyses. In this country, as in others, transport psychology remains very close to and influenced by industrial psychology. In France, this area of study was more closely linked to psychometry (as, e.g., Spain or Germany), since, as early as 1927, Lahy developed a psycho-physiological test for drivers (Pièron, 1954; Montoro *et al.*, 1991; Gundlach, 1998).

Fatigue, vigilance and monotony were first studied in professional rail and road drivers, then, as motorway accidentology developed, in all car drivers. In this field, scientific progress, to begin with at least, was closely linked to instrumentation and to experimental equipment since the level of vigilance in a given subject was calculated by an electro-physiological indicator. However, the outlook changed to more psychological perspectives as soon as it was considered that the electro-physiological indicator did not entirely explain the level of vigilance and in order to measure this level, the task performance level had to be measured too (Philipps-Bertin and Vallet, 1994). We must note that this research into vigilance will be revitalized by the arrival of information technology in cars (Brookhuis, 1995). European programs will attempt to build and promote devices, which measure the vigilance level in car drivers (SAVE, 1995; Montoro, 1997).

Publicity campaigns are part of the modernization of road safety, which began at the end of the 1960s, and the beginning of the 1970s. This modernization was mainly the result of the change in attitude of the authorities towards road users; it changed imperceptibly from an administrative approach – via the police and the law – to a social cause inspired by the negative effects of traffic. Nor must we forget the influence of the consumer movement and pressure groups who lobbied the authorities about the catastrophic accident rate.

So, during this period in the 1960s and 1970s, the psychologist's field of action, which had previously been confined to the selection of professional road or rail drivers, was extended to include all car drivers and all other road users such as pedestrians, cyclists, motorbike riders. The entire "traveling" population was first of all segmented according to types of transport, then according to age and sex. With this new segmented view, which was made possible by traffic and road safety statistics, the authorities attempted to address the users, to communicate with them. They now considered that citizens had to be made aware of the problems caused by accidents and that they had to contribute to their own safety by accepting and obeying safety regulations. For that is the question in safety matters – applying the law – given that road users cannot be constantly watched, nor watched everywhere. In other words, safety regulations have to be internalized and internalization is arrived at by persuasion. In addition, it is not possible to speak to each road user – pedestrian or driver – face to face and individually in order to persuade them to obey the rules. On the other hand, it is possible to address them as a collective being by using the mass media, particularly television, which, at that time, was expanding hugely.

That was the stage when social psychology was brought into the road transport area, chiefly in the form of opinion surveys designed to prepare advertising campaigns, but psychologists were also brought in to advise on communication and marketing. Laboratory experiments on methods for convincing audiences were used too. Thus investigation began into fear arousal, survey methodology, opinion measurement, attitudes to risk, speed, alcohol, and accidents, research into what determined these attitudes, their structure, their relation to behavior. At the same time as national transport research institutes were founded, small groups of psychologists in Great Britain, France, Germany, Denmark, Austria, Finland, Sweden, etc., carried out the first studies on road users' opinions for the advertisers preparing the publicity campaigns. It should be said that social psychologists did not design the messages, they pointed out the options in psycho-sociological terms, they described the social values attached to speed, risk, alcohol, they measured the gaps between social norms and legal norms, they described the motivation in risk taking, the individual and social representation of the car, its role as a status symbol... but they were not directly involved in the writing of copy, that is the province of advertising agencies.

Psycho-sociological studies, which paved the way for publicity campaigns, began at the end of the 1950s in Denmark at the Danish Council for Road Safety and in the 1960s in Britain (TRL) and France (ONSER), followed by nearly all the other countries in Western Europe. To begin with, the object of the studies was to prepare for new legislation, for example the changeover to the right in Sweden (1967). In 1970, Spoerer carried out an extensive study to prepare for changes in the traffic code in Germany. In 1967, a mass media campaign was undertaken in the UK to inform the public on the terms of the new road safety act regarding drinking and driving. In a more general way, the mass communications were used for the modification of road users behavior and attitudes (Wilde *et al.*, 1971).

The knowledge and measurement of opinions and attitudes toward safety, risk, speed limits, safety belts, drinking and driving are important elements in the effort to inform, persuade and train road users; they have been much in demand by the authorities and still have their preference. Let us note, however, that researchers have recently become interested in the relation between attitude and behavior and what can be predicted from attitude (Forward, 1994).

In the recent years, the trends, which have been noticed, have persisted more or less successfully. These trends are now deeply installed with some differences according countries due to scientific traditions, administrative concerns or political changes in Western or Eastern Europe. The recent years have seen what we call – with a critical eye – the "technical attendance" to car industry. The objectifying, mentalist outlook which stresses the importance of the cognitive functions was next to develop in another direction as a result of the new technology which was being installed in cars. Number of researchers in different countries have become specialists in the psycho-ergonomic assessment of devices such as anti-collision radar, automatic driving, radio-telephoning, etc. Psychologists have often taken part in wide-ranging European research projects such as Prometheus and Drive (research and technology development in advanced road transport telematics, 1991, 1992).

7. CONCLUSION

Transport psychology in Europe is an important and organized area of psychology. It covers nearly all the academic aspects with the notable exception of psychoanalysis, and theories of the unconscious and of the personality. It is closely linked to social and economic factors. Transport psychology has developed in relation to institutional demand, whether public or private. In contrast, it does not seem that scientific or academic institutions, apart from some notable exceptions, are particularly interested in this field or in trying to expand it. In most cases, it is the technical practitioners who decide, since academics do not often venture into the field. Yet, without a doubt, psychology applied to transport allow us to discuss, validate and put into practice the concepts and methods of general psychology.

8. REFERENCES

BAUMGARTEN, F. (1928), *Die Berufseignungsprüfungen, Theorie und Praxis,* München: R. Oldenbourg.

BAUMGARTEN, F. (Ed.) (1949), *Progrès de la Psychotechnique - Progress of Psychotechnics - Fortschritte der Psychotechnik 1939-1945,* Bern: A. Francke.

BAUMGARTEN, F. (1957), *Examenes de aptitud profesional,* Barcelona: Labor.

BROOKHUIS, K. (1995), "Driver Impairment Monitoring System", in VALLET, M., KHARDI, S. (Eds.), *Vigilance et transports*, Lyon: Presses Universitaires de Lyon.

CAPARROS, A. (1985), "Aspectos históricos de la psicología aplicada a la conducción", in Dirección General de Tráfico (Ed.), *I Reunión Internacional de Psicología de Tráfico y Seguridad Vial*, Madrid: Dirección General de Tráfico.

CARPINTERO, H., PEIRÓ, J., TORTOSA, F. (1988), *The Influence of European Thought on the Development of the American Psychology: The First Decades*, University of Valencia, Contract DAJA 45 87 M 0399, U.S. Army Research Institute, European Science Coordination Office.

CARPINTERO, H., TORTOSA, F. (1991), "Die Evolution psychologischer Forschungsfelder, Eine Betrachtung anhand der 'Psychological Abstracts'", in LÜCK, H. E., MILLER, R. (Eds.), *Theorien und Methoden psychologiegeschichtlicher Forschung*, Göttingen: Verlag für Psychologie – Dr. C.J.Hogrefe.

DANZIGER, K. (1990), *Constructing the Subject: Historical Origins of Psychological Research*, New York: Cambridge University Press.

DEHUE, T. (1995), *Changing the Rules: Psychology in the Netherlands 1900-1985*, Cambridge: Cambridge University Press.

DORSCH, F. (1963), *Geschichte und Probleme der Angewandten Psychologie*, Bern: Hans Huber.

ECHTERHOFF, W. (1991), *Verkehrspsychologie*, Bonn: Deutscher Psychologen Verlag.

FLEMING, D., BAILYN, B. (1969), *The Intellectual Migration in Europe and America, 1930-1960*, Cambridge (Massachusetts): Harvard University Press.

FORWARD, S. (1994), *Theoretical Models of Attitudes and the Prediction of Drivers' Behaviour*, Uppsala Psychological Reports, 434, Uppsala: University of Uppsala.

FRAISSE, P. (1965), "Histoire de la psychologie", in FRAISSE, P., PIAGET, J. (Eds.), *Traité de psychologie expérimentale*, Vol. 1, Paris: Presses Universitaires de France.

GARDNER, H. (1985), *The Mind's New Science: A History of the Cognitive Revolution*, New York: Basic Books.

GEUTER, U. (1985), "Polemos panton pater – Militär und Psychologie im Deutschen Reich 1914-1945", in ASH, M. G., GEUTER, U. (Eds.), *Geschichte der deutschen Psychologie im 20. Jahrhundert*, Opladen: Westdeutscher Verlag.

GIESE, F. (1921), *Psychotechnische Eignungsprüfungen an Erwachsenen*, Langensalza: Wendt & Klauwell.

GILGEN, A., GILGEN, C. (Eds.) (1987), *International Handbook of Psychology*, New York: Greenwood Press.

GUNDLACH, H. (1993), "Psychotechnische Untersuchungen bei der Deutschen Reichspost", in GOLD, H., KOCH, A. (Eds.), *Fräulein vom Amt*, München: Prestel Verlag.

GUNDLACH, H. (1996a), "The Hipp Chronoscope as Totem Pole and the Formation of a New Tribe – Applied Psychology, Psychotechnics and Rationality", *Teorie & Modelli, Rivista di Storia e Metodologia della Psicologia*, n. s., 1, 65-85.

GUNDLACH, H. (1996b), "Psychologie und Psychotechnik bei den Eisenbahnen", in GUNDLACH, H. (Ed.), *Untersuchungen zur Geschichte der Psychologie und der Psychotechnik*, München: Profil Verlag.

GUNDLACH, H. (1998), "El desarrollo de la psicología aplicada en Europa Central", in TORTOSA, F. (Ed.), *Una historia de la psicología moderna*, Madrid: McGraw-Hill (In print).

HALE, M. (1980), *Human Science and Social Order, Hugo Münsterberg and the Origins of Applied Psychology*, Philadelphia: Temple University Press.

HEBB, D. (1960), "The American Revolution", *American Psychologist*, 15, 735-745.

JAEGER, S., STAEUBLE, I. (1981), "Die Psychotechnik und ihre gesellschaftlichen Entwicklungsbedingungen", in STOLL, F. (Ed.), *Die Psychologie des 20. Jahrhunderts*, Vol. 13, *Anwendungen im Berufsleben*, Zürich: Kindler.

KLEBELSBERG, D. (1982), *Verkehrspsychologie*, Berlin: Springer.

KRÜGER, L., GIGERENZER, G., MORGAN, M. (Eds.) (1987), *The Probabilistic Revolution*, Vol. 2, *Ideas in the Sciences*, Cambridge, MA: MIT Press.

MALLART, J. (1942), *Organización científica del trabajo*, Barcelona: Labor.

MÉTRAUX, A. (1985),"Die angewandte Psychologie in Deutschland vor und nach 1933", in GRAUMANN, C. F. (Ed), *Psychologie im Nationalsozialismus*, Berlin: Springer-Verlag.

MISIAK, H., SEXTON, V. (1966), *History of Psychology: An Overview*, New York: Grune & Stratton.

MONTORO, L., TORTOSA, F., CARPINTERO, H. (1983), "Los origenes de los Congresos Internacionales de Psicología", *Revista de Historia de la Psicología*, 4(1), 43-57.

MONTORO, L., CARBONELL, E., TORTOSA, F. (1991), "Psicología y Seguridad Vial en España, 70 años de historia", *Revista de Historia de la Psicología*, 12(2), 157-174.

MONTORO, L., TORTOSA, F., CARPINTERO, H. (1992), "Brief History of International Congresses of Psychology (1889-1960)", in RICHELLE, M., CARPINTERO, H. (Eds.), *Contributions to the History of the International Congresses of Psychology*, Valencia, Brussels: Revista de Historia de la Psicología Monographs and Studia Psychologica, Leuven University Press.

MUELLER, F. (1960), *Histoire de la psychologie*, Paris: Payot.

MÜNSTERBERG, H. (1912), *Psychologie und Wirtschaftsleben, Ein Beitrag zur angewandten Experimental-Psychologie*, Leipzig: Johann Ambrosius Barth.

Nations Unies (1986), *Statistiques des accidents de la circulation routière en Europe (1986)*, Commission économique pour l'Europe, New York: Nations Unies.

OSIER, D., WOZNIAK, R. (1984), *A Century of Serial Publications in Psychology, 1850-1950, An International Bibliography*, Millwood, New York: Kraus International Publications.

PERLS, S. (Ed.) (1987), *Psychology, An International Perspective*, Alburquerque, New Mexico: Professional Seminar Consultants.

PHILIPPS-BERTIN, C., VALLET, M. (1994), *Le maintien de la vigilance des conducteurs de voitures: Des systèmes électroniques vers le génie cognitif?*, Recherche-Transports-Sécurité, Paris: INRETS.

PIERON, H. (1954), "L'orientation professionelle: Les problèmes généraux", in *Traité de psychologie appliquée*, Livre troisième, *L'utilisation des aptitudes*, Paris: Presses Universitaires de France.

Research and Technology Development in Advanced Road Transport Telematics (1991, 1992), Brussel: Commission of the European Communities.

ROTHENGATTER, J., DE BRUIN, R. (Eds.), *Road User Behaviour: Theory and Research,* Assen/Maastricht: Van Gorcum.

RÜEGSEGGER, R. (1986), *Die Geschichte der Angewandten Psychologie 1900-1940,* Ein internationaler Vergleich am Beispiel der Entwicklung in Zürich, Bern: Hans Huber.

SAARI, S. (1995), "The development of psychology in Finland", *News from EFPPA.*

SAVE (System for Effective Assessment of the driver state and Vehicle control in Emergency situations) (1995), *Project description,* Athens: Heldago.

SEXTON, V., MISIAK, H. (Eds.) (1976), *Psychology around the World,* Monterrey, CA: Brooks/Cole.

TORTOSA, F. (1998), *Una historia de la psicología moderna,* Madrid: McGraw-Hill.

TORTOSA, F., MONTORO, L., CARBONELL, E. (1989), *Psicología y seguridad vial en España, 60 años de historia,* Zaragoza: Libreria General.

TORTOSA, F., MAYOR, L., MONTORO, L., CARBONELL, E. (1989), "Motivación y tráfico", in MAYOR, L., TORTOSA, F. (Eds.), *Ambitos de aplicación de la psicología de la motivación,* Bilbao: Desclée de Brouwer, S.A..

TORTOSA, F., MAYOR, L., MONTORO, L., CARBONELL, E. (1990), "Motivación en la conducción y modelos de riesgo", in *II Congreso del Colegio Oficial de Psicólogos,* Comunicaciones, Area 8, *Psicología y Seguridad Vial,* Madrid: Colegio Oficial de Psicólogos.

TORTOSA, F., QUINTANILLA, I., CIVERA, C., DIAZ, R. (1993), "Psychologie et profession en Espagne, Profils historiques et situation actuelle, Special issue: Applied Psycholoy in Europe: The professionalization process", *European Review of Applied Psychology,* 43(2), 123-139.

TORTOSA, F., MAYOR, L., MONTORO, L., CARBONELL, E. (1993), "Motivazione e Traffico", in AUPI (Associazione Unitaria Psicologi Italiani), AUPI Dossier 0, *Psicologia e Sicurezza Viaria,* 99-136.

TORTOSA, F., MAYOR, L., MONTORO, L., CARBONELL, E. (1989), "Motivación y tráfico", in MAYOR, L., TORTOSA, F. (Eds.), *Ambitos de aplicación de la psicología de la motivación,* Bilbao: Desclée de Brouwer, S.A., (Enlarged edition).

VARII (1988), Special issue, *Ergonomics,* 31(4).

WILDE, G. (1995), *Target risk,* Toronto: PDE Publications.

YELA, M., PRIETO, J. M. (Eds.) (1993), "Applied Psychology in Europe: The professionalization process, Special issue", *European Review of Applied Psychology,* 43(2).

Chapter 3

MODELS IN TRAFFIC PSYCHOLOGY

Raphael D. Huguenin,
Swiss Council for Accident Prevention bfu, Berne
Kare Rumar,
Swedish Road and Transport Research Institute VTI, Linköping

1. INTRODUCTION

"There is nothing as practical as a good theory" (Lewin). This is true also within traffic psychology (OECD, 1997). Theories and models make it possible to sort our observations in a way which explains our observations, create ideas about how to make further studies, gives ideas about how to solve problems and predicts future development. Some decades ago there were ambitions to create a unified traffic science with its own theories and models. However, presently most researchers agree that the best research is carried out on the disciplinary level, that is, for example, traffic psychology (Rumar, 1990).

It is part of the function of useful scientific theories to provide an explanatory summary of certain facts or of any related phenomena and to predict events that are associated with them. Theories are at the same time overestimated and underestimated: On the one hand one expects them adequately to explain the diverse range of everyday phenomena, which often leads to irresponsible over-simplification; on the other hand they are assigned to the realm of the imaginary, which differs from practice not only in degree but also in principle. This can lead to a dichotomization between "theoretical" and "practical" and hence to non-observance of the theory by practical people (see Heckhausen, 1976).

The development of traffic psychology is characterized by the fact that, in the past 25 years, a large number of experiments and investigations were carried out to provide answers to individual questions and solutions for parts of a problem. In so doing, behavioral models or theories relating to road traffic were seldom used as the foundation for processing the data in an integrated way and for developing a theoretical basis for traffic psychology. Although more and more psychological models are being developed in this field (see, e.g., Michon, 1989, or Ranney, 1994), all too often individual results are placed alongside each other in an unrelated way and the benefits of a theory which would integrate this knowledge remain unexplored. Nevertheless, traffic psychology theories have not only been demanded (Hoyos and Pupka, 1977; Huguenin, 1988), driver behavior has also been the subject of well-founded explanatory hypotheses (e.g., Fuller, 1988; Klebelsberg, 1977; Koornstra, n.y.; Michon, 1985), theories (e.g., Van Der Molen and Bötticher, 1988; Näätänen and Summala, 1976), and systematic analytical approaches (e.g., Marek and Sten, 1977). Theoretical approaches have also been made concerning specific areas, such as risk, which is a key factor in the explanation of driving behavior (e.g., Wilde, 1988). Traffic psychology theory approaches range from empirically insufficiently comprehensible meta-theories to laboratory-tested models concerning certain aspects of the overall behavior of the driver.

Theories in general psychology are not able to adequately explain driver behavior. These theories are either so general that the special aspects of vehicle driving cannot be considered or are so specific that only certain aspects of driver behavior are touched upon. For example, learning theory can provide explanations of how certain abilities and skills necessary for driving a car can be acquired, or the reception and processing of information while driving can be explained using theories of perception. Assuming the sequence of a driver's behavior as shown in Figure 1, we need therefore a specific approach to be able to explain behavior in traffic, for example, despite adequate preparation – the risks cannot be avoided, thus making avoidance tactics necessary.

Based on this question the traffic psychology theories described below are those which refer to the general tasks in traffic and represent an overview regarding the psychological problem of individual risk, the strategy of the driver within the system and the driver's action in a problem and decision situation. There are many dimensions, which could provide a framework for the systematic presentation of the theories and models. Rothengatter (1997), for example, proposes three categories: driver task, functional control and motivational models. As, all in all, too many dimensions have to be taken into account, the following selection are not presented in an obvious structure or in a hierarchical order. The more specific theories are partly mentioned and readers are referred to other chapters where the models are connected to a relatively particular question and topic.

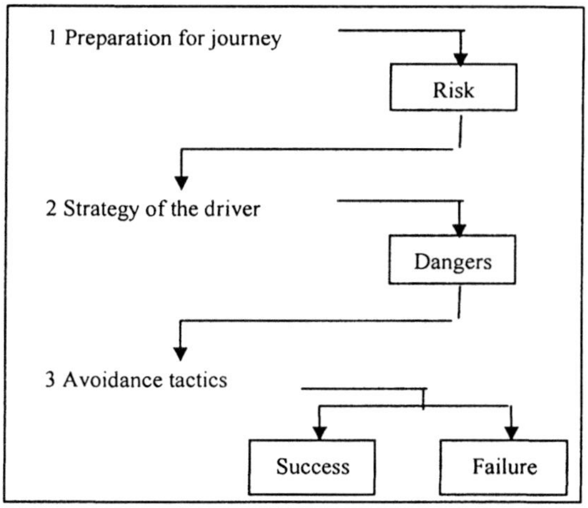

Figure 1. Sequence of driver behavior (Fleury, 1989)

2. THEORIES AND MODELS OF TRAFFIC PSYCHOLOGY

2.1. Points of Gravity of the Models: Perception and Proneness

There is no complete driver model. Most models focus on one or the other part of the driver task, the functional control carried out by the driver or the motivations of the driver. Each one of these approaches may take its starting point in the psychological characteristics of the driver or in the physical situation in which the driver is carrying out his task. The approach may be from the capacity of the driver or from the limitations of the driver. The models may focus on the basic perceptual point of view or one or more of the higher order cognitive, motivational, emotional or sociological influences.

Rumar (1985) presented an outline of a driver model which includes both the basic perceptual functions, the limited cognitive capacity, the higher order functions of memory and motivation, and what he calls three filters. In a later paper (Rumar, 1993), he developed these filters to represent human limitations of a physiological, psychological and sociological nature.

One of the best known perceptual psychologists, Gibson, argued that driving is in fact a human locomotion task of the same type as walking. The main differences are that the speed is a larger factor, the stopping distance considerably longer and the visual stimuli not as ecological as in walking. Walking is basically a perceptual task and so is driving. Most of the information necessary is available in the scene. The driver has problems if the information is too poor or inadequate. Gibson designed a driver model along these lines (Gibson and Crooks, 1938).

More than seventy years ago the trait-approach in traffic psychology was established (Shaw and Sichel, 1971). Based on the idea that accidents occur mainly because of human errors and that these errors are mostly due to specific personality traits, a relation between character variables and accidents was searched for. Certain models include up to a hundred variables, thus running the risk of missing out on an intelligible psychological interpretation of these factor combinations (Harano, Peck and McBride, 1975). Accident liability models (Maycock and Hall 1991; Forsyth, Maycock and Sexton, 1995), where predisposition (proneness) plays a role (see, e.g., Häkkinen, 1958), are examples. This concept has been criticized, especially with the argument that accidents occur statistically according to a Poisson distribution, but the relevance of the criticism has its limits (Kunkel, 1973). There is no doubt that during a relatively long period (for example, 20 years), a minority of drivers does not cause a majority of accidents. But some risks are unequally distributed among the driver population. Partisans of the "accident proneness" theory in its limited version maintain, without however really proving it, that poor drivers can be identified through a specific personality trait.

2.2. Adaptation and Objective versus Subjective Risk

During the last years, risk, specified as objective, subjective, compensated or homeostatically accepted risk, has been treated by using the concept of adaptation: Road safety measures do not always turn out positively or come up to the expectations of those who conceived them. Assuming that one is dealing with a basically efficient measure, the reason for its ineffectiveness can sometimes be found in the willingness and ability of the human being to adapt; he does not react passively to changes in the person-road-vehicle system but adaptively, which leads to variations in the effectiveness of safety measures and therefore to a variation in individual risk. The existence of such processes has been described in a series of traffic psychologists' studies (see OECD, 1990; Pfafferott and Huguenin, 1991).

Basically, adaptation can be interpreted in the light of the controversy in psychology between cognition of the objective and of the subjective risk. Objective risk can be defined as the product of occurrence probability and seriousness of an event (e.g., accident, police check). Subjective risk can be defined psychologically as the expectation of a dangerous event with a definite, but unknown seriousness and occurrence probability (Colbourn,

1978). Such expectations arise when a goal must be reached without the certainty that it can be reached. The more certain a driver feels, the less is his subjective risk.

The conflict between performance and safety tendencies is of special interest to traffic psychology. There is a tendency to avoid objective risk (but not to the extent that risk vanishes altogether), but success in avoiding risk will depend on the processing of cognitions associated with a situation. The absence of a conflict between performance and safety tendencies can be traced to at least three factors:

- Over-estimation of one's own abilities in the mastering of situations with objective risk;
- Conscious decision to take an objective risk;
- Defects in the perception of objective risks.

Tränkle, Gelau and Metker (1989) showed that the three "gaps" can be relevant for safety deficiencies, either uni- or multi-factorial. According to this, adaptation could be interpreted as a process which occurs when the perceived objective risk or the attitude towards objective risk changes and the adjustment leads to a new estimate of subjective risk.

After Klebelsberg (e.g., 1977, 1982), objective safety is determined by the physical safety conditions (e.g., friction coefficient), while subjective safety is perceived safety ("feeling of safety"). He makes two assumptions:

a) Objective and subjective safety qualify each other in the current situation, i.e., changing one of the two components can lead to a change in the relationship of the two components to one another;
b) "Behavior that is appropriate" to the situation (in road traffic) "requires that objective safety be at least as great or greater than subjective safety, because it is only then that the prevailing physical limit in a given situation will not be exceeded" (Klebelsberg, 1977; page 288).

From the model in Figure 2, one can see that safety at behavioral level increases if objective safety increases without subjective safety increasing to the same extent; it reduces if subjective safety increases without objective safety increasing at least by the same amount. Example: The improvement of visibility at a road crossing (greater objective safety) can lead to higher speeds at the crossing (because of an increase in subjective safety). The result may be less safety at the behavioral level.

Based on the risk adaptation concept (OECD, 1990), compensation mechanisms as responses to the introduction of safety measures have been described. It is obvious that compensation effects occur, but it is not necessarily always the case. Figure 4 depicts the different possibilities that have to be taken into consideration.

Following a usually factually sound definition of a problem, a measure is introduced with the purpose of improving safety in a certain part of the road traffic system. If the measure fulfills the expectations (1), it is referred to as a "primary effect."

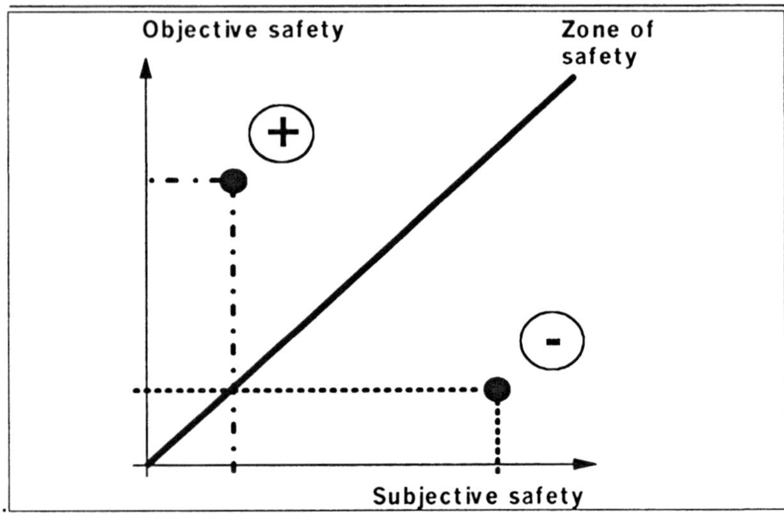

Figure 2. Paradigm of subjective-objective safety

If it does not result in adaptation processes on the part of road users and there are no other consequential effects within the system, the effect of the measure has permanence (2). In the event of an accident, the accident consequences are reduced according to the biomechanical parameters. However, if adaptation processes are involved, i.e., there is adaptive reaction by the road user.

The effect of the measure is usually influenced (3). This effect can be considered a "secondary effect." It will usually diminish the primary effect. In some cases, adaptation leads to an increase in effectiveness. According to the negative adaptation forms a, b, and c, different examples can be given:

a) For safety reasons a speed limit sign is erected at a motorway exit. However, this leads to an increase in vehicle speeds because many drivers believe from experience that such speed restrictions are usually set very low.
b) Following the installation of public lighting, it is noticed that, after a while, the speed on the relevant section of road increases and the safety dividend is, at least partially, cancelled. In place of or in addition to the intended effect, an unintended effect occurs. The road user changes his behavior in a way that was not intended when the measure was conceived.
c) The parents of children who receive road safety education may be more likely to allow their children to linger in the vicinity of the road.

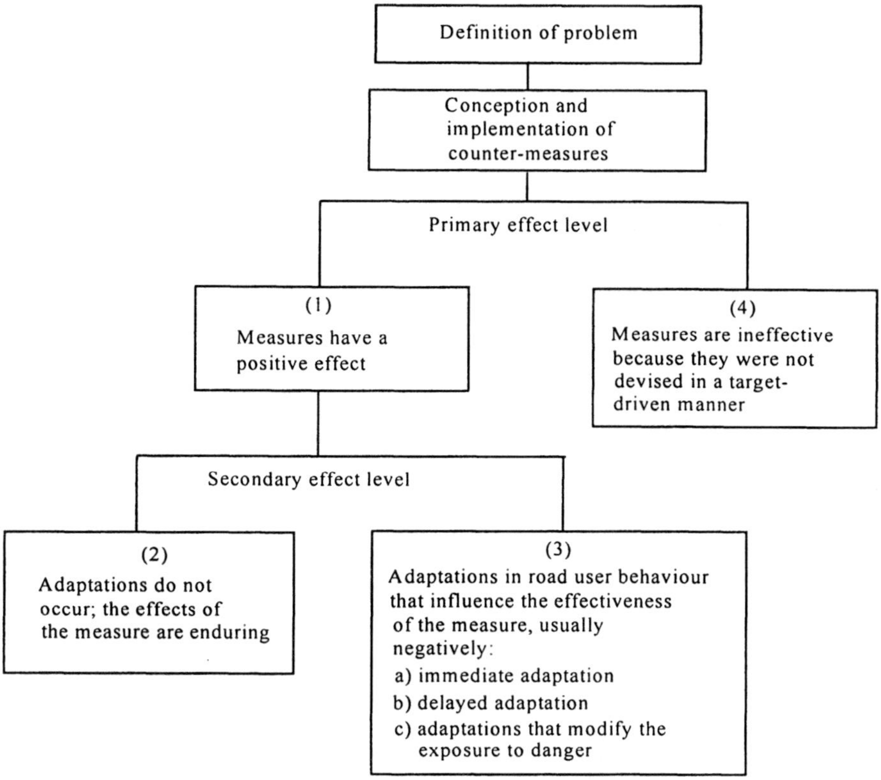

Figure 3. Differences in the adaptation process following the introduction of safety measures (schematic diagram; after Pfafferott and Huguenin, 1991)

Finally, it is important to separate the positive measures from those where proof of effectiveness is not possible because the measures are poorly devised and cannot therefore be effective (4). The studies into adaptation show that there was clear evidence for undesirable adaptation processes. Higher safety standards in a particular vehicle or model provide no guarantee for a more favorable accident ratio. In particular, when safety improvements are associated with or incorporated as an accompaniment to sporty or lively performance for opening up new speed horizons, the safety dividend can be offset or at least significantly limited (Bock, Brühning, Dilling, Ernst, Miese and Schmid, 1989). It can be shown that the type of safety measures incorporated in a vehicle has an influence on the reaction of the driver. The so-called compensatory or adaptive mechanisms are probably less marked in the case of passive safety measures and more pronounced in the case of active safety measures. The latter relate mainly to a "sporty" vehicle specification, the advantages of which are offset by a reduction in safety due to inappropriate driving. The assumption that passive safety installations bring about fewer negative adaptations than active measures can be articulated by the following propositions:

a) The less the safety dividend in vehicle design can be seen or felt, the greater its effect will be, i.e., the less its existence will be offset by risk-taking.
b) The more the safety dividend in vehicle design can be varied by driver decisions, the smaller the effect of the safety dividend will be, i.e., the more its existence will be offset by risk-taking.
c) The more the safety measures are aimed at protecting the vehicle passengers from the effects of a collision, rather than preventing accidents, the greater the net safety dividend will be, i.e., the less likely it is that a safety measure will be ruined by behavioral adaptation.

There is no doubt that the driver is often aware of the increased safety that derives from technical advances and that he sometimes transfers this awareness to the behavioral level. But, if so, this seldom occurs on a 1/1 basis, in accordance with Wilde's hypothesis (see below), and indeed does not always occur. That the phenomenon is very likely to occur following the implementation of safety measures in the driver-vehicle-environment system, however, is entirely plausible.

2.3. The Theory of Risk Homeostasis

The risk model which is most discussed in the field of motorway traffic is Wilde's (1988, 1994). The author poses the question: "Why certain drivers are willing to accept a degree of (objective) risk?" According to Wilde, knowledge of risk depends on the possibility of risk perception. The objective risk perceived is evaluated and compared to the accepted risk (comparer). The result is the optimal degree of attention required. The degree of attention actually exerted depends on supplementary factors such as the driver's capacity to decide or to manoeuvre. Wilde grouped these factors together in a detailed homeostatic model (Figure 4). Wilde's theory presents certain similarities with Klebelsberg's model. However, Wilde advances a precise dialogue: If we take measures to reduce objective risks, safety is improved as long as risk acceptance remains unaffected. In the same vein, when the objective risk remains constant, safety can improve by inciting the driver to take fewer chances. Nevertheless, these measures only remain efficient over a short period of time. Wilde believes that the equilibrium between estimated and accepted risks is maintained through risk homeostasis, and that safety only increases or decreases during the imbalance phase. Wilde formulates "the principle of maintaining the accident rate." "The number of accidents in any given country only depends on the accident rate which the population is willing to tolerate, and not on the specific measures taken in other sectors of the control system - at least not over a longer period of time" (Wilde, 1978; p. 142). This principle is not applicable at the individual level, but at the level of the social system constituted by the driver population. As such, it becomes delicate to declare the theory invalid, as it is no longer possible to formulate general predictions with exactitude. "We can wonder whether the theory is of any scientific interest whatsoever (as it cannot be tested), or if it is simply false" (Haight, 1986; p. 364). Wilde quotes a considerable number of studies (including his own) to support his theory. Many among them dwell on individual behavior after modification of one of the risk variables. All experiments and analyses quoted were not originally elaborated in order to test Wilde's theory.

The theory can be summarized on the basis of the following assumptions (Wilde, 1988):

- Traffic participants always compare the existing measure of subjective risk with the measure of accepted risk which corresponds to their personal level of activation (stimulation requirement);
- If there is a discrepancy between subjective and accepted risk, the individual tends to eliminate it;
- The probability of objective risk appears or already exists through the adaptation process;
- Risk willingness is regarded as an independent variable, which determines the accident rate. In its turn, it is stabilized through homeostatic regulation. The accident rate controls the direction and amount of the desired adaptation, but not the risk willingness itself (Wilde and Kunkel, 1984);
- The summing up of the objective risk over all traffic participants results in the accident rate and seriousness for a period of approximately one year;
- The accident rate, which is measured as a function of the exposure time, is constant;
- The model is based on system-theoretical reflection on theories of equilibrium.

Several studies and theoretical reflections have been developed on the basis of Wilde's theory. The best known is O'Neill's (1977) decision theoretical model of danger compensation. The theory is based on Taylor's (1964) observation that drivers tend to maintain the assumed risk per minute at a constant level.

2.4. Theory on Risk Behavior

Näätänen and Summala (1974, 1975, 1976) and Summala (1985 and 1986) established a theory of risk behavior in which one of the most important elements is subjective risk. They assume:

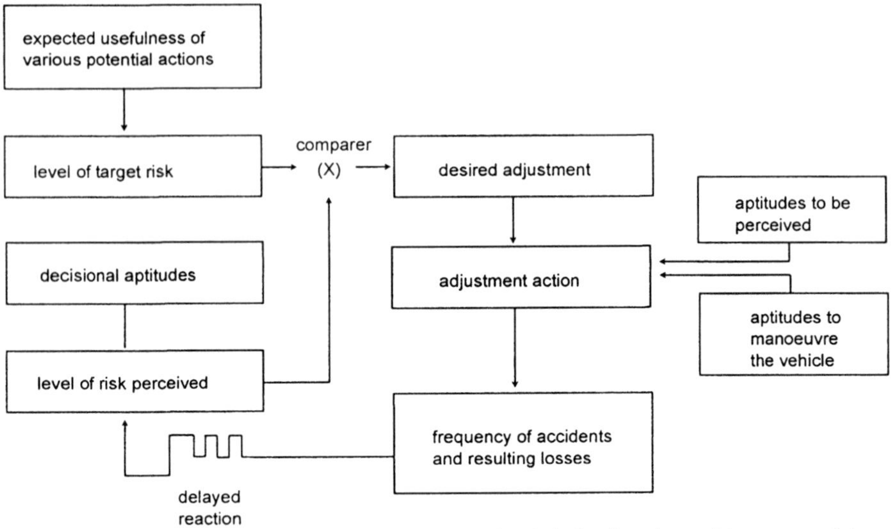

Figure 4. Homeostatic model comparing the driver's behavior, the accident rate and the level of target risk (Wilde, 1994).

- That subjective risk – defined as the cognition of danger – is a significant motivational factor (i.e., regulator) in the behavior pattern of the driver, and,
- That the subjective risk is insufficiently distinct.

The model postulates that accidents occur because the subjective risk is too low. The reason for this is that the driver overestimates his own ability. The model is not explicitly declared as a risk compensation theory but it implies that the lack of danger cognition leads to a correspondingly "more risky" driving style.

The model by Näätänen and Summala can be regarded as a relatively comprehensive cognitive structure-model of driver behavior. They describe the cognitive process by means of several constructs, which are controlled by the action caused by stimulation in a situation. Näätänen and Summala give a mainly cognitively orientated overall view of how driver behavior takes place and is controlled.

Näätänen and Summala's theory on risk behavior is also called the "zero-risk theory of driver behavior." In essence it proposes that drivers control risks on the basis of simple cues and features in traffic situations and normally avoid behavior, which elicits fear or anticipation of fear. The authors also postulate that drivers try to satisfy their motives regarding mobility by quick and dangerous driving on this activity so far as to avoid cognition of accident risk on the other. Driver behavior is determined to a large degree by habit, whereby the limits of safety and the avoidance of unpleasantness, as well as fear, play an important role. Reaching a certain destination by means of the chosen vehicle is referred to as the main motive, but other motivating components are also listed. At the same time, a distinction is made between extra, or excitatory, and inhibitory determinants of driver behavior. The following belong to the special (extra, sometimes also called excitatory) motives:

- Goals set for the trip or during it;
- Emotions stirred up in or before a traffic situation;
- Behavior models;
- Showing off and the need to prove oneself;
- Hedonic objective;
- Risk for risk's sake.

The most important of the inhibitory motives is the subjective risk and according to the authors, there is a lack of it. The lack of subjective risk can be substantiated by the following:

- Many forms of behavior in road traffic indicate a scarcity of subjective risk;
- In general, the individual does not reduce exposure in traffic to a minimum;
- Choice and maintenance of the vehicle do not demonstrate observation of the safety aspects;
- Safety measures based on the concept of subjective risk often miss their target;
- Experiences by traffic participants do not include elements of subjective risk with regard to accident expectation.

The control loop (Figure 5) starts from the stimulus situation to the motor responses. Perception is an active, selective process, which is controlled by drivers' motives as well as their experience. Perception triggers expectancies, and a decision is made as to what kind of behavioral change, if any, is needed.

It is postulated, by means of the model, that accidents take place because the subjective risk is too small, the reason being that the driver overestimates his own abilities. This has been substantiated by the results of several opinion polls in which 70-80 percent of the drivers classified themselves as good or very good drivers. Furthermore, it is indicated that most drivers believe that accidents only happen to others. In addition to these attitudes, perception also plays a role: The estimation of speed is often wrong, depending on the situation, too low, and the physical forces which come into effect in the case of impact wrongly assessed. This tendency is intensified by the subjectively easy task of driving. This is the reason for the driver's tendency to choose higher speeds, which should be eliminated. The model implies that the lack of danger cognition leads to a correspondingly riskier style of driving.

2.5. The Hierarchical Risk Model

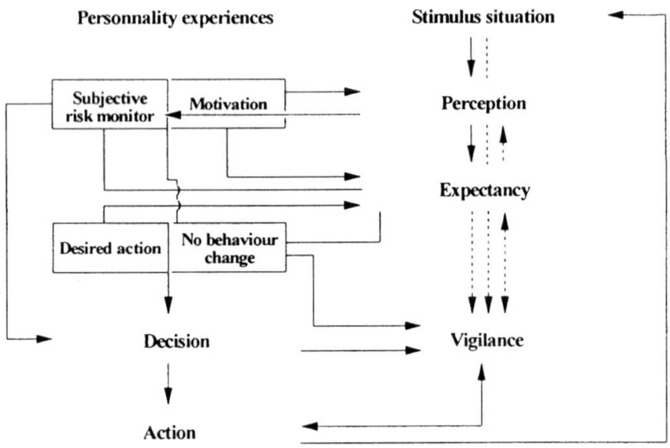

Figure 5. Flow diagram representing the most important factors and relationships in the decision process and behavior of the driver (Näätänen and Summala, 1976)

Van Der Molen and Bötticher (1988) developed a hierarchical risk model for traffic participants because of the contradictory results of empirical data of some risk theories described above. The main problem of driver behavior is that drivers adapt actions on different strategic levels, and on the basis of the environment and psychological processes.

The model is hierarchically structured in terms of a strategic, tactical and operational task level (Figure 6). The perception of the physical environment is, at each moment, influenced by the internal representation of similar situations. It also contains the knowledge of one's own limitations and abilities and the interactions with the environment. The internal representation determines judgmental processes by way of expectations. Motivations of the driver determine the subjective importance of the possible results of behavior alternatives. The model is a meta-model without explanations of the processes working within the described structure.

The interaction of safety motivation, expectation and judgment could be relevant in connection with adaptation processes. The authors postulate that, in equally attractive situations on the three different levels, a driver will not choose a more risky alternative when one with less risk is available.

Adaptation could also be explained under certain circumstances using balance theory. In such cases the need for equilibrium leads to a behavioral change as a reaction to an initial change in behavior. This could imply a demand for risk or similar phenomena. The adaptational change in risk acceptance in young drivers could be seen as an example in this context. After some experience with a positive adaptational behavior, young drivers are willing to take more risks and create a balance between the driving experience and the accepted risk.

3. ANALYTICAL, SYSTEM-ORIENTED APPROACHES

Analytical, system-oriented approaches aim to identify factors, which lead to failures in the functioning of the system and thereby enhance risk. This procedure allows the relevant mechanisms to be determined and enables (sub)models to be established which can then be generalized and thus increase the understanding of the system. Of course, this approach needs to be done in an interdisciplinary manner and has to take into account the interaction between the components, elements and factors. Because of the complexity, it is recommended to proceed in a hierarchical process. At each level, failures can be analyzed by modeling the way the vehicle-driver-road system works or fails. An example integrating the different components was suggested by Briggs in the late 1960's (Hobi, 1978, Fig7).

Evans (1984) interprets approaches to understanding traffic safety by placing them into three categories:

- "Engineering," meaning that physical changes to the system are considered without addressing possible induced user change;
- "Economic" or "danger compensation," meaning that users adjust behavior so that actual safety benefits are less than expected;
- "Risk homeostasis," meaning that users adjust behavior to re-establish prior levels of risk taking.

Evans formulates a general human behavior feedback model in which each of these approaches is included as a special case. The term feedback is chosen "to supersede such terms as danger compensation because feedback can, in principle, be of either sign and of any magnitude" (Evans, 1984; p. 4). Evans assumes that a component of a traffic system is altered in such a way that the engineering safety change is estimated to be SEng. Evans proposes that the actual safety benefit SAct. is not necessarily identical to the engineering change, but is rather related to it by:

SAct. = (l+f) SEng.

1. where f = degree of feedback in the system, and f of concerning engineering f = 1
2. economic $-1<f<0$
3. risk homeostasis f = -1

Evans associates the feedback parameter f with 28 examples of changes made to the road-vehicle-user system taken from the literature. In 15 cases changes were expected to increase safety, and they produced 4 different outcomes: Increase in safety more than expected, increase in safety less than expected, no increase in safety, or decrease in safety.

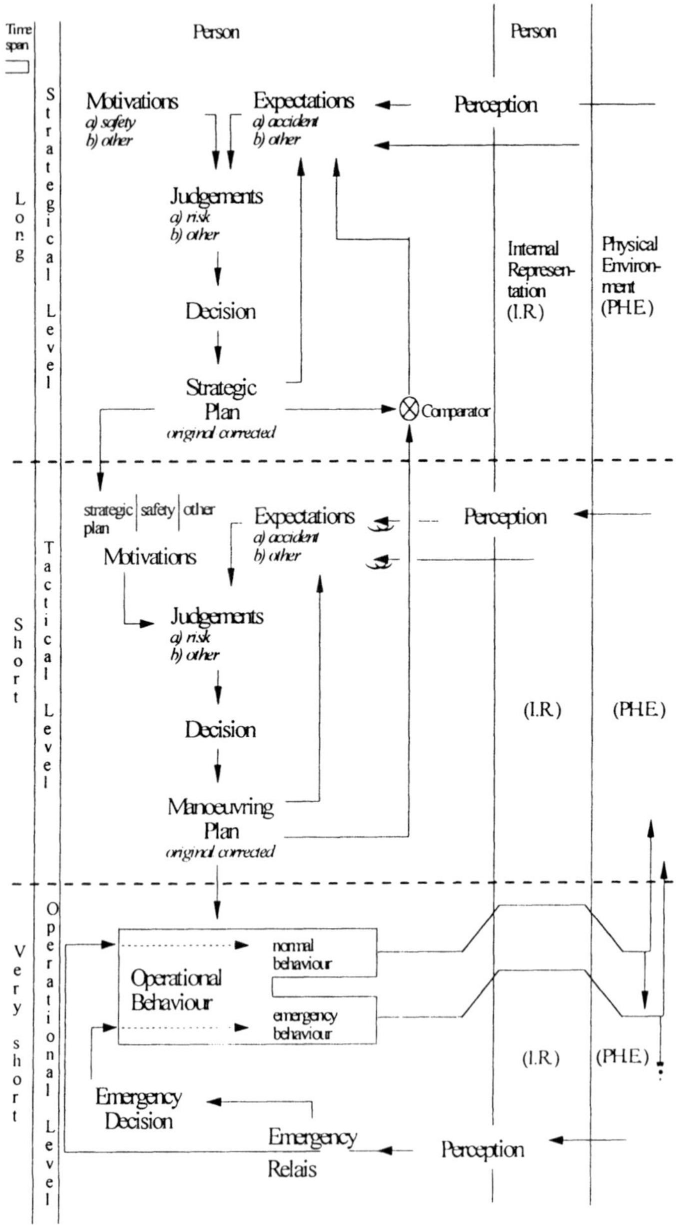

Figure 6. The hierarchical risk model for traffic participants of Van Der Molen and Bötticher (1988)

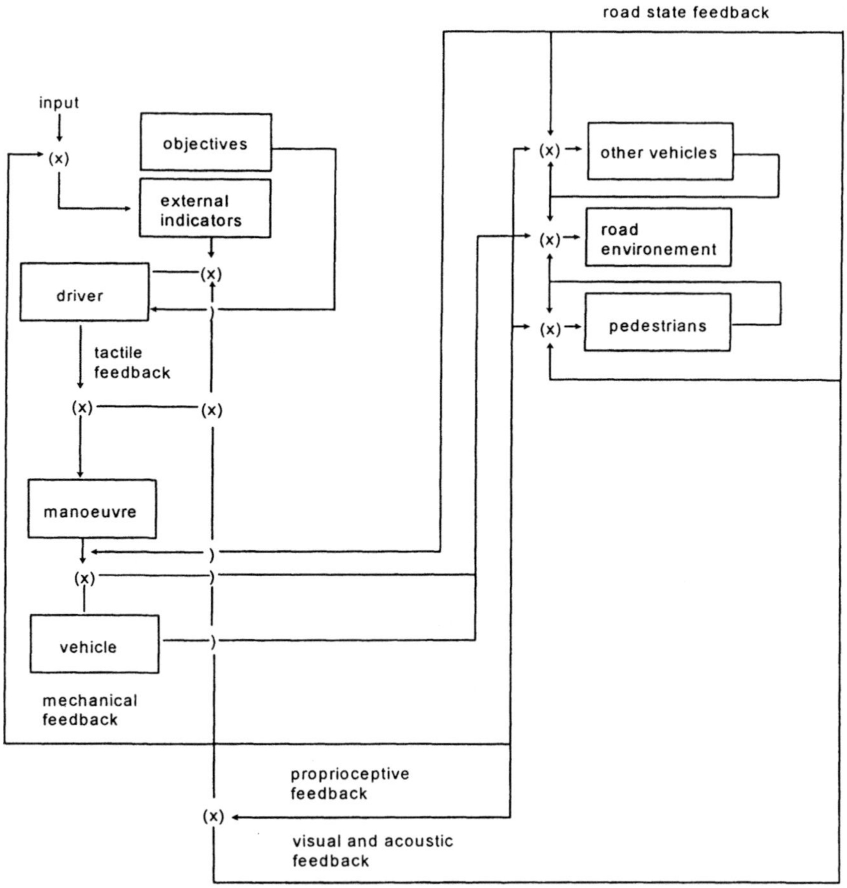

Figure 7. Functional diagram of a driver-vehicle-road system - taken from Briggs (Hobi, 1978)

In 13 cases changes were expected to reduce safety, providing analogous outcomes: Increase in safety, no safety reduction, or decrease in safety less than expected. Evans suggests that human behavior feedback is a pervasive phenomenon in traffic systems, which may greatly influence the outcome of safety measures. In some cases it may even generate effects opposite in sign to those intended. We need models that take into account the interactive nature of the process as represented by the parameter f.

4. MOTIVATION THEORIES

Various authors have tried to theoretically describe the motivational aspects of driving or of unsafe behavior. Under the influence of Vroom's

(1967) concept of expectancy and valence, Zink (1980) explains a model of unsafe behavior, which could also be of interest in road traffic. According to this model, the strength of the motivation for carrying out a certain activity depends on the valence (subjective notion of benefit) of an occurrence and on the expectation of the result of certain activities (behavior occurrence = valence x expected result; subjective benefit). Safety-oriented behavior can be expected when the valence for "accident-free" is high. Because of the low likelihood of an accident, however, unsafe behavior is of instrumental character, as it often leads to results of higher valence (e.g., the saving of time).

In order to make safety-oriented behavior more attractive, the valence for safety-oriented behavior would have to be increased (e.g., recognition) and/or positive results in the case of unsafe behavior would have to be eliminated. The reward (increase of valence) for the road user can take place intrinsically or extrinsically. An intrinsic reward is more promising because a closer connection between reward and action results.

The crucial question in a motivation theory for driver behavior is: "What valences become effective during driving or, in other words, what are the stimuli of driving?" (Hoyos and Pupka, 1977; p. 34). Apart from the fact that driving makes certain type of locomotion possible, the question whether - and if yes, how strongly - intrinsic motivation plays a role during driving is of special interest in traffic-psychological safety work. Based on an extensive study of literature, a general review of this theme was undertaken and the following observations on driving were made:

a) The driver seeks an optimal level of stimulation, so that the activation of the nervous system is neither too weak nor too strong, but an agreeable sensation.
b) In the sense of game behavior, driving often takes place without having an objective (compare also Nagayama, 1978).
c) Thrill can be generated especially well through driving. The motivating factor in generating fear-inducing situations is the expectation of overcoming this situation.

5. THE NEED FOR GENERAL STRUCTURES

Given the high number of variables, factors and micro models and the different approaches (which are often unrelated), it is recommended to find an adequate general structure as a basis for modeling behavioral factors in traffic. We could consider that road user actions are the result of more or less adequate behavioral controls. Along these lines, Rasmussen (1987) presented a hierarchical model including knowledge, rules and skills. Janssen developed a version of a model (Figure 8) that is based on

Rasmussen's who had probably the largest practical influence on modeling traffic behavior.

Reason (1994), also referring to Rassmussen's model, presented a Generic Error Modeling System (GEMS) differentiating between knowledge-, rule- and skill-based errors: Slips and lapses, deviating from the intended action (without being conscious) as well as unintended versus deliberate violations are integrated in the model. Based on these elements, Reason demonstrated how errors can be predicted and analyzed. Ranney (1994) adapted his classification of driving tasks according to Hale, Stoop and Hommels (1990; see Table 1). Traffic safety work is often based on models of the drivers' task and, from there, researchers move into the psychological area. Brown (1986) and Rumar (1985; Figure 9) are two examples. The latter stresses the self-paced task of the driver and hence the large importance of speed as the only regulating variable (Figure 10). Besides these elements, it is recognized that dispositional concepts, the capacity of the driver which determines his/her behavior in any given situation, and situational concepts – the influence the outside environment may have on behavior – certainly play an important role. It goes without saying that a global view of these concepts is indispensable. According to the researcher's perspective, the emphasis is on one or the other of these concepts. Thus, some concepts are focused on the overview and therefore relatively simple – as for example in Figure 11 – or more complex, as for example the model of Näätänen and Summala (1976).

More details between the different elements called human factors and their interrelationships are given in the overview metamodel of Figure 9. It is a "map" describing important psychological elements of road users, which explain the behavioral phenomena observed in certain traffic circumstances. The advantage of these models is that they take into consideration various major psychological dimensions, thus allowing for the analysis of behavioral determinants of the human sub-system. However, interactions between subjective variables (e.g., "driver") and objective variables (e.g., "road", "vehicle") are not integrated therein. It is possible to explain behavior as a function of the most important elements, taking different levels of behavioral observations as a general model. The driver's predispositions and his/her assimilation process should be taken into consideration (as well as his/her cognition of the situation). The model is nothing more than a program, but it allows the factors involved to be evaluated and the interactions within the sub-system "road user" to be determined. Based on these three observance levels, models of action can be established which take into account the characteristics of driving. The theoretical reference frame for the driver's action can be described at different levels. It can be applied for four or two wheel drivers but also to pedestrians, which encounter traffic when crossing the street.

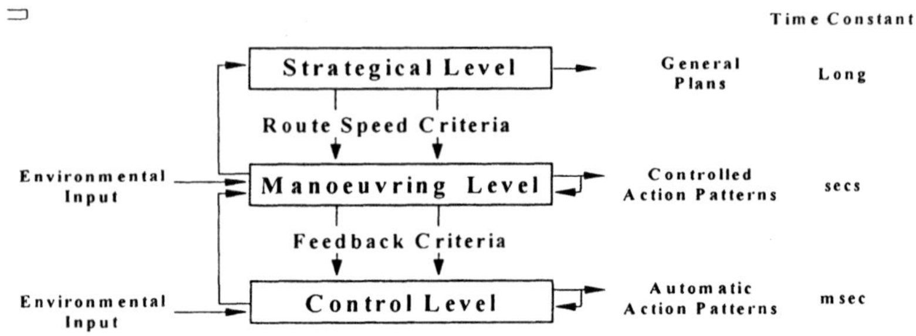

Figure 8. The hierarchical structure of the road user task. Performance is structured at three levels that are comparatively loosely coupled (after Janssen 1979).

Table 1: Classification of driving tasks (after Ranney, 1994)

	Strategic	Tactical/Maneuvering	Operational/Control
Knowledge	Navigating in unfamiliar area	Controlling skid	Novice on first lesson
Rule	Choice between familiar routes	Passing other vehicles	Driving unfamiliar vehicle
Skill	Route used for daily commute	Negotiating familiar intersection	Vehicle handling on curves

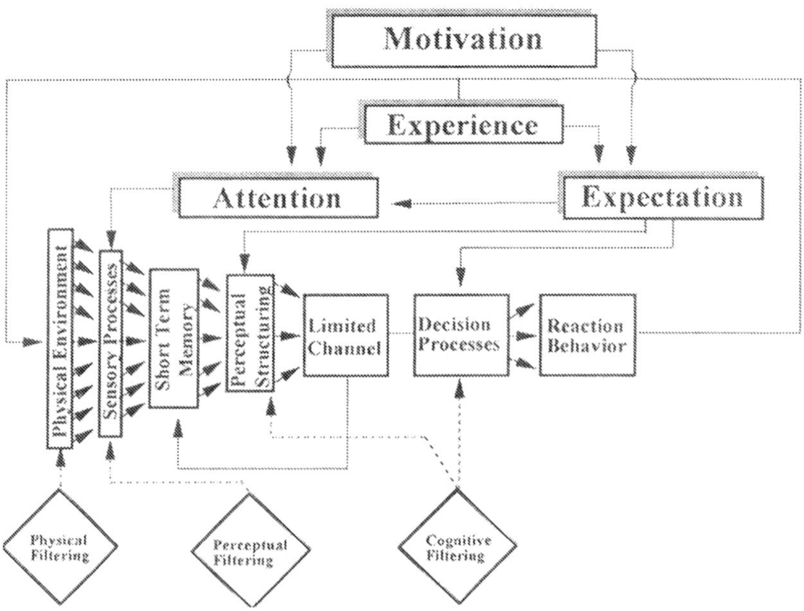

Figure 9. Psychological bases to resolve driver's task (Rumar, 1985)

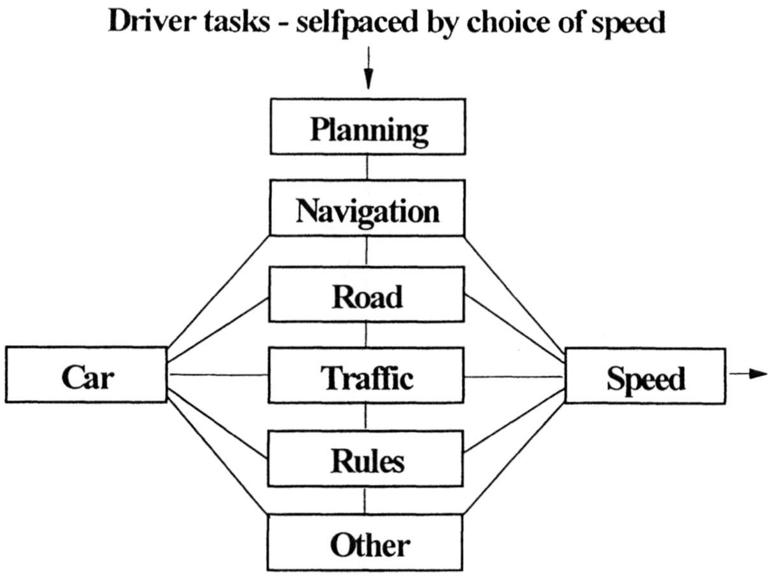

Figure 10. Driver's task model (Rumar, 1985)

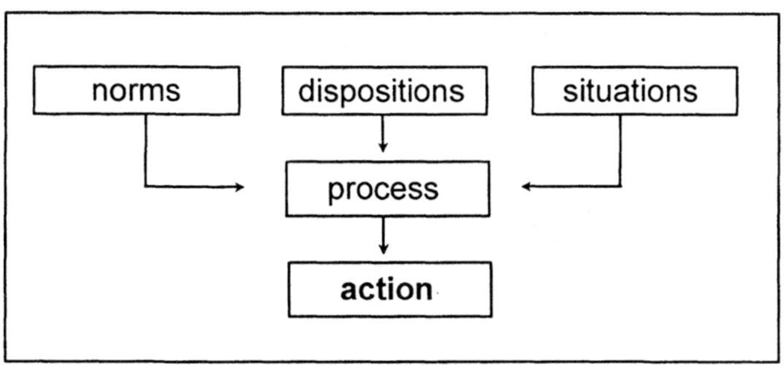

Figure 11. Determining elements on action

According to Figure 12, the three levels could be described as follows: The *predispositional level* consists of three elements:

a) *Driving suitability: Psychological and physical predisposition of the individual to drive a car:*
Here, theories of accident proneness could play a role for explaining errors, the relationship between a person's involvement in a traffic accident and the presence or absence of certain permanent factors of the individual (e.g., visual perception, selective attention, reaction time).

b) *Driving qualification: Acquired psychological and physical capabilities of the individual to drive a car:*
Models, which refer to road users' capacities, form the basis for explaining how adequate behavior in road traffic is acquired. This concept is of prime importance, as it allows for the integration of aspects of this process, which unfold in numerous circumstances. As such, the learning of driving skills, assimilation of experiences, acquisition concerning perception, anticipation and the role of memory, are included in this concept. Fuller (1984, 1988), for instance, analyzed the driving task from which he developed a driver behavior model directed at danger avoidance, based on learning theory. He presents a threat-avoidance model. With classical conditioning it can be shown that "a distinct preference for delayed avoidance as opposed to anticipatory avoidance responding, that is, a preference for the more risky behavior" exists. When confronted with a discriminative stimulus for a potential aversive event, a driver's actions depend in particular on the rewards and punishments for alternative responses.

c) *Driving capability: Momentary ability of the individual to drive a car:*
Laboratory studies have underlined the question of the reduction of a driver's suitability and qualification (i.e., driving skills, cognitive capacities). This deterioration of capability is very often due to the consumption of alcohol, but other drugs, fatigue and so on are important as well.

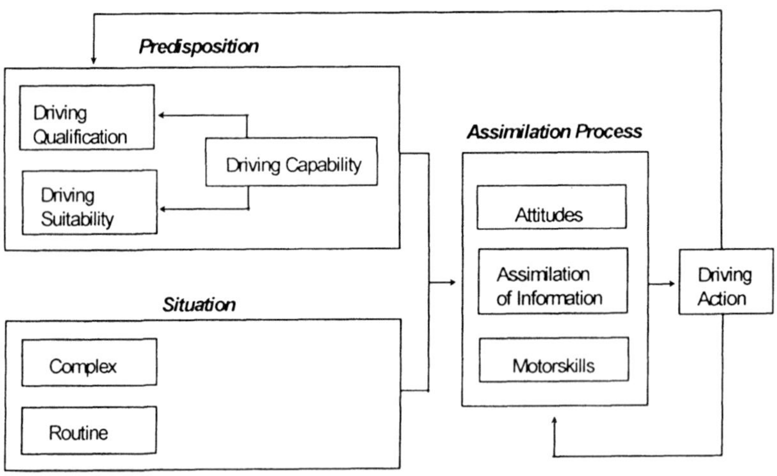

Figure 12. Connections between predisposition (P), situation (S), assimilation process (AP) and driving action (A) (Huguenin, 1988)

The factors at the *action level* could be interpreted as consisting of three elements:

a) *Attitude as a determinant of behavior: Tendency to react in the same way to certain stimuli or irritants.* As learning theory "is essentially based on a conceptualization of the task of driving which implies avoidance responses acquired with potentially dangerous stimuli, as well as an application of behavioral principles which are correctly chosen for the driving situation," Michon (1989; p. 145) observed that the model proves to be hardly applicable in complex learning situations. Therefore, numerous researchers have examined the concepts of "attitudes." Attitudes - based on values - partially determine behavior behind the wheel in a specific situation. This concept explains a given behavior pattern which does not correspond to acquired behavior which is normally observed. Often an attitude has exercised its influence and neutralized the expected action. Explanations of this phenomenon can be provided with the help of the concept of cognition and emotions. Based on Ajzen and Fishbein (1977), theories of planned behavior and explanations for intentional and unintentional (but erroneous) behavior have been given (Forward, 1997; Åberg, 1997). Another class of models seeks to explore the relationship between attitudes, behavior and individual accident rate (Parker, West, Stradling and Manstead, 1995).

b) *Assimilation of information: A cognitive process in which objects and situations are observed and judged*:
Perception is the most important element within this process.

c) *Motor skills: Movements of the body and parts of the body adapted to conditions of space and time.*

The factors at the *situational level* could be interpreted as consisting of two elements:

a) *Routine situations:* Situations which, due to predisposition and action determinants, can be dealt with using usual behavior patterns, i.e., without involving processes of choice and decision; actions take place in an automated way.

b) *Complex situations:* Situations, which include choice between alternative responses, in the course of which the preceding action is interrupted, because a decision involving alternative actions must be made. Complex situations involve conflict: for example where unexpected (emergency) situations suddenly occur.

Situative models concentrate on describing the situation in which the road user is found. Determining factors are analyzed according to the behavior, which occurs in the situation. The most well known approaches in this context are based on system theories, which allow the influence of a driver's environment on the way he/she drives to be evaluated. Behavior cannot be described in general terms, but only by reference to specific situations. A general form of description, which would ultimately permit predictions, can hardly be given.

Based on the above structures, an attempt can be made to sketch an action-theoretical model of driver behavior. In this sketch more importance is given to driver action than to driver behavior. There are several reasons for emphasizing the former. First, behavior is only consolidated in actions (action is goal-directed behavior). Second, behavior cannot be described in general terms, but only by reference to specific situations. Third, the multicausal type of observation can be applied to actions. Referring to Figure 9, a general view of the relationship between the three levels could be as follows:

The action *(A)* should be regarded as the function of situation *(S)*, predisposition *(P)* and assimilation process *(AP)*: [$A = f(S, P, AP)$];

The situation can be routine or complex. This is already a strong determinant of the action, which will be taken. The cognition of the situation *(CS)*, including to some extent its classification as routine or complex, depends primarily on the predispositions *(P)*, "driving suitability" *(DS)*, "driving qualification" *(DQ)*, and "driving capability" *(DC)*: [$CS = f(DS, DQ, DC)$];

The assimilation process *(AP)*, which is also based on predispositions *(P)* and which directly determines action, primarily takes place under participation of attitudes *(A)* processes assimilating information *(I)* and (often automatic) motor skills *(M)*: [$AP = f(A, I, M)$].

6. CONCLUSION

Although the models presented are in some cases vague, the definitions are not always clear, and most of the theories cannot be falsified in Popper's (1966) sense, there is evidence within the models for their heuristic value. It is therefore useful to consider all of them in the current context. One general problem with the models is that they are often not clear about whether the behavioral prediction or explanation applies on the individual or the collective level. Especially risk compensation theories assume that a balance between objective and subjective risk is required, resulting in constancy of risk, and this would necessitate corresponding cognitions. On the individual level there is little empirical evidence regarding this aspect.

Nevertheless, it cannot be denied that adaptation due to "risk compensation" exists (a better name would be "safety compensation"). It seems, however, questionable that these psychological mechanisms thwart the efforts to increase safety on a 1/1 scale. Wilde, for example, in this context, uses the principle of exhaustion in certain cases, having recourse to "disturbing conditions" which suggest the "total exploitation of a hypothesis," irrespective of deviating findings (Holzkamp, 1976). This is not always wrong, but it is sometimes problematic. After a theoretical analysis of the theory of risk homeostasis, Janssen and Tenkink (1987; p. 8) concluded: "There is no reason why exact homeostasis should be expected to occur as a response of road users to safety hardware improvements: partial compensation seems to be the rule instead."

Another basic question is: To what extent can theories of equilibrium be used as an adequate basis for the problem in question? Still unsolved is the problem of determining at what point equilibrium is reached. The theory may not define this on the basis of the accident occurrence, but should at least predict the accident level (danger to be circular).

Wilde's theory is the model, which has received the greatest amount of international attention, and provides the most complete explanation for the adaptation process. Consequently, the discussion concentrates often on this theory, which is in accordance with the demand for system-analytical investigation. Unfortunately, research cited in support of and against the theory has generally not had as its primary or original purpose the testing of the theory, and its hypotheses. Of course, on the empirical level, Wilde (1994) lists a number of results, which support his theory. Several authors have compiled corresponding counter-evidence in refutation of the model (Evans, 1986; Huguenin, 1984; McKenna, 1985), which, however, cannot be examined in detail here. Even Aschenbrenner, Biehl and Wurm's (1988) very interesting results have to be interpreted very carefully.

Generally there is no doubt that need for theories in traffic psychology exists. The models described above are interesting approaches, but there is still a lack, especially as the existing models are mostly based on a cognitive background. We need more concepts taking into account, on the one hand the basic perceptual aspects, and, on the other hand, the higher order, emotional and social psychological aspects.

7. REFERENCES

ÅBERG, L. (1997), "The Role of Perceived Risk of Detection in the Theory of Planned Behavior", in ROTHENGATTER, I., CARBONELL, E. (Eds), *Proceedings of the International Conference on Traffic and Transport Psychology*, Valencia (1996), Amsterdam: Elsevier (in print).

AJZEN, I., FISHBEIN, M. (1977), "Attitude-Behavior Relations: A Theoretical Analysis and Review of Empirical Research", *Psychology Bulletin*, (84), 888-918.

ASCHENBRENNER, M., BIEHL, B., WURM, G. (1988), *Mehr Verkehrssicherheit durch bessere Technik? Felduntersuchungen zur Risikokompensation am Beispiel des Antiblockiersystems (ABS)*, (Unpublished), Mannheim.

BOCK, O., BRÜHNING, E., DILLING, J., ERNST, G., MIESE, A., SCHMID, M. (1989), "Aufbereitung und Auswertung von Fahrzeug- und Unfalldaten", *Unfall- und*

Sicherheitsforschung Strassenverkehr. Heft 74. Bundesanstalt für Strassenwesen (BASt), Bergisch Gladbach.

BROWN, I. D. (1986), "Functional Requirements of Driving", paper presented at the *Berzelius Symposium Cars and Casualities,* Stockholm.

COLBOURN, C. J. (1978), "Perceived Risk as a Determinant of Driver Behavior", *Accident Analysis and Prevention,* (2), 131-141.

DEUTSCH, M., KRAUB, R. (1976), *Theorien der Sozialpsychologie, Fachbuchhandlung für Psychologie*, Frankfurt/M.

EVANS, L. (1984), *Human Behavior Feedback and Traffic Safety,* General Motors Research Laboratories, Warren.

EVANS, L. (1986), "Risk Homeostasis Theory and Traffic Accident Data", *Risk Analysis,* (1), 81-94.

FLEURY, D. (1989), "Accident Analysis Methodology", *Referat, gehalten anlässlich der First Round Table der International Scientific Initiatives on Road Traffic (ISIRT) de Wipselberg/NL,* Paris: INRETS.

FORSYTH, E., MAYCOCK, G., SEXTON, B. (1995), *Cohort Study of Learner and Novice Drivers: Part 3, Accidents, Offences and Driving Experience in the First Three Years of Driving*, TRRL Project Report 11, Crowthorne: Transport Research Laboratory.

FORWARD, S. (1997), "Subjective Norm and Who is the Significant Other?", in ROTHENGATTER, I. CARBONELL, E. (Eds) (1996), *Proceedings of the International Conference on Traffic and Transport Psychology,* Valencia, Amsterdam: Elsevier (in print).

FULLER, R. (1984), "A Conceptualization of Driving Behavior as Threat Avoidance", *Ergonomics,* (11), 1139-1155.

FULLER, R. (1988), "On Learning to Make Risky Decisions", *Ergonomics,* 31 (4), 519-526.

GIBSON, J. J., CROOKS, L. E. (1938), "A Theoretical Field-Analysis of Automobile Driving", *American Journal of Psychology,* 51, 453-471.

HAIGHT, F. A. (1986), "Risk - Especially Risk of Traffic Accident", *Accident Analysis and Prevention,* (5), 359-366.

HÄKKINEN, S. (1958), "Traffic Accidents and Driver Characteristics", *Finnland's Institute of Technology, Scientific Research Bulletin,* Helsinki.

HALE, A. R., STOOP, J., HOMMELS, J. (1990), "Human Error Models as Predictors of Accident Scenarios for Designers in Road Transport Systems, *Ergonomics*, (33), 1377-1388.

HARANO, R. M., PECK, R. C., MCBRIDE, R. S. (1975), "The Prediction of Accident Liability Through Biographical Date and Psychometric Tests", *Journal of Safety Research*, (7) (1), 16-52.

HECKHAUSEN, H. (1976), "Relevanz der Psychologie als Austausch zwischen naiver und wissenschaftlicher Verhaltenstheorie", *Psychol. Rundschau,* (1), 1-11.

HOBI, V. (1978), "Alkohol und Fahrverhalten", *Schweiz. Apothekerzeitung*, (22), 615-621.

HOLZKAMP, R. (1976), *Kritische Psychologie,* Frankfurt/M: Fischer.

HOYOS, G. C., KELLER, H., KANNHEISER, W. (1977), *Risikobezogene Entscheidungen in Mensch-Maschine-Systemen: Arbeitsplatzanalysen in Industriebetrieben,* Forschungsbericht zum Projekt HO 182/7, Institut für Psychologie und Erziehungswissenschaft, Technische Universität München, München: Lehrstuhl für Psychologie.

HOYOS, G. C., PUPKA, M. (1977), "Motivorientierte Aspekte der Verkehrspsychologie", *Unfall- und Sicherheitsforschung Strassenverkehr (7), Bundesanstalt für Strassenwesen (BASt)*, Bergisch Gladbach.

HUGUENIN, R. D. (1984), "Die Risikokompensationstheorie im Bereich des Strassenverkehrs - kritische Stellungnahme", *Referat anlässlich des V. GFS-Seminars Ingolstadt, Gesellschaft für Sicherheitsforschung*, Köln.

HUGUENIN, R. D. (1988), "The Concept of Risk and Behavior Models in Traffic Psychology", *Ergonomics*, 31 (4), 557-569.

JANSSEN, W. H., TENKINK, E. (1987), *Risk Homeostasis Theory and its Critics: Time for an Agreement*, TNO Institute for Perception, Soesterberg.

KLEBELSBERG, VON, D. (1969), *Risikoverhalten als Persönlichkeitsmerkmal*, Bern: Huber.

KLEBELSBERG, VON, D. (1977), "Das Modell der subjektiven und objektiven Sicherheit", *Schweiz. Z. Psychol*, (4), 285-294.

KLEBELSBERG, VON, D. (1977), "Psychologische Erklärungshypothesen für das Verkehrsverhalten", *Informationen und Mitteilungen, BdP, Sektion Verkehrspsychologie*, (9), 3-13, Bonn.

KLEBELSBERG, VON, D. (1982), "Die Bedeutung von subjektiver und objektiver Sicherheit: Fahrerverhalten als Risikoverhalten", in, *Verkehrssicherheit, Vorträge anlässlich des Seminars der Forschungsgruppe Berlin*, 5/6 November 1981, Daimler-Benz AG, Stuttgart: Forschung und Entwicklung.

KOORNSTRA, M. J. (n.y.), *Une approche systémique générale du risque collectif et individuel dans la circulation routière*, SWOV Institute for Road Safety Research, Leidschendam, The Netherlands.

KUNKEL, E. (1973), *Unfallneigung im Strassenverkehr: das persönliche Unfallrisiko unter dem Aspekt empirisch-statistischer Methoden*, TÜV Rheinland GmbH, Köln.

MAREK, J., STEN, T. (1977), "Traffic Environment and the Driver: Driver Behavior and Training in International Perspective", Springfield III: Charles C. Thomas.

MAYCOCK, G., LOCKWOOD, C. R., LESTER, J. F. (1991), *The Accident Liability of Car Drivers*, Research Report No. 315, Transport and Road Research Laboratory, TRRL, Crowthorne.

MCKENNA, F. P. (1985), "Do Safety Measures Really Work? An Examination of Risk Homeostasis Theory", *Ergonomics*, (2), 489-498.

MCKENNA, F. P. (1987), "Behavioral Compensation and Safety", *J. Occupational Accident*, (9), 107-121.

MEYER, W. U., SCHMALT, H. D. (1978), "Die Attributionstheorie", in FREY, D. (Ed), *Theorien der Sozialpsychologie*, Bern: Huber.

MICHON, J. A. (1985), "A Critical View of Driver Behavior Models: What Do we Know, What should we Do?", in EVANS, L., SCHWING, R. (Eds.), *Human Behavior and Traffic Safety*, New York.

MICHON, J. A. (1989), "Explanatory Pitfalls and Rule-Based Driver Models", *Accident Analysis and Prevention*, (21) (4), 341-353.

NÄÄTÄNEN, R., SUMMALA, H. (1974), "A Model for the Role of Motivational Factors in Drivers' Decision-Making", *Accident Analysis and Prevention*, (3/4), 243-261.

NÄÄTÄNEN, R., SUMMALA, H. (1975), "A Simple Method for Simulating Danger Related Aspects of Behavior in Hazardous Activities", *Accident Analysis and Prevention*, (1), 63-70.

NÄÄTÄNEN, R., SUMMALA, H. (1976), *Road User Behavior and Traffic Accidents*, Amsterdam: North-Holland Publishing Company.

NAGAYAMA, Y. (1978), "Characteristics of Excessively Car-Oriented People", in IATSS, *Mobility for Man and Society, Report of the Symposium on traffic science, International Association of Traffic and Safety Sciences*, Tokyo.

O'NEILL, B. (1977), "A Decision-Theory Model of Danger Compensation", *Accident Analysis and Prevention*, (3), 157-165.

OECD (1990), *Behavioral Adaptations to Changes in the Road Transport System*, Paris: OECD.

OECD (1997), *Road Safety Principles and Models*, Paris: OECD.

PARKER, D., WEST, R., STRADLING, S., MANSTEAD, A. S. R. (1995), "Behavioral Characteristics and Involvement in Different Types of Traffic Accident", *Accident Analysis and Prevention*, (27) (4), 571-581.

PFAFFEROTT, I., HUGUENIN, R. D. (1991), "Adaptation nach Einführung von Sicherheitsmassnahmen - Ergebnisse und Schlussfolgerungen aus einer OECD-Studie", *Zeitschrift für Verkehrssicherheit*, (37), 71-83.

POPPER, R. (1966), *Logik der Forschung*, Tübingen: Mohr.

RANNEY, T. A. (1994), "Models of Driving Behavior: A Review of their Evolution", *Accident Analysis and Prevention*, (26) (6), 733-750.

RASMUSSEN, J. (1987), "The Definition of Human Error and a Taxonomy for Technical System Design", in RASMUSSEN, J., DUNCAN, K., LEPLAT, J. (Eds), *New Technology and Human Error*, Chichester, U.K: Wiley.

REASON, J. (1994), *Menschliches Versagen, Psychologische Risikofaktoren und moderne Technologien*, Heidelberg/Berlin/Oxford: Spektrum Akademischer Verlag.

RUMAR, K. (1985), "The Role of Perceptual and Cognitive Filters in Obseved Behavior", in EVANS, L., SCHWING, R. C. (Eds), *Human Behavior and Traffic Safety*, London: Plenum.

RUMAR, K. (1990), "The Impossibility of a Unified Traffic Science", *IATSS Research*, 14, 1, 27-31.

RUMAR, K. (1993), "Road User Needs", in PARKES, A. M., FRANZÉN, S. (Eds), *Driving Future Vehicles*, London: Taylor & Francis.

SPOERER, E. (1979), *Einführung in die Verkehrspsychologie, Wissenschaftliche Buchgesellschaft*, Darmstadt.

SUMMALA, H. (1985), "Modeling Driver Behavior: A Pessimistic Prediction?", in EVANS, L., SCHWING, R. C. (Eds), *Human Behavior and Traffic Safety*, New York.

SUMMALA, H. (1986), *Risk Control is not Risk Adjustment: The Zero-risk Theory of Driver Behavior and its Implications*, University of Helsinki, Traffic Research Unit, Report 11, Helsinki.

SVENSON, O. (1978), "Risks of Road Transportation in a Psychological Perspective", *Accident Analysis and Prevention*, (4), 267-280.

TAYLOR, D. H. (1964), "Drivers' Galvanic Skin Response and the Risk of Accident", *Ergonomics*, (7), 439-451.

TRÄNKLE, U., GELAU, C., METKER, T. (1989), "Einflüsse von Alter und Geschlecht auf die Wahrnehmung situationsspezifischer Risiken im Strassenverkehr", *Zeitschrift für experimentelle und angewandte Psychologie*, Bd. XXXVI (2), 311-327.

VAN DER MOLEN, H. H., BÖTTICHER, A. M. T. (1988), "Risk Models for Traffic Participants: A Concerted Effort for Theoretical Operationalizations", in ROTHENGATTER, J. A., DE BRUIN, R. A. (Eds), *Road Users and Traffic Safety, 61-81*, Assen/Maastricht/Wolfeboro: van Gorcun.

VEILING, I. H. (1984), "A Laboratory Test of the Constant Risk Hypothesis", *Acta Psychologica*, (55), 281-294.

VOLLMER, G. R. (1974), Risikoverhalten im innerbetrieblichen Transportsystem Kranführer - Kran. Forschungsbericht No. 120, Bundesanstalt für Arbeitsschutz und Unfallforschung, Dortmund.

VROOM, V. H. (1967), *Work and Motivation*, New York/London: Wiley.

WILDE, G. J. S. (1978), Theorie der Risikokompensation der Unfallverursachung und praktische Folgerungen für die Unfallverhütung. *Hefte zur Unfallheilkunde*, (131), 134-156.

WILDE, G. J. S. (1982), "Critical Issues in Risk Homeostasis Theory", *Risk Analysis*, (4), 249-257.

WILDE, G. J. S. (1988), "Risk Homeostasis Theory and Traffic Accidents: Propositions, Deductions and Discussion of Recent Commentaries", *Ergonomics*, (31), 441-468, London.

WILDE, G. J. S. (1994), *Target Risk - Dealing with the Danger of Death, Disease and Damage in Everyday Decisions*, PDE Publications.

WILDE, G. J. S., KUNKEL, E. (1984), Die begriffliche und empirische Problematik der Risikokompensation: eine Erwiderung auf Dr. R. D HUGUENIN, *Zeitschrift für Verkehrssicherheit*, (2), 52-61, Darmstadt.

ZINK, R. J. (1980), "Arbeitssicherheit als Akzeptanzproblem aus motivationstheoretischer Sicht", *Zbl. Arbeitsmedizin*, (2), 39-48, Darmstadt.

8. RECOMMENDED REFERENCES TO READERS

ÅBERG, L. (1997), "The Role of Perceived Risk of Detection in the Theory of Planned Behavior", in ROTHENGATTER, I., CARBONELL, E. (Eds), *Proceedings of the International Conference on Traffic and Transport Psychology*, Valencia (1997), Amsterdam: Elsevier (in print).

ASCHENBRENNER, M., BIEHL, B., WURM, G. (1988), *Mehr Verkehrssicherheit durch bessere Technik? Felduntersuchungen zur Risikokompensation am Beispiel des Antiblockiersystems (ABS)*, (Unpublished), Mannheim.

COLBOURN, C. J. (1978), "Perceived Risk as a Determinant of Driver Behavior", *Accident Analysis and Prevention*, (2), 131-141.

EVANS, L. (1984), *Human Behavior Feedback and Traffic Safety*, General Motors Research Laboratories, Warren.

FORWARD, S. (1997), "Subjective Norm and Who is the Significant Other?", in ROTHENGATTER, I., CARBONELL, E. (Ed) (1996), *Proceedings of the International Conference on Traffic and Transport Psychology*, Valencia, Amsterdam: Elsevier, (in print).

FULLER, R. (1988), "On Learning to Make Risky Decisions", *Ergonomics*, 31 (4), 519-526.

HAIGHT, F. A. (1986), "Risk - Especially Risk of Traffic Accident", *Accident Analysis and Prevention*, (5), 359-366.

HALE, A. R., STOOP, J., HOMMELS, J. (1990), "Human Error Models as Predictors of Accident Scenarios for Designers in Road Transport Systems", *Ergonomics*, (33), 1377-1388.

HUGUENIN, R. D. (1988), "The Concept of Risk and Behavior Models in Traffic Psychology", *Ergonomics*, 31 (4), 557-569.

JANSSEN, W. H., TENKINK, E. (1987), *Risk Homeostasis Theory and its Critics: Time for an Agreement*, TNO Institute for Perception, Soesterberg.

KLEBELSBERG, VON, D. (1977), Psychologische Erklärungshypothesen für das Verkehrsverhalten, *Informationen und Mitteilungen, BdP, Sektion Verkehrspsychologie*, (9), 3-13, Bonn.

KLEBELSBERG, VON, D. (1982), "Die Bedeutung von subjektiver und objektiver Sicherheit: Fahrerverhalten als Risikoverhalten", in, *Verkehrssicherheit, Vorträge anlässlich des Seminars der Forschungsgruppe Berlin*, 5/6 November 1981, Daimler-Benz AG, Forschung und Entwicklung, Stuttgart.

MCKENNA, F. P. (1985), "Do Safety Measures Really Work? An Examination of Risk Homeostasis Theory", *Ergonomics*, (2), 489-498.

MICHON, J. A. (1985), "A Critical View of Driver Behavior Models: What Do we Know, What Should we Do?", in EVANS, L., SCHWING, R. (Eds.), *Human Behavior and Traffic Safety*, New York.

MICHON, J. A. (1989), "Explanatory Pitfalls and Rule-Based Driver Models", ***Accident Analysis and Prevention***, (21) (4), 341-353.

NÄÄTÄNEN, R., SUMMALA, H. (1976), *Road User Behavior and Traffic Accidents*, Amsterdam: North-Holland Publishing Company.

OECD (1990), *Behavioral Adaptations to Changes in the Road Transport System*, Paris: OECD.

PARKER, D., WEST, R., STRADLING, S., MANSTEAD, A. S. R. (1995), "Behavioral Characteristics and Involvement in Different Types of Traffic Accident", *Accident Analysis and Prevention*, (27) (4), 571-581.

PFAFFEROTT, I., HUGUENIN, R. D. (1991), Adaptation nach Einführung von Sicherheitsmassnahmen - Ergebnisse und Schlussfolgerungen aus einer OECD-Studie, *Zeitschrift für Verkehrssicherheit*, (37), 71-83.

RANNEY, T. A. (1994), "Models of Driving Behavior: A Review of their Evolution", ***Accident Analysis and Prevention***, (26) (6), 733-750.

RASMUSSEN, J. (1987), "The Definition of Human Error and a Taxonomy for Technical System Design", in RASMUSSEN, J., DUNCAN, K., LEPLAT, J. (Eds), *New Technology and Human Error*, Chichester, U.K: Wiley.

REASON, J. (1994), Menschliches Versagen, Psychologische Risikofaktoren und moderne Technologien, Heidelberg/Berlin/Oxford: Spektrum Akademischer Verlag.

ROTHENGATTER, T. (1997), "Psychological Aspects of Road User Behavior", *Applied Psychology: An International Review*, 46 (3), 223-234.

SHAW, L., SICHEL, H. S. (1971), *Accident Proneness: Research in the Occurrence, Causation and Prevention of Road Accidents*, New York: Pergamon Press.

SUMMALA, H. (1985), "Modeling Driver Behavior: A Pessimistic Prediction?", in EVANS, L., SCHWING, R. C. (Eds), *Human Behavior and Traffic Safety*, New York.

SUMMALA, H. (1986), *Risk Control is not Risk Adjustment: The Zero-risk Theory of Driver Behavior and its Implications*, University of Helsinki, Traffic Research unit, Report 11, Helsinki.

VAN DER MOLEN, H. H., BÖTTICHER, A. M. T. (1988), "Risk Models for Traffic Participants: A Concerted Effort for Theoretical Operationalizations", in ROTHENGATTER, J. A., DE BRUIN, R. A. (Eds), *Road Users and Traffic Safety, 61-81*, Assen/Maastricht/Wolfeboro: van Gorcum.

WILDE, G. J. S. (1994), *Target Risk - Dealing with the Danger of Death, Disease and Damage in Everyday Decisions*, PDE Publications.

PART TWO
MIND AND BEHAVIOR IN TRAFFIC

Chapter 4

PERCEPTION

Viola E. Cavallo,
INRETS, Laboratoire de Psychologie de la Conduite, Arcueil, France
Amos S. Cohen,
Universität Zürich, Psychologisches Institut, Zürich, Switzerland

1. INTRODUCTION

Perception can be defined as the processing of information gathered by the various sensory systems. It enables individuals to acquire knowledge of their environment and interact therewith.

Perception is an integral part of most human activities, but it plays a particularly important role in driving, which may be considered as being "predominantly a perceptual task" (Gibson and Crooks, 1938).

Although all sensory modalities work together as a perceptual system (Gibson, 1950, 1966), vision plays the most prominent role in driving. Driving, according to Gibson and Crooks (1938), is locomotion through a terrain by means of a tool (the automobile), with the aim of reaching a destination. The basic activity consists of achieving a path (a trajectory), determined by its speed and direction, so as to respect the spatial constraints (road infrastructure) and to avoid a collision with the obstacles encountered (other vehicles, pedestrians). Driving is therefore guided mainly by vision, which enables the driver to acquire environmental information from a sufficient distance, so as to anticipate the future driving situation and to take regulatory action in time. Emphasis will therefore be placed on visual perception, without under-estimating the contribution of the other sensory modalities (Sivak, 1996).

Perception is a decisive factor of safety when traveling, as it forms a basis for the driver's decisions and actions. In-depth accident studies indeed suggest that perceptual errors are a major contributory factor to accidents: they are estimated to represent between 40% to 50% of the total amount of driver errors (Nagayama, 1978; Hills, 1980). These errors concern failure to detect obstacles as well as incorrect interpretations and misjudgments. Under-estimating the curvature of a bend, over-estimating the distance of the vehicle ahead or over-estimating the time available for overtaking are all examples of characteristic perceptual errors when driving.

The apparent ease of driving and the speed with which the basic skills are acquired often lead to under-estimating the complexity of the perceptual processes involved. Perception is not simply a reflection of the surrounding environment, but a representation resulting from a construction. According to the cognitive approach, which is our main reference, perception is essentially the function of interpreting sensory data, which are incomplete and ambiguous. In this way, perceptual activity is not limited to "bottom-up" processing (also called "data-driven" or "stimulus-driven" processing), that occurs in part automatically and depends largely on the characteristics of the environment and the sensory system; it also includes "top-down" processing guided by representations and concepts ("concept-driven" or "hypothesis-driven" processing), calling on previously acquired knowledge and experience. Both aspects are vital when driving: Because of bottom-up processes the driver is able to perceive unexpected and new events and to react in emergencies, whereas top-down processes enable him to anticipate future events and actively search for the relevant information.

Driver perception is not primarily a contemplative function, but it has to be operational, as it is questioned by the demands of the driving situation and its outputs contribute to behavior control. The driver's perceptual activity therefore also depends on the objectives and the conditions of the driving activity. Top-down processing, in particular, is limited by the temporal constraints of the driving task, which involve only a few seconds, whereas interrogating stored knowledge requires time. On the other hand, this coupling with the action provides feedback information, which can be used to check the accuracy of perception and, if necessary, correct it. This feedback loop enables the driver to learn and to adapt remarkably well to the environment.

Traditionally, road safety research and application work has placed great importance on bottom-up mechanisms. The underlying behaviorist concept is that of a passive driver whose reactions are simply triggered by a stimulus. Seen from this point of view it would suffice to provide the driver with the appropriate information and insure his visual system is operating correctly, to obtain the desired behavior. For a long time, attempts were made to establish a relationship between a driver's visual abilities and his driving behavior (Burg, 1967, 1968). The correlation found between accidents and various ophthalmologic indicators (e.g., static and dynamic acuity, photopic, mesopic and scotopic acuity, color vision, movement vision, etc.) was extremely low, despite the size of the samples. The apparent paradox for drivers under 25, with the best visual acuity and highest accident rate, shows the limits of such a simplistic approach. Today, these tests are considered to be useful to prevent people with significant visual defects from driving, but they do not provide a way of predicting behavior or accident risk for an individual driver.

The main research focus has consequently shifted to analyzing the way drivers pick up and process visual information, in relation to the multiple determinants of the perceptual activity. The volume of work and diversity of approaches prevented us from providing an exhaustive presentation, but led us to choose a couple of themes which enable us to address significant aspects of the driver's perceptual activity.

2. SPATIO-TEMPORAL PERCEPTION

The driving activity includes basic spatio-temporal skills which play an important role, for example, when controlling speed and trajectory, car following, overtaking, negotiating a bend, etc. In sum, all the situations where the driver has to manage the interaction of his vehicle with the road infrastructure and traffic. These activities require elementary processes regarding the perception of space and motion, including the perception of speed, distance and time-to-contact, which will be discussed below.

2.1. Perception of Ego-Speed

A driver's speed, and the way he adapts it to the driving situation constraints is closely linked to traffic safety. To choose an appropriate speed, a driver must be able to correctly evaluate it. For this the tachometer is of limited use, as the driver usually bases his estimations on visual, auditive and vestibular sensations. The importance of vision in this complex system may be illustrated by a well known illusion, i.e., the sensation of movement when a train moves off from an adjacent track: the visual stimulation alone can set off a sensation of self-motion (called "vection"), although there is no actual movement.

The visual basis for this sensation of speed consists of the "optical flow", the importance of which has been emphasized by Gibson (1950). The optical flow results from the movement of an observer in a textured environment and contains information on the speed and direction of movement (its "propio-specific function") together with information on the environmental structure (its "extero-specific function"). Optical flow consists in an apparent movement of the texture elements in the observer's visual field, which can be described by velocity vectors (Figure. 1). The sensation of speed depends on the amplitude of the velocity vectors: the longer the vectors, the greater the sensation of speed. However, this source of information is ambiguous, as the amplitude of the velocity vectors is not determined in a univocal way by the speed of the observer, but also by the distance of the texture elements: the angular velocities of distant elements are less than those of nearer elements. Because of this, we do not feel a sensation of speed in a plane (except when taking off and landing), whereas we do in a car, despite actual speeds being much lower. The sole processing of angular velocities, and the resulting sensation of speed, is therefore not a reliable basis when evaluating speed. Only the integration of distance information can provide a way of weighting the angular velocities of the environmental elements and accurately estimating the actual speed. The level of the eye in relation to an object with known dimensions may be used as a frame of reference (Gibson, 1979).

Figure 1. Representing the optical flow in terms of velocity vectors (Gibson, 1979) for an observer who moves parallel to the ground and is looking in the direction in which he is traveling. The focus of expansion, an indicator of the direction of movement, is on the horizon.

An important discovery concerning the perception of ego-speed was to note that it depends more specifically on the stimulation of the peripheral visual field. Using selective visual field occlusion techniques it has been shown (Brandt, Dichgans and Koenig, 1973) that a stimulation covering 30° of the central field produces practically no sensation of speed. On the contrary, to mask the central field at an angle of 120° reduces the sensation of speed only slightly, when compared to the stimulation of the whole field. These laboratory findings were corroborated by road experiments

(Salvatore, 1968): Passengers in a vehicle were exposed to either a stimulation of 25° of the central field, or a bilateral stimulation of 25° of the peripheral visual field. Speed estimations were more accurate with peripheral vision, whereas speed was underestimated in central vision. These results indicate that the size of the visual field is a factor, which significantly affects speed perception. Restricting the visual field to 40-45°, which can often be seen in driving simulators, raises the issue of the similarity of the processes of speed perception in real and simulated conditions, and therefore the question of whether the simulation is valid.

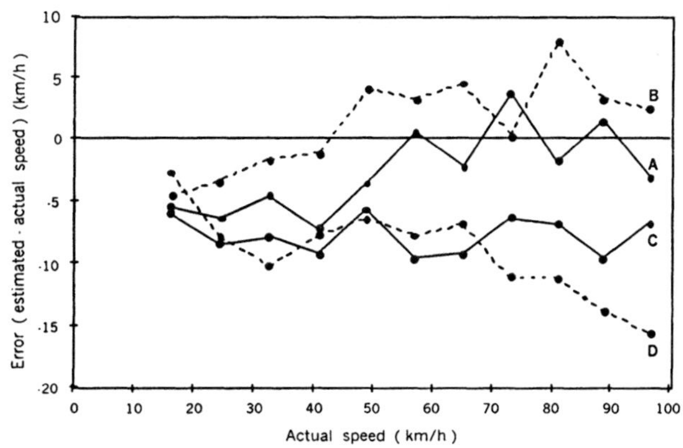

Figure 2. Speed estimation errors as a function of actual speed for car passengers in (A) normal visual conditions, (B) blindfolded, (C) with diminished hearing, (D) with diminished hearing and blindfolded (Evans, 1970).

In actual driving conditions, speed perception is not based on visual information alone, but it interacts with auditive and vestibular information. Verbal speed estimations of passengers in an actual driving situation (Evans, 1970) show significant differences, depending on the contribution of the different sensorial modalities (Figure. 2). In normal conditions when visual, auditive and vestibular information is simultaneously available, low speeds are under-estimated and estimations for average speeds of between 50 and 80 km/h are relatively accurate. However, when the observer is not provided with auditive information, under-estimations are systematic and bear no relation to speed. It would therefore appear that efforts by manufacturers to reduce engine noise is problematic in terms of safety, as it may encourage drivers to drive faster.

Distortions when estimating speed are also produced by the interference of vestibular information when changing speed: speed estimations are more accurate with low acceleration (or deceleration), whereas speeds are underestimated when acceleration is greater (Salvatore, 1968). This result draws the attention to transit situations, where drivers must significantly reduce speed, for example, approaching built-up areas, leaving motorways and entering bends, in which speed is typically under-estimated and the resulting deceleration insufficient.

Speed perception is also influenced by a negative adaptation effect. This comprises a reduction in the amplitude of the sensation of speed during a lengthy stimulation period. To maintain the sensation of a constant speed, vehicle speed must be increased. The higher the initial speed and greater the increase in speed: Irving (1973) noted in an actual driving situation an increase of approximately 5 km/h at 50 km/h, 15 km/h at 75 km/h, and 25 km/h at 110 km/h. In this study speed was stabilized after about one minute. Other experiments have shown a significantly longer adaptation time: Schmidt and Tiffin (1967) consider that after 30 min the curve plateau is far from being reached. However, for the most accurately estimated speeds (50 to 60 km/h), adaptation is low, even for significant exposure times (Schmidt and Tiffin, 1967). Adaptation is observed more particularly on motorways, during a relatively uniform and lengthy stimulation, which leads drivers to imperceptibly increase their speed to maintain a constant sensation of speed. Significant braking difficulties may occur in these conditions, particularly for novice drivers.

To a certain extent, the adaptation effect can be offset by the visual structure of the environment. The variation of the sensation of speed with the environmental structure is easy to observe: Driving along a tree-lined road produces a greater sensation of speed than driving through essentially featureless countryside. This phenomenon, linked to the spatial frequency of the stimulation, was illustrated by a simulator study (Denton, 1980). Denton (1980) projected a succession of patterns comprising either a uniform texture or transversal strips onto a simulated roadway. The spacing between the strips was progressively reduced according to an exponential function, creating an impression of increased speed. Drivers were required to reduce their initial speeds by half when entering the test zone with one of these different visual patterns. Transversal strips produced greater reductions in speed than the uniform texture (Figure. 3).

This type of pattern was introduced on approaches to roundabouts (Denton, 1973, 1980), on motorway exits (Malaterre, 1977) and when approaching high-accident rural curves (Shinar, Rockwell and Malecki, 1980), but the long-term effectiveness of this measure is still open to question. Evaluating the effect of these devices generally shows a reduction in speeds following their installation, but this effect fades with time. Shinar *et al.* (1980) suggest that this type of layout may however be effective for drivers who are unfamiliar with the location.

The illusion of increased speed produced by these markings does not therefore resist perceptive learning (refer to 3.3.), as drivers familiar with the surroundings draw up a new representation, by establishing a more appropriate relationship between the visual stimulation and the actual speed. This example illustrates that perception is indeed more than bottom-up processing, i.e., taking into account sensorial data from the receptors. It also shows the difficulty of providing effective ergonomic measures over long periods.

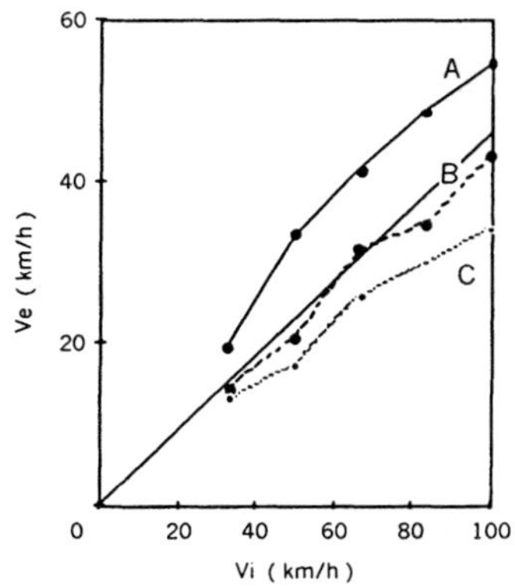

Figure 3. Speed estimated at half the initial speed, in relation to three types of visual patterns (A: uniform texture with no transversal strips; B: transversal strips where inter-strip distance is reduced from 6 m to 4.6 m over a distance of 800 m; C: transversal strips where the distance varies from 6 m to 3 m over 400 m).

2.2. The Perception of Distance

Estimating distance in relation to fixed or mobile obstacles is a factor involved in many driving activities. Combined to speed information, it is required, for example, to adjust headway with the preceding vehicle, to control deceleration or to overtake. Although distance information is of little use on its own, it can be regarded as a basis to be considered in many spatio-temporal estimations (see also 2.3).

Distance perception is based on a great number of distance cues (cf. Schiff, 1980). These include monocular (pictorial) cues, such as texture gradient, linear perspective, aerial perspective, occlusion and the relative size of an object together with its height in the visual field. Binocular vision, due to binocular disparity, enables stereopsis and a mobile observer also has motion parallax. Only oculomotor sources (convergence, accommodation) are of little value in a driving situation, as they operate only over small distances (up to 3-5 m). It should be emphasized that among these cues, the familiar size of an object is the only cue that provides information regarding absolute distance: knowledge of the actual size of an object (e.g., passenger car or truck) can be used to estimate one's distance by using the object's angular size.

In normal driving conditions, most of these cues are available simultaneously and may have a cumulative effect. However, their respective weight in a given situation, and the way they interact are not known with any degree of accuracy.

In view of the significant redundancy of the distance cues available to the driver in a large range of driving conditions, it can be estimated that distance perception does not raise major problems for traffic safety. This may perhaps explain why there is so little work dealing with this issue available.

The distance perception mechanism only seems to fail when the depth cues are dramatically reduced, as in thick fog. In thick fog, only the rear lights of the preceding vehicle can be seen, and distance perception is based on their visible characteristics, such as luminance, the angular size of lights and the angular distance between lights. All other distance indicators cease to exist or are considerably diminished. In addition to this reduction of distance cues, fog also modifies the actual characteristics of the depth cue of aerial perspective. Aerial perspective results from the fact that distant objects are seen through more atmosphere than nearer objects. Light is scattered by the particles in the atmosphere, resulting in a reduction in contrasts, the occurrence of blur and a change of color with increased distance. In fog or mist, as in the case of atmospheric pollution, these effects are accentuated and it can be assumed that objects in fog are perceived to be further away than in clearer conditions.

It has indeed been shown that objects of an unknown size seem twice as far away in fog than in clear weather (Ross, 1975). The distance of cars, which are seen by their lights only, is overestimated by about 60% (Cavallo, Doré, Colomb, and Legoueix, 1997, Figure 4). The lesser overestimation in the case of cars is probably due to the presence of the familiar size cue. It can be assumed that the overestimation of distance contributes to the modification of car-following behavior insofar as it is likely to reduce temporal headways and thus increase the danger of collision. The misperception of distance in fog is particularly fallacious in that the driver is unaware of it. It is therefore difficult to counteract.

It has however been shown that the characteristics of the rear fog light configuration have a decisive effect on distance perception (Cavallo, Doré, Colomb, and Legoueix, 1997). The perception of vehicle distance in fog can be improved by the presence of two fog lights, as the overestimation of distance is reduced by 35% and estimates are more consistent than with only one fog light. Furthermore, distance overestimation is significantly reduced when the width between fog lights is as great as possible. This demonstrates that distance perception in fog can be improved by relatively simple ergonomic means that take into account the perceptual mechanisms of the driver.

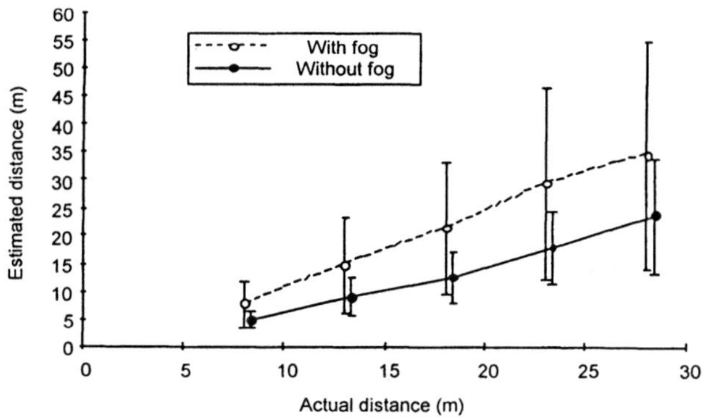

Figure 4. Estimated distance of vehicles in fog and in normal visibility conditions.

2.3. The Perception of Time-to-Contact

A fundamental aspect of driving is predicting critical events in the next few seconds. A driver on a collision course, for example, has to accurately evaluate how close he is to the impending collision in order to brake or swerve at the right time. In other words he has to judge time-to-contact (TTC), which refers to the time that remains before reaching an obstacle, and thus to the time available for taking action. TTC is considered to be a crucial parameter in controlling avoidance behavior.

TTC is also likely to be involved in more complex judgment tasks such as overtaking or left-turn maneuvers, where the driver has to determine whether there is enough time for the planned action. The underlying concept of temporal action control is being used increasingly often in the study of driving behavior, which addresses the anticipatory aspects of actions, such as braking, trajectory control, car following, traffic merging decisions, curve taking, stop-or-go decisions at intersections, and so on. Various equivalent terms have been employed, depending on the situation under investigation, including "time-to-collision", "time-to-arrival", and "time-to-go".

A considerable amount of work has been devoted to determining the factors that influence the driver's temporal predictions and, in particular, identifying the visual information required to correctly perceive TTC. This work frequently uses TTC estimation tasks (e.g., showing an observer a situation where he is approaching an obstacle; the stimulation is stopped some seconds before impact and the observer is asked to indicate the moment when he expects the collision to take place), or studies TTC at a specific time in the action sequence (e.g., when starting to brake).

Two main hypotheses have been formulated as to the nature of visual information used to perceive TTC. According to the first of these, which is in line with the ecological optics approach (Gibson, 1950, 1966, 1979; Lee, 1976) TTC might be based on specific characteristics of the optical flow (see Figure 1). For example, in direct collision situations between a driver and an obstacle (e.g., when catching up the preceding vehicle) TTC is visually specified by the expansion of the retinal image of the obstacle. Lee has shown mathematically that TTC can be obtained by the angular extent of the obstacle image (θ) and the instantaneous rate of expansion of that visual angle (θ'): $TTC=\theta/\theta'$). The optic variable defined by this ratio has been termed "local tau" (Tresilian, 1991). The advantage of using such a tau-type variable is argued to be the simplicity and economy of the method: TTC may be derived purely from two-dimensional optical variables, which are directly available on the retina of the driver, without the need for three-dimensional parameters such as speed, obstacle size and distance. Studies have shown that TTC can, in fact, be estimated solely on the basis of tau information when spatial information is lacking (Todd, 1981; Kaiser and Mowafy, 1993).

Another potentially usable means of obtaining TTC information is based on spatial variables such as speed (s) and distance (d): $TTC=d/s$. The use of this method assumes that both distance and speed information are integrated to obtain TTC. For this reason, this method is referred to as a cognitive or computational strategy. Several studies suggest that information regarding ego-speed and distance to the obstacle are taken into account when available, which is the case in natural driving situations (Cavallo and Laurent, 1988; Groeger and Brown, 1988; Cavallo, Mestre, and Berthelon, 1997; Kappé and Korteling, in print). Moreover, it has been shown that distance information provided by the relative size of the obstacle is used in passive judgment tasks (DeLucia, 1991; Caird and Hancock, 1994) as well as in active collision avoidance tasks (DeLucia and Warren, 1994). Relative size information even overrides local tau information under certain conditions, in particular in poor environments (Kappé and Korteling, in print), and sometimes induces significant judgment errors. For instance, it has been shown that for small-sized objects (motorcycles, pedestrians) the impending collision is perceived as occurring later than for large objects (Caird and Hancock, 1994; Stewart, Cudworth and Lishman, 1993). This means that the time available for action is overestimated when the obstacle is small.

Although the visual origin of TTC remains a much debated theoretical question, it can be assumed that several sources of information are involved in TTC perception. Their respective weight and combination rules are likely to depend on the task to be accomplished and still have to be determined with accuracy.

One of the persistent characteristics of TTC estimation is a systematic underestimation of about 20 to 30% (Figure 5). This underestimation bias is likely to reflect a kind of "safety margin" due to a "built-in or learned tendency to err in the safe direction" that allows drivers more time to avoid unwanted contacts in potentially dangerous situations (Schiff and Oldak, 1990). This interpretation is consistent with the observed effects of various internal and external factors on temporal estimates.

Regarding the internal factors, it has been shown that women give more conservative judgments than men (McLeod and Ross, 1983; Caird and Hancock, 1994). A higher safety margin is also noted for novice compared to experienced drivers (Cavallo and Laurent, 1988) and for field-dependent drivers compared to field-independent drivers (Cavallo, Berthelon, Mestre, and Pottier, 1998).

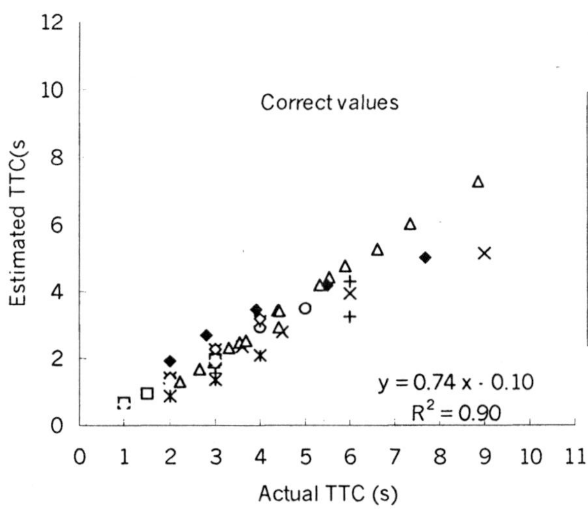

Figure 5. Estimated time-to-collision in ego-motion experiments.

One of the external factors influencing TTC perception is speed. This applies in situations of both ego-motion (McLeod and Ross, 1983; Cavallo and Laurent, 1988; Kappé and Korteling, in print) and object-motion (Schiff and Detwiler, 1979; Schiff and Oldak, 1990; Caird and Hancock, 1994): higher speeds generally produce higher estimates, indicating that the collision is perceived as further away in time. In this case the driver is likely to overestimate his available time. Evidence provided by a study on left-turn maneuvers (Hancock, Caird, Shekhar, and Vercruyssen, 1991) indeed showed shorter accepted gaps and a higher collision frequency at greater speeds. A second external factor influencing TTC estimation is the structure of the environment. Results of studies have indicated that safety margins are reduced in impoverished environments (Cavallo, Mestre, and Berthelon, 1997; Kappé and Korteling, 1998). However, the positive effect of a rich environment is observed only in ego-motion situations, due to the generation of global optical flow allowing for the perception of ego-speed. The environmental structure bears no influence on object-motion situations where only local optical flow is generated. Finally, the size of the vehicle-obstacle is a third factor influencing temporal estimates. Safety margins are lower with smaller than with larger obstacles (DeLucia, 1991; Caird and Hancock, 1994). Accepted gaps at left-turns are also lower for smaller than for larger vehicles.

These findings draw attention to typical situations where TTC may be misperceived: at high speeds, in impoverished visual conditions (driving at night, in fog, in tunnels, in simulators, etc.) and in the case of smaller obstacles (motorcycles, pedestrians). In sum, this involves situations where it is difficult to estimate speed and distance and where tau variables cannot be used. Traditional ergonomic measures designed to improve vehicle conspicuity (third brake light, fog lights, increased light intensities, etc.), are not therefore sufficient in these situations, where the main objective should be to improve conditions for speed and distance perception.

3. SELECTIVE ATTENTION

The reliability of steering operations may be related to driver perception regarding, for example, his own locomotion, the movement of other objects and target recognition. When perceiving the world, it is usual to consider more than a single dimension of the stimulation and even integrate input from different modalities (e.g., Bubb, 1977), including proprioception. The perception of visual scenery controls a driver's behavior. Consequently, there is a close link between a perspective view of the road and the driver's steering behavior (Otten and Schroiff, 1988). When a perceptual error occurs, as in the case of illusive curves (Shinar, 1977), driver action is inappropriate. Similarly, if approaching a junction produces an impression of right of way contrary to the traffic regulations in force, then accidents have to be expected (Leutzbach, 1987, Figure. 6).

The possibility of misjudgment easily can be reduced. This is why speed perception should be supplemented in relation to object recognition, as reflected in a driver's information input. This is a precondition when estimating traffic environment semantics. The issue to be discussed is whether a driver can recognize all the targets in the environment or, inversely, whether there are any limitations to information input.

Visual orientation is facilitated by information input through the retina as a whole. This can, nevertheless, be divided into the foveal and parafoveal fields. This differentiation should only indicate that these two regions satisfy different functions to varying degrees (Table 1). They must, however, be considered as integral parts of the same system. In general, parafoveal vision monitors the environment, which includes locomotion, whereas foveal vision is better suited to target recognition.

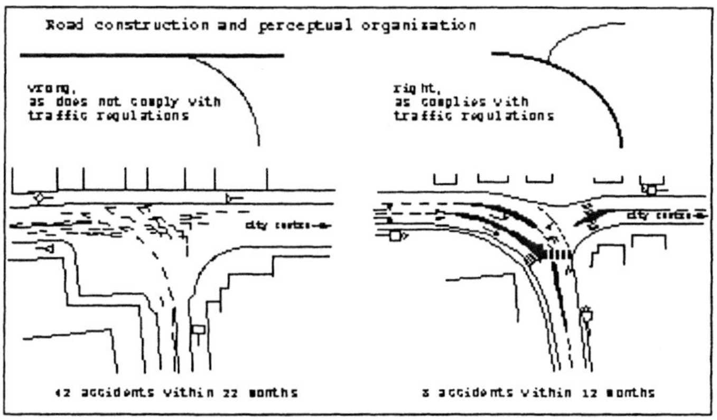

Figure 6. Frequency of road accidents at a junction. Their probability is significantly higher when the Gestalt-Psychology-Principle of continuation (above) is disregarded, i.e., the driver's perceptual organization does not comply with traffic regulations (left). The accident rate decreases when the perceptual organization complies with traffic regulations (right) (Leutzbach, 1987).

The two retinal regions function together. Extrafoveal information even controls the process of selecting a single target for the next fixation from all the available targets. This must be ascertained by parafoveal vision before initiating a goal-oriented eye movement, as search patterns are not distributed at random unless the observer has no advance information about the target (Haber and Hershenson, 1973). In conditions of exogenous cueing, attention moves rapidly, before the eye, to the cued location detected by parafoveal vision (Stelmach, Campsall, and Herdman, 1997). This pre-attentive process is of crucial importance when traveling at relatively high speeds, i.e., when it is vital for the driver to select the required input from all the objects available at the same time. The recognition of only the most relevant objects from the excessively high amount of information available is essential, as the driver has a relatively limited capacity for input and information processing. Distinguishing between redundancy and information and subsequently choosing the most important input at any given time is a somewhat difficult task which involves divided attention, as revealed in the way the driver's eye behaves.

Table 1. Comparison between foveal and parafoveal vision in relation to their functions.

	retinal region	
demand	foveal field	parafoveal field
detailed input	appropriate	inadequate
rapid input	appropriate	inadequate
visual attention	usually corresponds to fixation	possible for limited period, unfavorable
monitoring the environment	inadequate due to limited viewing angle	adequate
form identification	appropriate	inadequate
own speed of travel	not specialized	specialized
movement of targets in environment	slow speeds (<15°/s)	high speeds
color vision		
photopic conditions	appropriate	appropriate - within limited range
scotopic conditions	achromatic: cones inactive	achromatic: cones inactive

3.1. Eye Movement Behavior

Good eyesight, as defined by legislation, is a pre-requisite for driving a car. How a driver uses his eyes, as indicated by his oculomotor activity, is of even greater importance in terms of object recognition. The fovea is the luxury region of the retina for photopic vision. The arguments in favor of foveal vision state that fixation targets provide not only accurate input but also reflect the driver's focus of visual attention in conditions of exogenous cueing (Festinger, 1971; Schneider and Deubel, 1995; Stelmach *et al.*, 1997). Moreover, the time required to pick up visual information must be as short as possible when fixing an object, and gradually increase with the eccentricity of the retinal image (Webb, 1977). Foveal input is over-represented in the cortex, as each foveal cone is connected to the visual cortex through a single fiber of the optical nerve, in contrast to the parafoveal region. In the distant periphery, some dozens or even hundreds of photoreceptors are connected to a single fiber. This architecture favors the processing of foveal input in terms of object recognition. In sum, the probability of object recognition is highest when it is fixated. It decreases with increased eccentricity towards the periphery, depending on stimulus characteristics (Engel, 1974, 1977), the subject's prior knowledge and expectancy (Ikeda and Takeuchi, 1975), his actual work-load (Leibowitz, 1973) or road characteristics, as well as driver experience (Cohen, 1984). Consequently, eye movement research uses oculomotor activity as a dependent variable as well as a technique for studying a driver's perceptual capability. His gaze is not distributed at random, unless he is driving along redundant routes (e.g., McDowell and Rockwell, 1978). On the other hand, when driving a car in conditions involving a substantial work-load, the driver selects his input (1) in relation to the relevance of the object to his actual task requirements, (2) depending on the relevance of the adverse object in his field of vision and (3) according to prior input. This relationship can be accurately described in the form of time discrete process models (Cohen and Hirsig, 1991). The temporal and spatial limitations of information input, which can be indicated as peripheral capacity, requires highly selective attention, divided between the most important objects and events. This does not refer to the ability of the brain to process information, but simply to the limits of its input, as illustrated by the selective eye fixations.

3.1.1. Temporal Limits of Peripheral Capacity

Traffic regulations require a driver to perceive the traffic constellation without interruption. Can he recognize every object in his environment, as required and indicated by his gaze? There are some temporal limits to central information processing. This will not be dealt with in this paper. Fixation times last on average about 1/3 s in daytime conditions (Figure. 7). This value is relatively constant. It is not dependent upon either road characteristics (Cohen, 1987) or traveling speed (McDowell and Rockwell, 1978). However, in night-time conditions and when a driver is tired or under the influence of alcohol, this is extended to approximately 0.5 s. Generally, there is significant inter-and intra-individual variability. Nevertheless, if day-time conditions are considered as an optimistic frame of reference to estimate the upper limit of the peripheral capacity, then it is highly probable that a driver might recognize three objects per second due to focused attention, as shown by eye fixations. This frequency of three fixations per second represents the average input rate of discrete "packages of information" (Gaarder, 1975). A specific time is presumably needed to extract the required information. If the fixation time is too short, then object recognition is impeded (Mori and Abdel-Halim, 1981). If the fixation time is too long, then the driver stares. This should be avoided. However, the distribution of fixation times represents an unimodal curve. The observed durations do not allow for a distinction between the types of fixation, which have long been categorized by Purkinje (1825) as being devoted to exploring, processing and staring according to their respective duration.

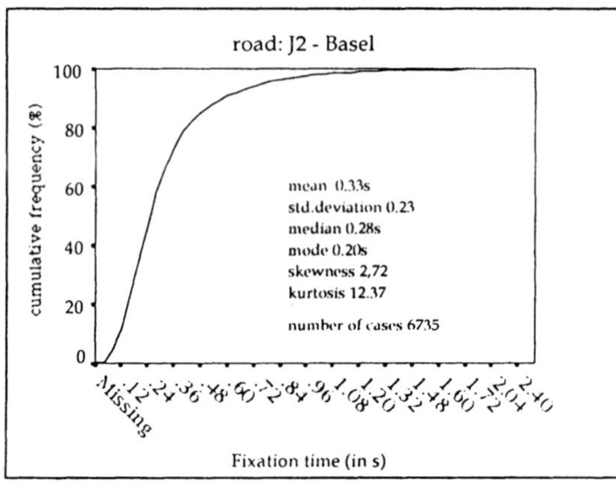

Figure 7. Cumulative frequency of the fixation times observed on a rural road (Cohen, 1996).

The driver's peripheral capacity for information input represents his temporal limitation and refers to the greatest number of objects he can clearly recognize during a time interval, and which can influence his steering behavior. As to whether the average frequency of three fixations per second is sufficient to recognize the most relevant objects, essential for accurate decision-making, also depends on their density along the road in relation to traveling speed. If a driver passes on the average over more than three relevant targets per second but can deal with only three of them, then it is highly likely he will overlook the remaining targets. The consequence is a deficit in the information required for perfectly adaptive behavior. To avoid such situations, further research is needed to accurately estimate the environmental information load in order to adjust traffic conditions in relation to human resources. The driver's temporal capacity for inputting information seems to be far too limited to perceive his whole environment, at least in urban traffic. The consequences are firstly that drivers must carefully learn to use their capacity as efficiently as possible. Secondly, excessive information must be avoided when designing the environment. Finally, the assumption that drivers can perceive an entire traffic constellation is too optimistic a view by far. The consequences of considering a driver's spatial limitations in identifying the presence of an object in his environment are indicated below.

3.1.2. Spatial Limits of Peripheral Capacity

Foveal and parafoveal vision both contribute to an individual's visual orientation in terms of a well-coordinated system. Peripheral vision monitors mainly the environment. When an element attracts a driver's attention and contains information to be objectively, or in the opinion of the observer, of use when perceiving his environment, he is likely to gaze at this element. The information already acquired determines what he will pick up next (Neisser, 1976) in relation to (1) the actual goal of observation (Yarbus, 1967), (2) the relevance of available information with regard to his actual task and (3) the driver's individual characteristics (Cohen and Hirsig, 1991). The next fixation target must then be detected in advance, using parafoveal vision, to program the ballistic eye movement (Carpenter, 1977).

The angular range around the current fixation point, where an object can be detected as a possible target for the next fixation, will be indicated as the spatial limit of the peripheral capacity or as the "useful field of vision." Its actual size depends on a variety of environmental factors, stimuli variables and driver's status, such as his actual workload. If it exceeds his capacity to process the information, then the size of the useful field of vision decreases when driving (Cohen, 1984) and, in extremely unfavorable conditions, reaches only the extension of the fovea (Mackworth, 1976).

The variety of factors regulating the size of the useful field of vision allows for only an approximate estimation of the spatial limitations for target detection in the parafoveal field of vision. An estimation can be derived from the actual size of the eye movements, indicated in amplitude. Figure. 8 shows the cumulative distribution of the amplitudes observed while driving on a rural roads, which total on average approximately 4.2 ° (median approximately 2.9 °). In 90 % of the cases amplitude totals approximately 8.0 ° or less. These data simply indicate that most of the targets fixated are within a relatively limited range around the current fixation. Complementary findings indicate that increasing the eccentricity of the target decreases the probability of it being detected and that, if detected, reaction time increases accordingly (Cohen, 1984). The consequence of the gradient of reduced visual capability towards the periphery is that the driver is unable to detect the targets in the far periphery of his field of vision, except within a limited range around his current fixation point. The reason for this is related to central as well as peripheral factors (e.g., cortical representation and the density of photoreceptors). This view contradicts the assumption that a driver can perceive his environment as a whole. Inversely, it can be assumed that only part of the relevant information is considered when making decisions.

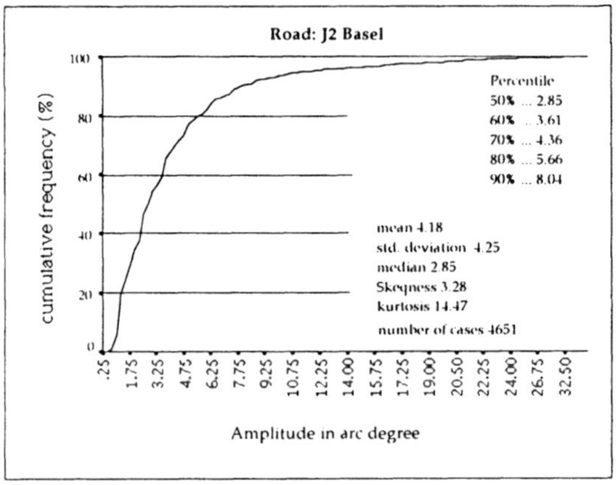

Figure 8. Cumulative frequency of the amplitudes observed on a rural road (Cohen, 1996).

3.2. Traffic Relevant Information in Curves

The perceptual process represents an interaction between "bottom-up" and "top-down" information processing facilitated by a mechanism linking between the both, as Gregory (1980) describes in his model or Neisser (1976) in his schema theory. Even thought the outcome of the perceptual process depends on a variety of variables it represents a reconstruction of the environmental conditions according to the input.

The temporal as well as the spatial limitations of information input are the reason for requiring a highly selective process of attentional shifts during the course of time, always toward the most important object or occurrence in the same moment, as manifested by the fixation of the eye, respectively in their succession. Each "package of information" must enrich the knowledge about the environment in a goal-oriented manner, i.e., facilitating accurate decision-making.

The general requirement of highly selective divided attention cannot be expressed in the form of simple rules, prescribing the essential targets of fixation, because of the heterogeneity of the traffic conditions, which are, furthermore, unpredictable. Beyond this, the information a driver needs in a certain moment depends on several factors, like his prior input, which is still stored, the information available or his individual characteristics. Generally it can be stated, that the driver always should prefer an object for the next fixation which contains a rather high amount of relevant information facilitating to extend his knowledge about the environment in a goal oriented manner.

Another approach, a more specific one, is derived from the task analysis and regards just free driving conditions. When considering only the subtasks of the vehicle's guidance and its control while traveling around sharp curves, the spatial locations containing relevant information can be identified (Figure. 9). Guidance information is located in relatively long distances, meaning around the road's vanishing point. The driver must fixate his eye in this area for inputting information required for planning the adequate speed or path of driving in relation to the environmental condition and subsequently to set up appropriate feed-forward motor programs for the next few seconds. These programs should always be understood only as a modifiable basis for any alternation, if they are necessary. Therefore, the driver has to monitor his environment continuously by parafoveal vision for detecting any alternation, if it occurs, and then shift his attention toward it for estimating as to whether any consequence is required.

Simultaneously the driver has also to deal with the vehicle's actual movement parameters. Speed of traveling is perceived due to parafoveal vision whereas object recognition or controlling the lane position is favorably facilitated when fixating either the right or the left lane marking in rather short distances. Because the driver has to fulfill all the subtasks simultaneously but can input the required "packages of information" only in succession, the minimum requirement is to divide his attention between the objects containing the respective information and to store it for a while. In other words, the driver has to fixate ideally the road's vanishing point for inputting guidance information and then the lane marking in his surrounding for inputting control information, but without prescribing how to fetch the input.

When fulfilling the attentional shifts between input for guidance and control, then the required oculomotor activity must differ in a right handed as compared to a left handed curve, because of the unequal spatial relationships. The road's vanishing point is located in a right-handed curve to the right of the actual heading angle whereas it is located to the left in a left-handed curve (Figure. 9). When shifting attention from guidance to control information and vice versa, then the driver's eye can be moved with smaller amplitudes in a right handed curve, as compared to a left handed curve. Experienced drivers do shift their attention in the described manner.

Newcomers, on the other hand, do not adapt their visual search to the curve's direction (Cohen and Studach, 1977; Cohen, 1985).

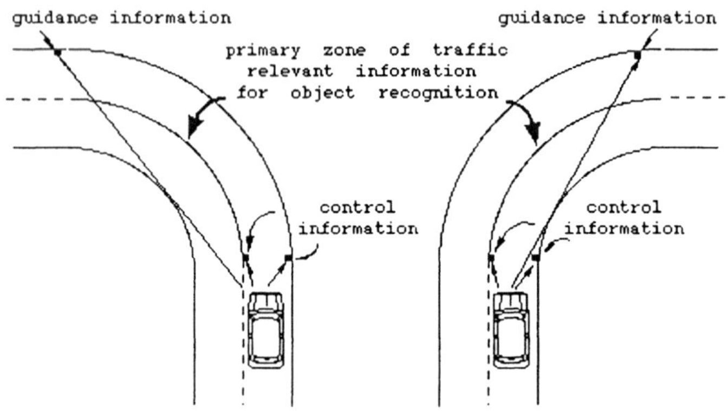

Figure 9. Schematic representation of environmental locations containing information for the vehicle's guidance and control (e.g., lane position) in relation to the curve's direction.

3.3. Perceptual Learning

The temporal and spatial restrictions of inputting information indicate that a driver picks up a very limited part of the targets available, but nevertheless has the impression that he perceives his entire environment. This discrepancy between input and perception indicates that he sees more than his sensory data includes. Neisser (1976) explains this phenomenon in terms of a schema theory as being due to perceptual enrichment, which results from the assimilation of two kinds of information. These are (1) recalled stored information and (2) actual sensory input, which are integrated using an interactive information processing mechanism. Gregory (1989) describes this process in relation to artificial intelligence as an algorithm, controlling the integration of sensory data and perceptual knowledge, which is derived from the individual's conceptual knowledge.

Assuming that conceptual and perceptual knowledge regulate information processing, then it must be expected that driving experience, which corresponds to perceptual learning, must influence the information a driver needs and is seeking. This hypothesis is supported by experimental findings. Drivers with rather limited experience (1) frequently fixate their eyes on targets in nearer distances, probably because they are more likely to be occupied with the vehicle's lane position than with planning their future path and traveling speed. Inexperienced drivers (2) obviously consider the environmental conditions less carefully than their experienced counterparts, e.g., in curves. Once again, inexperienced drivers are more frequently occupied with the car's lane position, while more mature drivers input more guidance information from greater temporal distances and divide their attention better than the novice motorists. Moreover, (3) drivers with insufficient experience move their eyes with greater amplitude but (4) fixate, in general, on less informative targets. They also (5) rarely anticipate the difficulties ahead and are consequently suddenly confronted with a greater workload, for example, when entering a sharp curve. The role of perceptual learning is of increased importance when steering a car in complex traffic conditions, i.e., under conditions involving significant time restraints. This is not so much the case when traveling along a less complex straight road (Cohen and Studach, 1977; Shinar, 1977; Cohen, 1985, 1987). The driver's visual orientation is obviously regulated by a long-term process of perceptual learning. This period might supposedly be reduced by adequate training, like any sensory-motor activity, which is result of proper training. Appropriate methods must still be developed (Lamszus, 1998) in accordance with the knowledge already accumulated (Cohen, 1998).

Modifying a driver's visual search pattern reflects a cortical process governing the movements of the eye, which illustrates the highly selective mechanism of input required for target recognition (Yarbus, 1967). Perceptual learning might increase the efficiency of stimuli discrimination, meaning the differentiation between information and redundancy. Bruner (1964) suggests that learning also facilitates the development of a set of category codes for stimuli integration in the form of relationships, redundancy and spatial relationships. Yarbus (1967) even suggests a close relationship between gaze behavior and thinking. It can therefore be supposed that while acquiring experience, the driver improves in task performance, determining the information needed, etc. and is therefore able to input the information more selectively and goal-orient the information required. In this way a fixation pattern simply reflects a peripheral manifestation of cortical processes. If so, then the evaluation of information must also be revealed in the driver's motor behavior.

This issue was investigated by Cavallo, Brun-Dei, Laya, and Neboit (1988) using the occlusion method when driving through curves. The results showed that both beginners and experienced drivers were able to estimate with precision the right moment to change direction at the entrance to curves, with or without occlusion, but only experienced drivers were able to produce the correct steering wheel rotation. On the other hand, on leaving the curve, visual contact was essential to decide the right moment for realignment. In general, the modes of regulation were dependant upon perceptual learning. Novice drivers used a more retroactive regulation while experienced drivers used a more proactive one. This suggestion was backed up by the drivers' visual input. Novice drivers do not change their visual search in a curve-approaching zone, when compared to a straight road. On entering a curve they pick up information from relatively short distances, obviously to control the lane position of the vehicle. Mature drivers, however, increase their control of lane regulation when approaching a curve, when compared to a straight road. When entering a curve, they can fixate their eyes on greater distances, without disregarding their lane, whereas novice drivers consider essentially the lane position of the car (Cohen, 1985). The combination of these two rather different experiments suggests a close relationship between a driver's input and his behavior. They depend on his abilities, which become more elaborate as perceptual learning progresses.

4. CONCLUSIONS

Perception, especially visual perception, is very closely related to traffic safety. It influences the way drivers adapt to the actual circumstances or, in the case of inadequate adaptation, their steering failures. Whether a driver is able to manage a crucial situation under time pressure depends on two factors. The first concerns the vehicle's movement parameters, particularly its traveling speed. The second refers to object recognition in the driver's environment, which cannot be shorter than his stopping distance in terms of longitude. As the driver cannot recognize every object, an efficient distribution of his controlled attention is of vital importance, as indicated by his eye movement behavior.

A driver's peripheral capacity restricts the number of objects or events he can deal with by subsequently fixating them within a given time period. On the other hand, there is spatial limitation regarding the current fixation as a potential target for a subsequent fixation, the size of which depends on a variety of factors. The relevant findings explain why a significant proportion of accidents can be attributed to delayed or impeded hazard finding. The driver can recognize only part, but not all, of the information available. He nevertheless has the impression he has recognized everything. This illusion is the result of cognitive enrichment, which refers to the synthesis of current input and recalled information. As a result, there is no immediate feedback signal for behavioral adjustment to information density. A perceptual error, like the phenomenon of illusive curves, defined as one which is physically sharper or more dangerous than it is perceived by the approaching driver (Shinar, 1977), can be detected just with a temporal delay. Furthermore, adaptation to speed of traveling causes the underestimation of the actual velocity. This deviation can be detected only when the driver considers his tachometer. The perceptual limitations suggest that own locomotion as well as the environment can hardly be perceived completely or with perfect accuracy. A new approach is therefore required. A driver's abilities and limitations should be considered when building road networks.

An exact knowledge of the environmental conditions is essential for perfectly adjusting the vehicle's movement to the actual conditions as well as for the anticipated circumstances. However, this essential precondition can hardly be fulfilled, if at all, because of the driver's limited capacity for inputting the exhaustively great amount of information available entirely. The consequence is that decision-making is governed just by the selected input, in contrast to a perfect representation of the external conditions. The existing deficiency of input remains unrecognized, as a consequence of cognitive enrichment. Therefore, perception must be understood as an assimilation of "bottom-up" up sensory data and "top-down" knowledge. The compound of these two kinds of information represents the driver's basis for generating his adaptive behavior, which includes three kinds of possible failures. First, the driver can make a decision without considering an obstacle at the right time or at all, just because of his limited capacity for input as well as information processing. Secondly, he might make a decision according to "top-down" information, i.e., with regard to his perceptual knowledge. Thirdly, the result of the perceptual process might deviate from environmental characteristics, like under-estimating a curve's bent or the own speed of traveling, or over-estimating the distance to oncoming traffic. Each of these perceptual failures impedes the driver's adaptive behavior as a single factor and especially in their combination. On the other hand, no direct sensory feed back loop exists for signaling a deviation between the subjective perceptual errors described and the actual traffic conditions. As a consequence handling errors might occur to a certain degree. Because the driver has no signal for behavioral adjustment, he needs some spatio-temporal space for being able to compensate handling errors, when recognized. Therefore, road construction and traffic regulation must consider the driver's need for a tolerance required for delayed adjustments.

Finally, visual perception is a crucial factor, which modulates traffic safety. The visual modality provides the input from sufficiently great distances for enabling one to drive a car at rather high speeds. Nevertheless, the driver's capacity to input the available information and to process it is limited. If perceptual errors occur, lead to inappropriate steering behavior combined with unfavorable environmental conditions, then hazards and in extreme cases even accidents might be expected. For avoiding the commencement of collision there is a need to take the driver's perceptual possibilities and limitations for avoiding risky maneuvers in advance. We need drivers, who are able to perceive their environment within the framework of human limitation, which can be increased due to perceptual learning, as the differences between beginners and mature drivers indicate. On the other hand, we need a network of roads, which takes the driver's possibilities into account.

5. REFERENCES

BRANDT, T., DICHGANS, J., KÖNIG, E. (1973), "Differential Effects of Central versus Peripheral Vision on Egocentric and Exocentric Motion Perception", *Experimental Brain Research,* 16, 476-491.

BRUNER, J. (1964), "The Course of Cognitive Growth", *American Psychologist,* 19, 1-15.

BUBB, H. (1977), "Analyse der Geschwindigkeitswahrnehmung im Kraftfahrzeug", *Zeitschrift für Arbeitswissenschaft,* 31, 103-112.

BURG, A. (1967), *The Relationship between Vision Test Scores and Driving Record: General Findings,* Report 67-24, Department of Engineering, University of California, Los Angeles.

BURG, A. (1968), *Vision Test Scores and Driving Record: Additional Findings,* Report 68-27, Department of Engineering, University of California, Los Angeles.

CAIRD, J. K., HANCOCK, P. A. (1994), "The Perception of Arrival Time for Different Oncoming Vehicles at an Intersection", *Ecological Psychology,* 6, 83-109.

CARPENTER, R. H. S. (1977), *Movements of the Eye,* London: Pion.

CAVALLO, V., LAURENT, M. (1988), "Visual Information and Skill Level in Time-to-Collision Estimation", *Perception ,* 17, 623-632.

CAVALLO, V., MESTRE, D., BERTHELON, C. (1997), "Time-to-Collision Judgments: Visual and Spatio-Temporal Factors", in ROTHENGATTER, J. A., CARBONELL VAYÁ, E. J. (Eds.), *Traffic and Transport Psychology: Theory and Application,* Amsterdam: Pergamon.

CAVALLO, V., BERTHELON, C., MESTRE, D., POTTIER, A. (1998), "Visual Information and Perceptual Style in Time-to-Collision Estimation", in GALE, A. G. *et al.* (Eds.), *Vision in Vehicles VI,* Amsterdam: Elsevier.

CAVALLO, V., BRUN-DEI, M., LAYA, O., NEBOIT, M. (1988), "Perception and Anticipation in Negotiating Curves: The Role of Driving Experience", in GALE , A. G. *et al.* (Eds.), *Vision in Vehicles II,* Amsterdam: Elsevier.

CAVALLO, V., DORÉ, J., COLOMB, M., LEGOUEIX, G. (1997), "Probleme der Wahrnehmung der Entfernung von Fahrzeugen im Nebel", in SCHLAG, B. (Ed.), *Fortschritte der Verkehrspsychologie,* Bonn: Deutscher Psychologen-Verlag.

COHEN, A. S. (1984), *Einflussgrössen auf das nutzbare Sehfeld,* Forschungsbericht N° 100, Bergisch Gladbach: Bundesanstalt für Strassenwesen.

COHEN, A. S. (1985), "Visuelle Informationsaufnahme während der Fahrzeuglenkung in Abhängigkeit der Umwelt und der Fahrpraxis", *Schweizerische Zeitschrift für Psychologie und ihre Anwendungen,* 44, 249-288.

COHEN, A. S. (1987), *Blickverhalten und Informationsaufnahme von Kraftfahrern,* Forschungsbericht N° 168, Bergisch Gladbach: Bundesanstalt für Strassenwesen.

COHEN, A. S. (1996), Lärmschutzwände und Sicherheitsverhalten, Unveröffentlichter Forschungsbericht, Zürich: Psychologisches Institut der Universität Zürich.

COHEN, A. S. (1998), *Visuelle Orientierung im Strassenverkehr: Eine empirische Untersuchung zur Theorie des visuellen Abtastens,* Bern: bfu.

COHEN, A. S., HIRSIG, R. (1991), "The Role of Foveal Vision in the Process of Information Input", in GALE , A. G. *et al.* (Eds.), *Vision in Vehicles III,* Amsterdam: Elsevier.

DELUCIA, P. R., (1991), "Pictorial Depth Cues and Motion-Based Information for Depth Perception", *Journal of Experimental Psychology: Human Perception and Performance*, 17, 738-748.

DELUCIA, P. R., WARREN, R. (1994), "Pictorial and Motion-Based Depth Information during Active Control of Self-Motion: Size-Arrival Effects on Collision Avoidance", *Journal of Experimental Psychology: Human Perception and Performance*, 20, 783-798.

DENTON, G. G. (1971), *The Influence of Visual Pattern on Perceived Speed*, Report LR 409, Crowthorne, England: Department of the Environment, TRRL.

DENTON, G. G. (1980), "The Influence of Visual Pattern on Perceived Speed", *Perception,* 9, 393-402.

ENGEL, F. L. (1974), "Visual Conspicuity and Selective Background Interference in Eccentric Vision", *Vision Research*, 14, 459-471.

ENGEL, F. L. (1977), "Visual Conspicuity, Visual Search and Fixation Tendencies of the Eye", *Vision Research*, 17, 98-108.

EVANS, L. (1970), "Speed Estimation from a Moving Automobile", *Ergonomics,* 13, 219-230.

FESTINGER, L. (1971), "Eye Movements and Visual Perception", in BACH-Y-RITA, P., COLLINS, C. C. (Eds.), *The Control of Eye Movements,* New York: Academic Press.

GAARDER, K. R. (1975), *Eye Movements, Vision and Behaviour,* New York: Halstead Press, Division of John Wiley.

GIBSON, J. J., CROOKS, L. E. (1938), "A Theoretical Field-Analysis of Automobile Driving", *The American Journal of Psychology,* LI, 453-471.

GIBSON, J. J. (1950), *The Perception of the Visual World,* Boston, MA: Houghton Mifflin.

GIBSON, J. J. (1966), *The Senses Considered as Perceptual Systems,* Boston, MA: Houghton Mifflin.

GIBSON, J. J. (1979), *The Ecological Approach to Visual Perception,* Boston, MA: Houghton Mifflin.

GREGORY, R. L. (1989), "Light on Black Boxes", *Perception,* 18, 281-284.

GROEGER, J. A., BROWN, I. D. (1988), "Motion Perception is not Direct with Indirect Viewing Systems", in Gale, A. G. *et al.* (Eds.), *Vision in Vehicles II,* Amsterdam: Elsevier.

HABER, R. N., HERSHENSON, M. (1973) *The Psychology of Visual Perception,* New York: Holt, Reinhart & Winston.

HILLS, B. L. (1980), "Vision, Visibility, and Perception in Driving", *Perception,* 9, 183-216.

HANCOCK, P. A., CAIRD, J. K., SHEKHAR, S., VERCRUYSSEN, M. (1991), "Factors Influencing Drivers' Left-Turn Decisions", *Proceedings of the Human Factors Society*, 35, 1139-1143.

IKEDA, M., TAKEUCHI, T. (1975), "Influence of Foveal Load on the Functional Visual Field", *Perception and Psychophysics*, 18, 255-260.

IRVING, A. (1973), "The Perceptual Problems of the Driver", *Report of the 1st Congress of I.D.B.R.A.*, Zürich.

KAISER, M. K., MOWAFY, L. (1993), "Optical Specification of Time-to-Passage: Observers' Sensitivity to Global Tau", *Journal of Experimental Psychology: Human Perception and Performance*, 19, 1028-1040.

KAPPÉ, B., KORTELING, H. (in print), "Time to Contact Estimation is Based upon Inferential and Direct Visual Information", *Perception*.

LAMSZUS, H. (1998), "Blickschulung in der Ausbildung von Kraftfahrern - ein unabgeklärtes Aufgabengebiet mit Forschungsbedarf", *Zeitschrift für Verkehrssicherheit*, 44, 2-11.

LEE, D. N. (1976), "A Theory of Visual Control of Braking Based on Information about Time-to-Collision", *Perception*, 5, 437-459.

LEIBOWITZ, H. W. (1973), "Detection of Peripheral Stimuli under Psychological and Physiological Stress", in *Visual Search*, Washington: National Academy of Science.

LEUTZBACH, W. (1987), "Wahrnehmung und Fahrverhalten", in *Wahrnehmung und Verkehrssicherheit*, München: ADAC.

MACKWORTH, N. H. (1976), "Stimulus Density Limits the Useful Field of View", in MONTY, R. A., SENDERS, J. W. (Eds.), *Eye Movements and Psychological Processes*, New York: Wiley.

MALATERRE, G. (1977), *Régulation de la vitesse à l'approche de points singuliers*, Rapport de recherche, Arcueil: ONSER.

MCDOWELL, E. D., ROCKWELL, T. H. (1978), "An Exploratory Investigation of the Stochastic Nature of Drivers' Eye Movements and their Relationship to the Roadway Geometry", in SENDERS, J. W., FISHER, D. F., MONTY, R. A. (Eds.), *Eye Movements and the Higher Psychological Functions*, Hillsdale, New Jersey: Erlbaum.

MCLEOD, R. W., ROSS, H. E. (1983), "Optic-Flow and Cognitive Factors in Time-to-Collision Estimates", *Perception*, 12, 417-423.

MORI, M., ABDEL-HALIM, M. H. (1981), "Road Sign Recognition and Non-Recognition", *Accident Analysis and Prevention*, 13, 101-115.

NAGAYAMA, Y. (1978), "Role of Visual Perception in Driving", *IATSS Research*, 2, 64-73.

NEISSER, U. (1976), *Cognition and Reality: Principles and Implications of Cognitive Psychology*, San Francisco: Freeman.

PURKINJE, J. (1825), *Beobachtungen und Versuche zur Physiologie der Sinne*, Berlin: Reiner.

ROSS, H. E. (1975), "Water, Fog and the Size-Distance Hypothesis", *British Journal of Psychology*, 58, 301-313.

SALVATORE, S. (1968), "The Estimation of Vehicular Velocity as a Function of Visual Stimulation", *Human Factors*, 10, 27-32.

SCHIFF, W. (1980), *Perception: An Applied Approach*, Boston, MA: Houghton Mifflin.

SCHIFF, W., DETWILER, M. L. (1979), "Information Used in Judging Impending Collision", *Perception*, 8, 647-658.

SCHIFF, W., OLDAK, R. (1990), "Accuracy of Judging Time to Arrival: Effects of Modality, Trajectory and Gender", *Journal of Experimental Psychology*, 16, 303-316.

SCHMIDT, F., TIFFIN, J. (1967), "Distortion of Drivers' Estimates of Automobile Speed as a Function of Speed Adaptation", *Journal of Applied Psychology*, 53, 536-539.

SCHNEIDER, W. X., DEUBEL, H. (1995), "Visual Attention and Saccadic Eye Movements: Evidence for Obligatory and Selective Spatial Coupling", in FINDLAY, J. M., WALKER, R., KENTRIDGE (Eds.), *Eye movement research*, Amsterdam: Elsevier.

SHINAR, D. (1977), "Curve Perception and Accidents in Curves: An Illusive Curve Phenomenon?", *Zeitschrift für Verkehrssicherheit*, 23, 16-21.

SHINAR, D., ROCKWELL, T. H., MALECKI, J. A. (1980), "The Effects of Changes in Driver Perception on Rural Curve Negotiation", *Ergonomics*, 23, 263-275.

SIVAK, M. (1996), "The Information that Drivers Use: Is it Indeed 90% visual?", *Perception*, 25, 1081-1089.

STELMACH, L. B., CAMPSALL, J. M., HERDMAN, C. M. (1997), "Attention and Ocular Movements", *Journal of Experimental Psychology: Human Perception and Performance*, 23, 823-844.

STEWART, D., CUDWORTH, C. J., LISHMAN, J. R. (1993), "Misperception of Time-to-Collision by Drivers in Pedestrian Accidents", *Perception*, 22, 1227-1244.

TODD, J. T. (1981), "Visual Information about Moving Objects", *Journal of Experimental Psychology: Human Perception and Performance*, 7, 795-810.

TRESILIAN, J. R. (1991), "Empirical and Theoretical Issues in the Perception of Time to Contact", *Journal of Experimental Psychology: Human Perception and Performance*, 17, 865-876.

WEBB, N. G. (1977), "Orientation of Retinal Rod Photoreceptor Membranes in the Intact Eye Using X-Ray Diffraction", *Vision Research*, 17, 625-631.

YARBUS, A. L. (1967), *Eye Movements and Vision*, New York: Plenum.

Chapter 5

FACTORS INFLUENCING DRIVING PERFORMANCE

Michel Vallet,
INRETS, National Institute for Transport and Safety Research, France

1. INTRODUCTION

The act of driving is a highly complex endeavor, partly due to time constraints for detecting, perceiving, and processing information before making a decision and reacting on a road environment full of unpredictable events. It is the driver who has the sole responsibility for how actions are carried out, and reactions to incidents depend wholly on his/her capacity to manage the situation.

An important part of the driving task includes play or fun activities. In the book, *Les jeux et les hommes – The Games and the Men*, Caillois (1958) underlined that the excitement is a "voluptuous panic... typical from industrial societies" which make available for people machines able to increase tenfold the pleasure leading to "a destructive madness." These principles, contributing to the driving activity in a more or less important manner according to the personality of the driver, are clearly in contradiction with the concept of a pragmatic and sensible way of running an efficient performance. Nevertheless, according to Caillois (1958), the game "makes [one] discover [...] life in general by increasing the ability to overcome obstacles and to face difficulties." This point of view evokes the principle of learning the driving task on a driving simulator, where the risk of a poor performance has no consequences.

These activities are mainly based on visual perception of the dynamic environment, the relative position of the vehicle, and the instruments on board.

Some part of the driving task might be automated through practice, that is to say the level of attentional demand to run this activity is quite low. Most of these sub-tasks belong to the *operational level* of the driving task, involved in the general control of the vehicle, consisting of shifting gears for example. The level of performance for these activities is quite independent from contextual events and is more connected with the level of the driver's experience.

The tasks related to the *tactical level* usually require driver's attention, at a level that can be very different according to the degree of complexity induced by external events, bringing about a global mental workload that can highly influence the driving performance.

The level of attention required for the *strategical level*, and, consequently, the level of performance to run the sub-tasks related to this stage, are more dependent upon the driver's familiarity with the area, and the usability of guiding and navigation information available to run this activity related to orientation processes.

The concept of driving performance might be considered from several points of view: From a global point of view, efficiency in reaching a destination quickly; from a behavioral point of view, efficiency in controlling a vehicle comfortably; from a safety point of view, ability to avoid road accidents...

The evaluation of driving performance can be conducted by using subjective and objective criteria and/or by recording any dysfunctional driving activity.

In the first case, various criteria can be taken into account: Evaluation of the driver's knowledge concerning driving regulations; evaluation of the level of his past experience in the driving activity through questionnaires; quantification of the amount of transgressions to road regulations; observation of the driver's behavior in order to quantify the reaction speed to detect critical situations, the correct control of the vehicle trajectory, the efficiency in perceiving relevant information from the external environment, and so forth.

In the second case, the evaluation of the driving performance can be made through the record of accidents, with several difficulties raised by this approach: Firstly, a road accident is fortunately considered as a very rare event in terms of statistics; secondly, there might be a bias linked to the type of source used such as police or assurance record; and finally, an accident does not necessarily happen because of deficiencies of the driver and can be due to technical problems of the vehicle, or uncontrolled external reasons.

The factors influencing the driver's performance might be distinguished according to their origins:

- "Internal factors" influencing functional abilities of the driver considered on a long term scale such as age, and experience, or on a short term scale, such as vigilance, stress, and fatigue, not to mention the important part played by motivations and emotions, as taken into consideration in the concept of Risk Homeostasis Theory (1995).
- "External factors" affecting the ability of the driver in his general capacities such as noise, pollution, discomfort, and disturbance of the attention processes required for correct management of the driving task itself, due to high level of complexity of the overall road context.

The driving performance is difficult to define precisely and complex to quantify; nevertheless, some of the factors that can influence and modify driving efficiency are well known.

2. THE PSYCHOPHYSIOLOGICAL BASIS OF DRIVING PERFORMANCE

Taillard and Mouret (1990) consider that there is a relationship between performance and vigilance: "Vigilance is a necessary condition for the survival of an individual, his autonomy, his communication and his interaction with the external environment. It is from the quality of this vigilance that depends the quality of the performance during the work."

The numerous amounts of works on vigilance allow a good understanding of the phenomenon and of its variations. Vigilance was described in 1923 by the neurologist Head in order to define "the state of high efficiency for the central nervous system," inducing the quickness of adaptive answers. Macworth (1957) used this term to precise that the vigilance is "the state of preparation to detect and to answer to short modifications of the environment, that happen in a random way." The term *sustained attention* is used more in contemporary psychology (Warm, 1984; Parasumaran, 1984). Basic vigilance from the nervous system is distinguished from operational vigilance preparing to the behavior. In this second area, the studies from Lecret-Grillon (1976-1985) have been devoted to the driver's fatigue.

According to the author, the notion of vigilance "fits well to the road issue when considered under the specific view of accidents prevention;" she recommended we conduct research by using electrophysiological techniques rather than monotonous performance tasks.

Nevertheless, Mackie (1987) thinks that research on vigilance in real conditions or simply in realistic conditions is not a sufficient enough or relevant way to maintain driver's vigilance.

Vigilance varies slowly (24 hours cycle) or quickly during the day (90 minutes cycle) in relation to the biological rhythms.

The circadian variations of vigilance (close to 24 hours) have been studied in detail: vigilance follows a circadian rhythm with a diurnal maximum at the end of the day. It has been shown that the "output" is actually optimum during the evening, but the tasks requiring mnesic functions or intellectual activities are better conducted in the morning (Baddeley, 1970). On the contrary, vigilance evaluated subjectively goes by a maximum in the middle of the day (12-13 hours) (Monk, 1983) and then decreases.

Already, in 1966, in a driving simulator study, Janssen showed that subjects drove quicker at midday than at midnight, but their ability to avoid the roadsides is more important at midnight than at 8 o'clock. The rhythm of vigilance shows important inter-individual differences; age, gender, and the required amount of sleep (low and high level of requirement according to the individual) modified its variation greatly; but a parameter which is usually neglected is the behavioral typology morning/evening defined by Horne and Ostberg (1976) as a function of the temperature curve, the preferred hours of getting up and going to bed, and vigilance.

In terms of vehicle accidents, Folkard (1997) underlined the temporal aspects of road safety, by analyzing the critical moments of the day in the works conducted by Langlois, 1985; Hamelin, 1987; Lavie, 1991; Horne, 1995. Most of these studies take into account other important factors such as the traffic density, the accident frequency involving an isolated vehicle, the duration of the driving activity previously to the accident, and the involvement of alcohol for the night accidents.

The part played by these factors can be controlled during driving simulator sessions (Lenné, 1997). The most important risk happens at 3 o'clock in the morning (range from 2 to 4 o'clock) with a second maximum at 3 o'clock.

The task duration, combined with a tendency to sleepiness, has been studied through more sophisticated approaches. Akerstadt set up a model for the vigilance level including a third dimension (1995): sleep inertia, happening after waking at the end of the night or after naps during the day. Age is related to natural variations of the vigilance. The way of living also plays an additional part: For example, the young drivers easily drinking alcohol or taking drugs at the end of the week-end, with also important lack of sleep, and a clear consequence on the driving performance.

3. EXTERNAL FACTORS OF MODIFICATION IN THE VIGILANCE AND IN THE PERFORMANCE

3.1 The Awakening Drugs

Three categories of awakening drugs are considered as counter-measures to sleepiness (Roussel, 1995):

- Caffeine is able to keep driver alert and awake, but its effect decreases quickly and it can induce some undesirable after-effects, linked to this quick decrease leading to take in a repetitive way more or less important quantity of it. From 200 mg to 400 mg (2 to 4 cups) of caffeine can induce anxiety, nervousness, restlessness and even tightness.
- Amphetamines have a great awaking power, verified on many occasions during the Second World War. It is dope for the sportsmen and also a genuine drug in that increases in dosage become necessary bringing about physical and psychological dependency: Deprivation syndrome, high appetence for the product. Their secondary effects are fearsome on the cardiovascular system (hypertension and tachycardia), the gastrointestinal (nauseas, organs), and, most importantly, in the neuropsychiatric area (evolution toward delirium).
- Eugregoric drugs, of which the more efficient is the Modafinil, are as powerful as amphetamines, in spite of a very different biochemical process. The lack of important secondary effects and of induction in drug addiction made these molecules very valuable in order to compensate sleepiness due to illness, but it is not ideal to use them in the same common way like the caffeine.

The role of caffeine, used in a common way in coffee, tea, and cocoa, has been studied by Horne (1995). On the English motorway rest areas, a cup of coffee (250 ml) has 100 mg of caffeine (between 72 and 137 mg). It is the most acceptable of the uppers, and efficient on tired drivers but with a low efficiency on awaked one.

Horne's experimentation compares the consequences of a nap of 15 minutes to ingesting 150 mg of caffeine or of a placebo.

The results happened to be convincing:

- The number of driving incidents went from 10 during the first hour (7 minor incidents and 3 average) to 12 during the second hour for the placebo group, 6 for those who napped, and at 2 for those who consumed caffeine. These results show significant differences;
- The sleepiness subjectively evaluated on a scale varied in a similar way as the number of incidents according to the time and the groups;
- The changes between the first and the second hour are similar to those concerning the incidents, and the differences between conditions indicated again a maximum efficiency for the caffeine taking. The authors cleverly underlined that since, all of the drivers did not have the chance to take a nap, caffeine was the more efficient way to improve vigilance.

The analysis of the visual strategies does not seem to be relevant, as they do not seem to be well related to the other sleepiness criteria.

3.2. The Medicines

The potential danger represented by medicines and illegal drugs for vigilance while driving concerns authorities, who include this topic in the examination for the driving license (Lagier, 1995).

Tranquillizers and sedative medicines are highly consumed by French people, 15% for the 15 years old and over, with 30% of chronic users (6.7% in Canada), which put the France at the 7th position on 45 countries.

The percentage of drivers involved in accidents while under sedative drugs in comparison with drivers having no accidents is from 5 to 10%, sometimes more. The same statistic of 5 to 10% is found for hospital drivers who had taken illegal drugs. This statistic has also been observed during non-obligatory examinations recommended by a big company to its drivers, through the working medical examination. Lagier underlined the difficulty in identifying the precise impact of drug-taking on accidents occurrence.

The analysis of motorists statements for drunk driving indicated a very high percentage of low or no alcohol level in the blood: it was medicines and/or illegal drugs that were responsible for this non alcoholic intoxication.

Lagier indicated that the processes inducing the effects are numerous:

- Medicines overdosed;
- Unspecified effect, especially at the beginning of the cure;
- Rare but unpredictable effects;
- Effect of deprivation/interaction with other medicines;
- Interaction with alcohol.

In these two last cases, the effects can be synergetic, that is to say multiplied.

Visual disorders can also be induced by some medicines, especially antidepressants.

3.3. The Effect of Alcohol on the Vigilance

In the middle of 1980, medical doctors (Chiron) considered that 10% of the lorry drivers were extreme drinkers, that is to say chronic alcoholics, in comparison with 7.4% in a random sample and 10.4% in workers as control sample. "We could have expected to have less extreme drinkers at safety post such as lorry driving than in the general population of the employees."

The regulations for alcohol level in the blood on one hand, and society's effort, including motorway companies to reduce, or even to eliminate, any possibility of driving after alcohol drinking on the other hand, have certainly been very efficient. Indeed, in a recent survey from INRETS (Biecheler, 1994), a positive alcoholic level has been detected for 19.2% of the 714 drivers, that is to say 15.1% with an alcoholic level < 0.24 mg/l, 2.4% with an AL between 0.25 and 0.39 mg/l and 1.7% > 0.4 mg/l.

Surprisingly, one fourth of the drivers in exceeding the speed limits were in the group of slight alcoholics in comparison with 17% among non-alcoholic drivers or illegal alcoholic drivers. The survey revealed that among lorry drivers (3.9% of the total sample), 6.9% presented an alcoholic level > 0.25 mg/l and 0.8% were exceeding the speed limit, while, in the other group of road professionals representing 5.4% in total, there is 0% with an alcoholic level > 0.25 mg/l, but 8,7% were exceeding the speed limit. These figures revealed that light alcoholic level can be at the origin of accidents. The demonstration of the causal relationship between alcohol level and road accidents have been made since 1964 in the summer study of Grand Rapids.

Several other works conducted since then have shown that the number of accidents and the risk of accidents increases regularly with alcohol level. It has to be noted that young drivers are more exposed to risks and that this risk happen around 0.5 mg/l, while the effect of alcohol level < 0.25 mg/l on speed has been demonstrated.

Very few published data are available concerning the accidents risks of lorry as a function of the alcohol level: a Finish study (Summala, 1994) indicated that the lorry drivers are very sober and that they resist in a better way than the private car drivers to the fatigue.

3.4. The Physical Factors from the Environment

It is relevant to distinguish the environment inside the vehicle cabin, from which the physical factors which might influence the driver's vigilance and fatigue, from the environment where drivers sleep, such as the rest areas of the motorway.

If some data are available concerning physical conditions from the cabin, there are no scientific results about the effects of conditions of rest and sleep during the trip.

The experimental or real conditions of observation concerning the effects of noise on the mental performance and attention are extremely diversified. It is necessary to consider in a separate way the results about the performance studied in standardized cases, where the activity is strongly activated, from the one obtained after long and monotonous tasks, in the framework of vigilance studies affected by noise while driving.

A careful study of the important literature indicates the lack of homogeneity, even the contradictory aspects of the results. This leads the authors, especially Koelega (1986), to underline that the disparate aspect in the definition of the tasks and the noise explain why, finally, "nothing is really known concerning the effect of changing noise on the sustained attention."

The significance of a review of this type seems to be quite poor according to the fact that, since 1980, Loeb already judged that the studies on the relationships between noise and performance only suggested that the effects of noise depend upon the type of noise and the nature of the task. This suspicion toward the possibilities to really understand was shared by Cohen (1980) who indicated that it was necessary to take into account the significance of the sound, the context and the social situation of tested people.

The analysis of the results of various authors' studies led Hancock (1984) to conclude that continuous white noise induces a decrease in performance if cognitive demand is high, but that it induces no clear effect if the task of sensory vigilance present a low level of demand. In this last case, an improvement in performance can been observed, especially when noise varies according to its volume or its frequency. Indeed, the detection of a signal is better in a noise, which is modulated in its volume around an average of 64.5 dB rather than in a noise of an equivalent intensity but continuous and stable (Kirk and Hecht, 1963). This is similar to the results of Tarrière (1964) showing, in a task of visual vigilance, an improvement of the performance due to significant noise (music, speech).

In fact, we should not neglect the origin and the frequency of signals being detected because these factors seem to modulate the effect of noise on the performance. In a review on this topic, Hockey (1970) concluded that the decrease in vigilance under noise effect happens especially when the task required the detection of signals coming from various sources. In the case of a unique source of signals, the performance in the noise decreases when the frequency of the signals is high, while it is improved or not modified when the frequency of the noise is low.

So, noise would decrease performance when the sources are multiple, or, in the case of a unique source, when the frequency of the signal is high.

According to Hockey (1984), the relationships between task complexity and noise effects would depend upon the "activating" effect of the noise, optimum and so positive in the first case, and in excess and so negative in the second. Another interpretation considered by the same author (Hockey, 1970) suggests that the selectivity of the attention increases in relation to the time spent working in the noise. This is based upon an experiment in which the main task is to track a mobile target moving vertically in the central visual field controlled by a lever hold in the right hand, while the secondary task is to detect six luminous signals spotted around the target, two of them being close from the center and four of them being in peripheral vision. The experiments last for 40 minutes in a sound ambience of 70 dB and of 100 dB.

From a neurophysiological point of view, it is obvious that, whatever the results of well conducted experiments, working in the noise induce an unquestionable biological workload. Either the noise is filtered and so hidden by a central process, which implies an increase activity of some of the structures not available anymore for other functions, or the noise is constantly analyzed at a central level. In the second hypothesis, it implies a time-sharing in the information processing directly related to the task and to those due to the noise. Furthermore, it is well known that it is impossible to pay attention simultaneously to two types of information coming from different sensory modalities. Visual or mental attention (mental computation) is able to stop the information at a sub-cortical level. The limited duration of these experiments and the conditions themselves do not correspond at all to the problems encountered in everyday working conditions, where the sources of pollution are not limited to the only sound environment, which, we know, can be perceived in a different manner according to the noise exposure outside the working place. Finally, the inter-individual differences were not taken into account in these approaches from which the results are far away able to answer to the raised questions.

3.5. The Vigilance of Truck Drivers and the Noise

Inside the vehicle, the level of noise varies according to numerous parameters: Type of vehicle, speed, windows opened or closed, quality of the road surface, etc. Nevertheless, the noise spectrum inside the vehicle presents a lot of low frequencies (2 to 63.5 Hz), with levels reaching 100 dB in this part of the spectrum. An increase in the speed induces an increase of the level centered on the octave range to 32 Hz. The opening of a window induces an increase of about 15 dB between 4 and 20 Hz for a vehicle going at around 100 km/h and of 8 dB, between 4 and 16 Hz, at 80 km/h. The use of the weighting A allows to present the level of noise by only one value, but it neglects the low frequencies. The levels, in Leq, vary on a road trip from 66 dB (A) for an expensive automatic car to 77 dB (A) for a small vehicle.

The variation of the noise and the messages from the radio plays an important part in maintaining vigilance compatible with the driving task. Germain observed that truck drivers increased their radio volume when they felt tired in order to stay awake. Tarrière (1985) studied the effect of noise in the cabin of a lorry, under driving simulator conditions, on 17 men of 34 years old in average, exposed during two hours at 80 dB (A). The main task was driving in central vision and the noise did not modify it. On the contrary, the secondary task, consisting of detecting signals by peripheral vision, is modified by the noise; indeed, the number of errors increases in the noisy conditions, but with a great inter-individual variability. From a biological point of view, the recording of the electroencephalogram indicated a decrease in vigilance during the two driving hours, with alpha rhythm happening (relaxed awakening or closed eyes awakening) less important in the noise. The FC and its variability are clearly increased by noise, which also causes more blinking of the eyelids and peripheral vasomotor reactions. During the two hours, the progressive decrease in the muscular tone of the nape of the neck is less obvious in a noisy background.

Fakhar and Vallet (1993) showed that a level of noise of 80 dB during a trip of four hours on the motorway induces a decrease in the vigilance of the driver after three hours; however when this noise is slightly increased during the two first hours, a level of 76 dB does not induce these variations. This research studied the combined effects of noise and vibrations on the vigilance of car drivers: Vigilance is used as a criterion of physical environment quality inside the vehicles.

The noise inside the car at 130 km/h represented the first experimental condition. This noise of 76 dB (A) is amplified and increased to 4 dB (A) in order to get a second, more nosy condition.

The normal vibrations are 0.7 m s-2 between 0 and 50 Hz. The use of a second vehicle, with hard shock absorbers, has been necessary in order to get strong vibrations of 1.52 m s-2.

The driving task has been observed for 15 000 km on motorway: 9 subjects (6 men, 3 women) averaging 35 years in age, drove 400 kilometers 4 times on the motorway, each route corresponding to a combination of 2 levels of noise and vibration.

After the setting up of a global indicator for vigilance by merging data based upon the optimal filtering of Kalman (EEG = ratio X/B; EOG = duration of the eye blinking; ECG = cardiac variability), we also set up a vigilance scale varying from 0 to 10.

An ANOVA has been conducted on the global indicator for vigilance. This analysis has several factors:

- Two modalities of vibrations;
- Two modalities of noise;
- Time duration with three modalities (1st, 2d and 3rd hour);
- Driving dynamic with two modalities.

We observe that vigilance decreased during the third hour, when noise is higher than 80 dB (A). The vibrations have no effect at this level and during time duration. There are no synergetic effects between noise and vibrations for these levels and these driving durations.

Our study indicated that there is a significant effect of noise on vigilance. On the contrary, no significant effect of the vibrations have been demonstrated, neither by themselves, nor when combined with the noise.

Our main results concerning noise are the following:

- Strong noise induces a decrease of the vigilance;
- The decrease of the vigilance level under noise appears only after a given amount of driving time (after two hours in our experiment, we notice that most of the subjects were not aware of this phenomenon). This result underlines the importance of the recommendation from ASSECAR (association for the safety on the motorway) concerning the necessity to take a break every two hours.

3.6. The Role of Vibrations and Infrasonic Vibrations

These two physical phenomena are actually linked but their effects on the drivers seem to be different. The frequency of car cabin resonance is at 2 Hz, one of the wheels is at 10 Hz, while the internal combustion engine induces a maximal level in a range from 10 to 20 Hz. Concerning the aerodynamics infrasonic vibrations, they are related to the vibrations of the vehicle body under the turbulences of external and internal air.

For long duration exposures, corresponding to driving situations for which vigilance might be weak, vibrations would bring about visual problems. According to Roure *et al.* (1983), a subject under vertical vibrations presents a decrease in visual acuity around 5 Hz, a frequency corresponding to the resonance of an important body mass, in addition to an effect of 10-30 Hz, because of the resonance located to the eyeballs. The works of Griffin and Lewis (1978) and of Jex *et al.* (1981), showed that about half of all road drivers have more or less visual problems, particularly with peripheral vision.

More fundamental works support the fact that there is a decrement in sensory-motor performance due to vibrations exposure. For the human being, it induces a disturbance in the manual control (Lewis and Griffin, 1976) with consequences on the visuo-motor coordination (Berthoz, 1971; Gauthier *et al.*, 1985). These phenomena would be related to the interference between physiological signals, coming from muscular sensors, and the disturbing messages coming from the vibrations.

In the lorry cabins, the influence of vibrations on the driving safety on motorway, studied by Gillepsie (1982), did not bring any clear conclusion concerning the relationships between these two phenomena, while, for Arnberg (1979), only random vibrations of big magnitude might modify the performance and the vigilance.

The synergetic effects of noise, vibrations and heat on vigilance and on the performance of the driver have been studied by Petit and Tarrière (1991). They conclude that, among the recorded physiological parameters, the frequency of the eyelid-blinking reveals slight decrease in vigilance as a function of time, but does not distinguish the respective effects due to noise, to vibrations and to heat. It is the same for the EMG: Even if the muscular fatigue is modified by heat and noise, the "direct" effect of vibrations on the EMG hides any possible effect of fatigue. On the contrary, the three other physiological parameters and the driving performance of the subjects (answering time and errors made) are in correspondence:

- Heat induces a negative effect on the vigilance level of the driver because it increases the ratio of alpha waves, increases the number of errors and the answering time for the detection task, and finally, disturbs the tracking task. It has also a negative effect on the cardiac rhythm, inducing a drastic increase of the number of heart beats.

- Noise has a small impact on vigilance (there is a slight increase of the alpha/beta ratio), does not disturb the cardiac rhythm and improves the overall driving performance (tracking and detection).
- Vibrations diminish slightly the level of vigilance, increase in an important manner the cardiac rhythm, improve the main tracking task, but at the detriment of the secondary task of detection, which is deteriorated at the same pace.
- The combination of vibration-noise-heat always induces the same type of effect by lowering vigilance: Either it tends to amplify the negative effects of each parameter considered separately, or it decreases the positive effects of the noise on the physiological parameters and, especially, on the task performance.

4. REFERENCES

CAILLOIS, R. (1958), *Les jeux et les hommes*, Paris: Gallimard.

COBLENTZ, A. (1989), "Vigilance and Performance in Automatized Systems", *NATO ASI serie D, Behavioural and Social Sciences*, vol. 49, Netherlands: Kluwer Academic Publishers.

EVANS, L. (1991), *Traffic Safety and the Driver*, New York: Van Nostrand Reinhold Book.

FAKHAR, S. (1991), *Effet du bruit et des vibrations sur la vigilance des conducteurs de véhicules légers en situation de conduite monotone*, Thèse de doctorat, Lyon.

GERMAIN, C., HAMELIN, P., NIERAT, P. (1992), "Le transport au quotidien : Logiques de production", *Synthèse INRETS n° 18*, Paris: Bron.

HAMELIN, P. (1995), "Le travail des conducteurs routiers : Structures de production, conditions de travail et risque", *Actes Congrès SIA 1995. 07.16.*

HAMELIN, P. (1990), "Les enquêtes sur les conducteurs de poids lourds", *3rd International Conference on Survey Methods in Transportation*, Washington D.C.

HARTLEY, L. R. (1997), "Fatigue and Transport, Second International Conference on Fatigue in Transportation", *Accident Analysis and Prevention*, vol. 29, n°4.

KUIKEN, M. J. (1995), *Instructional Support to Drivers: The Role of In-Vehicle Feedback in Improving Driving Performance of Qualified Motorists*, Traffic Research Centre, Netherlands.

MCDONALD, N. (1984), *Fatigue, Safety and the Truck Driver*, London: Taylor and Francis.

MACKIE, R. R. (1977), *Vigilance: Theory, Operational Performance, and Physiological Correlates*, New York: Plenum Press.

MERCIER-GUYON, C., "Alcohol, Drugs and Traffic Safety", *Proceedings of the 14th International Conference*, Annecy, 3 vol.

PHILIPPS-BERTIN, C. (1996), *Apport d'une activité cognitive dans le maintien de la vigilance des conducteurs automobiles*, Thèse de doctorat, Université Lumière Lyon 2.

Transportation Research Circular (1994), *Driver Performance Data Book Update: Older Drivers and IVHS*, n°419, Transportation Research Board, National research Council.

ROTHENGATTER, T., CARBONELL, E. (1997), *Traffic and Transport Psychology: Theory and Application*, Pergamon, Elsevier Science Ltd.

VALLET, M. (1991), *Le maintien de la vigilance dans les transports*, Caen: Paradigme.

VALLET, M., KHARDI, S. (1995), *Vigilance et transports – Aspects fondamentaux, dégradation et prévention*, Lyon: Presses Universitaires de Lyon.

DE WAARD, D. (1996), *The Measurement of Drivers' Mental Workload*, Traffic Research Centre, Netherlands.

Acknowledgments are due to Dr. Annie Pauziè for the English translation and constructive comments. M.V.

Chapter 6

LEARNING TO DRIVE

Ray Fuller,
Department of Psychology, Trinity College, Dublin, Ireland

1. INTRODUCTION

The job of machine-operator is not everyone's cup of tea. It conjures up images of control wheels and levers, oily rags, water-cooling, noise and macho males in blue overalls. And yet there is one machine that most people of both sexes yearn to operate, the car. To late-adolescents this promises liberation from the speed limits of their own physiology, from the geographical limits imposed by their immediate neighborhood, from the restricted destinations and time-synchronizing demands of public transport services, from the attentions of an overweening parent. It also provides a powerful statement of adulthood, perhaps no less than the loss of one's virginity, an event to which it may indeed contribute. But not only this. The modern developed world has been progressively designed around the mobility provided by the car, with services widely distributed over urban, rural and residential areas. Ownership of a car has for many become almost a necessity. Not surprisingly, most young people are strongly motivated to learn to drive.

2. WHAT IS INVOLVED IN LEARNING TO DRIVE?

Learning to drive is no easy, simple matter, however. The human brain needs to go through some fairly radical changes before it can be entrusted to guide safely perhaps a ton of metal hurtling at high speed towards a similar object approaching from the opposite direction and before it can control the vehicle like an extension of its own body, like a "wheeled exo-skeleton." What then has to be learned in learning to drive? A useful starting point is the conceptualization provided by the SHEL model (Hawkins, 1987), which describes the various interfaces between the human operator, the car and its operating environment (see Figure 1).

Figure 1. The SHEL model: interfaces with the human operator

SHEL is an acronym for Software, Hardware, Environment and Liveware, where the latter term, Liveware, is a rather inelegant (and unintentionally almost dismissive) term for the human in the system. The liveware in the center of the figure represents the driver. The interface between the driver and the Hardware, the machine or car in this instance, focuses on controls and displays. Effective, efficient and safe control of a car requires a skilled performance at this interface and the development of this skill typically follows a pattern characteristic of coordinated visual-motor skills in general:

- Gradual transition from conscious rule-based and discrete acts to relatively unconscious, automatic coordinated skill-based acts (from declarative to procedural knowledge);
- Associated decreases in the requirement for working memory capacity and release of capacity for higher-order perceptual and judgmental processes;
- Associated expansion of the window of planned behavior.

Research on visual-motor skill acquisition has identified important roles for feedback and for practice. Feedback has a vital corrective function but also a motivational effect on performance (Holding, 1989), while practice facilitates the development of automatic routines for control operations such as changing gear, steering and even activating appropriate switches. Practice effects typically follow a power law (see Newell and Rosenbloom, 1981 and Anderson, 1983).

In computing, Software represents the set of instructions, which tell a computer what to do. In the SHEL model the interface between the driver and Software is concerned with those instructions, which tell a driver what to do, which control his/her behavior on the road. These might (and almost certainly will) include:

- Regulations relating to roadway use by drivers;
- Standard procedures – specifications of what behavior is required under what conditions;
- Informal rules for behavior;
- Codes of social behavior (see later).

Elements of all of these types of "controlling instruction" will have been assimilated during childhood and adolescence by the driver, even before s/he first takes the wheel of a vehicle, but very many remain to be learned through instruction, training and experience. The seeming universal over-representation of young drivers in crashes, in spite of huge variation in requirements for, and experience of, formal training testifies to the often painful role of experience in this process. Learning what behavior is required under what conditions to achieve both satisfactory mobility and a safe outcome is a major challenge and is not helped by a number of factors such as the probabilistic (as opposed to certain) relationship between the key variables involved (taking a particular corner at 100 km/h may lead to loss of control... or it may not) and the relatively low frequency of occurrence of many hazardous contingencies (such as the roadway beyond a bend being completely blocked by an obstruction).

The Environment. In this context, the interface between the driver and the road and traffic environment, can at times, seriously undermine the ability of the driver to carry out the task safely and the driver needs to learn to recognize particular conditions and their properties. While performance problems caused by such general environmental factors as vibration, noise and extremes of temperature and airflow have been eliminated from the modern car, there remain the problems of reduced visibility during hours of darkness and in weather conditions such as fog and heavy precipitation and conditions associated with a decrease in road surface adhesion (rain after a dry spell, loose gravel, wet leaves, frost, ice). More specific environmental learning has to do with interpreting the evolving road and traffic scenario in front of the driver in terms of required control actions (mainly speed and direction), a skill sometimes referred to as "reading the road" or "hazard recognition," where potentially serious aversive outcomes are possible, such as crashing, hitting a vulnerable road user or losing control of the vehicle. A subtle part of this learning has to do with the management of the driver's vigilance or attention, not only maintaining a level necessary for the task in hand but also deploying it effectively (to sources, and potential sources, of task-relevant information) and efficiently (avoiding monitoring of irrelevant sources or excessive dwell-time on particular sources). These processes are in part under external control, dictated by characteristics of the road and traffic environment.

The fourth interface about which the driver has to learn is Liveware, the interface with other road users such as drivers, cyclists and pedestrians. It needs to be recognized that driver behavior has a strong social dimension, a dimension that expresses how we relate to other people. Do we show off to them, threaten them, bully them, ignore them, compete with them, punish them? Do we show deference or assertiveness, rage or revenge? For a safe and reliable performance, the driver needs to learn to separate the way s/he drives from emotional needs and assimilate the social code of the road of mutual care, consideration and courtesy. We expect this from professional drivers and airline pilots for example. We don't expect Captain Jones who has had a breakfast row with his wife to take it out on his passengers or other aircraft in his vicinity. But because roadway use is so unexceptional, so well stitched into the fabric of daily life, we tend to carry over into it ongoing emotional and social needs and fail to recognize the inappropriateness of their expression there. Indeed the very anonymity and protection offered by the metal capsule of the car may well disinhibit expression of feeling previously restrained for fear of social censure. Reflecting this point, there is clear evidence for a distinct association between low levels of socialization in everyday life and poor social behavior on the road (Evans, 1991).

So learning to drive is in part also about learning how to relate to other people on the road. But beyond this basic social learning there is the further dimension of communication, knowing what, how and when to communicate to other road users and how to interpret the messages being transmitted from them. Communication elements include indicator lights, brake lights, headlamp flashes, horn, eye-contact and gestures such as waving. More subtly they include the "posture" of other vehicles, whether nudging out from a parking spot or in terms of their apparent trajectory when moving. The driver needs to learn to use these cues appropriately, as well as interpret them.

Learning and experience of the interfaces described above ultimately enable the development in the driver of:

- Skilled routines, which operate in an automatic, unconscious way as part of attention, search, decision and control sequences (skill-based learning);
- A large body of rules, which specify appropriate responses given particular situation or scenarios (rule-based learning);
- Knowledge structures, which provide a long-term memory, based organization of knowledge relevant to the driving task and its sub-domains (knowledge-based learning).

At any moment in time these mental models provide:

- A dynamic representation of the trajectories of the driver's own vehicle relative to the road environment and that of other road users (vehicles, cyclists, pedestrians) in the environment;
- A prediction of the imminent outcomes of those trajectories (rather like an internal video sequence which runs ahead of real time);
- Response options linked to the various road and traffic configurations.

For the interested reader, further discussion of these features may be found in, for example, Brown and Groeger (1988), Camerer and Johnson (1991), Brown (1993), and Redding and Seamster (1994).

3. LEARNING AS A CONTINUOUS PROCESS

Of course, as indicated earlier, the task of learning to drive is situated in a life history of the individual in which many relevant elements have already been assimilated. The inexperienced driver will have had varying degrees of direct exposure to the contingencies of driving through his or her active participation in traffic as a pedestrian and perhaps cyclist. S/he will also have observed events both as a vehicle passenger and perhaps also as a viewer of road safety educational films and regular fictional movies. S/he will also bring to the driving task an acquired knowledge-base derived from experience in environments other than the roadway, such as the effects of impact with hard objects, characteristics of the trajectories of moving objects, expectations about other people's behavior and so on. Such knowledge is important for the development of safe driving because it enables the driver to rule out certain contingencies as improbable or impossible (e.g., "if I apply the brakes I will go faster") and facilitates the identification of valid contingencies (e.g., directing attention to the relationship between road-surface conditions and the effects of hard braking).

How we behave as drivers on the roadway is also continuously subject to control by its consequences. Responses, which are followed by rewarding consequences, will tend to be strengthened, those followed by punishing (or non-rewarding) consequences will tend to become weakened. Thus if risk-taking achieves desirable rewards (such as the saving of time) this behavior will be reinforced, a phenomenon described elsewhere as "learned riskiness" (Fuller, 1992). Similarly, rule-following will tend to be abandoned where experience shows the consequences of rule-following are not what is expected. Hence drivers are more likely to ignore speed limits (and even traffic signals) when streets are deserted (such as in the early hours of the morning). The implications of this analysis have been discussed fully elsewhere (Fuller, 1991). The fundamental point however is that driver behavior like any other is continuously being remolded by its consequences and that this process may undermine as well as strengthen safe procedures.

4. WHAT FORMS DOES DRIVER TRAINING CURRENTLY TAKE?

To be competent and safe, in principle, we do not have to learn through direct trial-and-error experience how to act in every possible circumstance. We can be taught both general and specific rules and procedures for correct and safe behavior. After basic vehicle control skills, such rule learning is a fundamental component, if not the fundamental component, of most training programs: learning what the rules and procedures are for dealing with situations we are likely to meet. So, we can reasonably ask, are training programs successful in reducing young driver accidents?

In the different European countries, varying requirements for driver training and different training methodologies have emerged (Lynam and Twisk, 1995). Variations relate to features such as:

- The age at which a driver may drive different types of vehicle (for a car, 16 in Norway, 17 in Ireland and the UK, 18 most other EU countries but 16 in Sweden and France as an accompanied driver under a contracted apprenticing or apprentice system);
- The mandatory requirement for theory and practical training (most countries but not the UK, Ireland or the Netherlands);
- The opportunity to practice with laypersons (not permitted in Germany, Denmark, the Netherlands, Portugal, Greece, Luxembourg);
- The requirement for practical training and certification of driving instructors (all countries except Ireland);
- The requirement of a probationary period for newly qualified drivers (Germany, Spain, Portugal, the Netherlands).

What is perhaps surprising is that despite these wide variations in training experience, Lynam and Twisk (1995) concluded that there was no evidence that the differences in national systems produced major differences at the level of national casualty totals. Now although this conclusion should be treated with caution, because a number of recent developments have not yet been fully evaluated, what it may underline is:

- The observation that the traditional system of training aims mainly to prepare learners to pass the driver licensing examination, after which drivers are left on their own (ETSC, 1996);
- The importance of the experience of driving on the roadway as a key learning process (see Hall and West, 1994). My guess is that the fundamental problem is that appropriate safe behavior simply cannot be completely taught – it has to be learned by doing and that inevitably means making mistakes on the roadway, and
- The effect of driver testing providing a common gateway through which only equivalently competent drivers pass, irrespective of the nature of their previous training.

Thus, in the EU, young qualified drivers are more or less as good, or bad, as each other. But does this mean then that driver training can have no substantial part to play in helping to reduce the inordinately high level of young driver casualties? I think not. It has to be recognized that, universally in Europe, driver training courses have developed on a piece-meal basis and are not based on evaluative research. Typically they have not been designed through the application of established principles of learning or skill acquisition. Thus, to take an example, in the UK, training is controlled by a Government Driving Standards Agency, which maintains a register of Approved Driving Instructors who are periodically tested for ability and fitness to give instruction. A recent in-depth study of this system (Groeger and Clegg, 1994) found that less than 5% of training was given to more complex and hazardous maneuvers such as overtaking and lane changing and less than 2% to elements such as hazard recognition and anticipation. And as we shall see below, inadequate ability in this area distinguishes young from older drivers and is typical as a contributory factor in young driver accidents.

5. YOUNG DRIVER CHARACTERISTICS

We do know that young drivers tend to have a range of general characteristics, which have implications for their safety (Twisk, 1995). These characteristics relate to risky life-style, risk exposure, risk-seeking and overestimation of competence.

5.1. Risk Life-Style

The first noticeable feature is that, within the group of young drivers, there is undoubtedly a subgroup of drivers who deliberately take risks and as a consequence are over involved in both traffic violations and accidents. Their risk behavior is not just confined to roadway use but characterizes their life-style as a whole. They are more likely to engage in drug use, heavy drinking and petty crime. Their poor socialization and risk seeking may well be the result of both personality and environmental factors. However they are by no means the entire problem. Only about 15-20% of young male drivers fall into this category. There is no evidence that the majority of young driver accidents arises from intentional high risk or antisocial driver behavior.

5.2. Risk Exposure

A second characteristic of young drivers is that they are more vulnerable because of the conditions under which they drive. Compared with others, they are more likely to drive vehicles, which offer less protection in the event of a collision. They are more likely to drive with a car full of friends, thereby increasing the number of casualties in the event of a crash. They are more likely to drive during darkness and particularly on weekend nights, periods typically associated with a higher incidence of driving with elevated BALs. Late at night there are of course the further risk-contributing effects of drowsiness and the opportunity provided by lower density traffic to drive faster, especially in urban environments.

It is also worth noting that the alcohol-related accident risk of young drivers starts at much lower BALs (20 mg per 100 ml) and rises more steeply than in older drivers. Driving with a BAL limit of 50 mg per 100 ml, the involvement in a fatal accident of an 18-19 year old is estimated to be almost five times higher than for a 30-34 year old. Despite this, young drivers are not generally over-represented in alcohol-related accidents, except on weekend nights. Although such factors as these may selectively influence young driver accident rates they do not account entirely for young driver over involvement in accidents (Hodgson *et al.*, 1981). In reviewing the evidence, Jonah (1986) concluded that "even when one controls for the quantity and quality of exposure to risk, young drivers are still at greatest risk of casualty accident involvement" (p. 257).

5.3. Risk Seeking

A third aspect of the young driver relates to attitudes to driving and, in addition, the driver's vulnerability to peer pressure. Evidence suggests that for some young adults, high-risk driving, which they see as a demonstration of superior control skills, is associated with social status and is reinforced by peer influences. It can also be unintentionally reinforced by the high-risk driving behavior modeled by heroes in fictional drama on film and TV (Evans, 1991).

5.4. Overestimation of Competence

Fourth, young male drivers typically overestimate their own competence and the possibilities for correction in a developing accident scenario (Gregersen, 1996).

To summarize then, young drivers:

- Include a sub-group of high-risk lifestyle individuals;
- Tend to drive under more vulnerable conditions;
- Are subject to peer pressures to adopt high-risk driving styles;
- Overestimate their ability to drive safely.

This propensity to take risks on the road declines with greater maturity and major life events such as getting married. But increasing age is not the only saving factor, the other, and perhaps the most important, is the accumulation of experience. If we look at the competence of young drivers compared with older ones we typically find the symptoms of inadequate experience of the contingencies found in the road and traffic environment. Thus, as summarized in Fuller (1988) they:

- Are poor at identifying distant hazards (Laidlaw, 1975 in Brown, 1982);
- Take longer to perceive and respond to simulated potentially dangerous traffic conditions (Quimby and Watts, 1981);
- Underestimate the risk of accident in video traffic sequences involving vehicle handling and so-called driving reflex skills (Matthews and Moran, 1986);
- See less risk in specific driving situations (Finn and Bragg, 1986);
- Are more likely to be in driving situations where they may come into conflict with other drivers (Jonah, 1986);
- Are less likely to be able to stop within the limits of forward visibility (Quimby and Watts, 1981);
- Give lower ratings of the seriousness of the consequences of their own "offensive" driving (Brown and Copeman, 1975).

These characteristics have implications for the content of driver training and for training methods.

6. DRIVER TRAINING

6.1. Hazard Perception

Because of their inexperience young drivers have had relatively little opportunity to develop reliable predictive models of the outcomes of potentially hazardous situations, whether these relate to the behavior of other road users or to the speed, trajectory and attentional options selected by the driver himself or herself. Not surprisingly young driver accidents are characterized by driving too fast for the prevailing conditions (e.g., see Michiels and Schneider, 1984). This is also a problem for older drivers and motorcyclists, but it is especially true for young novice drivers. This lack of an ability to relate speed to conditions (such as negotiating a bend) has also shown up in the relatively high involvement of young drivers in single vehicle accidents. Thus driver training needs to pay more attention to the development of hazard perception skills, including perception of those hazards involved in fast driving. Recognition of this is becoming realized in the form of extensions to driver testing which incorporate hazard perception elements (usually written scenarios) and modifications to training curricula. But given the apparent requirement for experience for the development of hazard perception skills (see e.g., Lund and Williams – 1985 – who concluded that post-license defensive driving courses do not reliably decrease the likelihood of crashing), the problem remains of how to pack several years of driving experience of hazards into a time-limited, commercially viable and safe training program. The obvious technologies to support such learning include video, multimedia and, perhaps potentially most important of all, simulations of various degrees of fidelity up to virtual-reality driving. The advantages of simulations are of course that they enable active learning in an error-tolerant learning environment. But for the moment, the availability of the appropriate technology for driver training is effectively non-existent and so consequently is the research which could evaluate its effectiveness (see Hoskovec and Stikar, 1992). We do know, however, that in other domains such as pilot training, the use of various levels of simulation realism for learning hazard recognition and management has been demonstrably successful (see review by Johnston, 1995). What is needed is a similar investment in the exploration of training needs and methodologies for the novice and continuing vehicle driver.

6.2. Self Representation

A further way in which the learning of safe driving might be facilitated is by aiding the driver to formulate self-representations of the contingencies of the traffic environment with which s/he interacts. Brewer (1974) assembled considerable evidence in support of the contention that the learning of contingencies is markedly enhanced by subjects' awareness of those contingencies. Unfortunately we still don't know very much about the conditions under which an awareness of contingencies arises and comes to affect behavior, although Resnick and Ford (1978) conclude that verbalizing goals and strategies before attempting to solve a problem greatly enhances the likelihood of solution. This is an area, which clearly merits further research attention. In the meantime it is worth noting that the act of keeping an "event" diary has helped individuals identify contingencies in health-related areas such as cigarette smoking behavior. Perhaps a "driving-event" diary might help the novice driver in the same way. In an early exploratory study, Lewin (1982) found that self-recording of near accidents and mental imagery practice improved driving skills. Hatakka, Keskinen, Katila and Laapotti (1997) found a strong association between self-evaluations and safety-related aspects of driver behavior. And Gregersen (1996) has shown that in dealing with the problem of overestimation of their competence, "insight" training for young drivers is more effective than skill training.

6.3. Group Training

In-car training typically follows the one-on-one pattern of pupil and instructor, which limits feedback to the trainee to elements of his or her performance. A development from this of potential merit is the concept of training in pairs or even groups of three learners at a time, once individuals have mastered basic vehicle control skills. Each learner takes a turn at driving but all observe the performance of their peers and hear feedback regarding it. This training scenario provides not only exposure to a potentially much wider range of errors and how to deal with them but enables the learner to develop an active role as an evaluator of performance and provides an opportunity for learners to share and discuss problems and experiences (see conclusion by Gregersen, Brehmer and Moren, 1996, p.306). This technique has been tried experimentally in the training of commercial airline pilots with promising results (Johnston, Regian and Shebilske, 1995). Perhaps it is time to transfer that success to the driver training environment. It may even stimulate the design and development of vehicle models dedicated to effective training, perhaps providing three front seats for trainees with the instructor supplied with parallel controls and independent communication to all trainees from an elevated position above the trainee driver.

6.4. Human Factors

Lastly a notable omission in current driver training relates to the opportunity and requirement for systematic learning about human factors. What people do on the roadway is ultimately constrained by their competence; this competence always sets the limiting conditions for behavior. But human performance, though highly adaptive, is unreliable. It is unreliable because it is vulnerable to a wide range of influences such as:

- Biological individual differences (e.g., age);
- Learned individual differences (e.g., experience);
- Task-generated factors (e.g., noise, emotion, fatigue, sleepiness);
- Non-task generated factors (e.g., sleep-loss, aggression, stress, alcohol).

Although competence may be undermined by such factors, this is not the whole of the picture. Behavior is in turn a function of individual values (which provide motivation for particular acts), social norms (influencing behavior through group dynamic processes) and the culture within which the road user operates, a culture which in turn transmits values, rewards and punishments (e.g., through enforcement). Thus quite independent of the quality of training the driver has received, what s/he actually does on the roadway at any moment of time is vulnerable to a vast array of influences. It would seem appropriate for drivers to be made aware of these influences and processes, to become familiar with what might be called the human factors of driving. Such knowledge does not, of course, guarantee that the driver will become safer, but recognizing the problem must be the first step towards a solution. What is fundamentally needed, perhaps, is a shift in emphasis in driver training from preparation to pass the license test to preparation for safe, effective and efficient driving for life.

7. REFERENCES

ANDERSON, J. R. (1983), *The Architecture of Cognition*, Cambridge, Mass.: Harvard University Press.

BREWER, W. F. (1974), "There is no Convincing Evidence for Operant or Classical Conditioning in Adult Humans", in WEIMER, W. B., PALERMO, D. S. (Eds), *Cognition and the Symbolic Processes*, New York: Wiley.

BROWN, I. D. (1993), "How is Driver Behaviour Acquired?", in GRAYSON, G. B. (Ed.) *Behavioural Research in Road Safety IV*, Crowthorne: Transport Research Laboratory.

BROWN, I. D., COPEMAN, A. K. (1975), "Driver's Attitudes towards the Seriousness of Traffic Offences, Considered in Relation to the Design of Sanctions", *Accident Analysis and Prevention*, 7, 15-26.

BROWN, I. D., GROEGER, J. A. (1988), "Risk Perception and Decision Taking during the Transition between Novice and Experienced Driver Status", *Ergonomics*, 31, 585-597.

BROWN, I. D. (1982), "Exposure and Experience are a Confounded Nuisance in Research on Driver Behaviour", *Accident Analysis and Prevention*, 14, 345-352.

CAMERER, C. F., JOHNSON, E. J. (1991), "The Process-Performance Paradox in Expert Judgement: How Can the Experts Know so Much and Predict so Badly?", in ERICSSON, K. A., SMITH, J. (Eds), *Toward a General Theory of Expertise: Prospects and Limits,* Cambridge: Cambridge University Press.

ETSC (1996), *Driver Training and Testing: The Need for Improvement,* Brussels: European Transport Safety Council Briefing.

EVANS, L. (1991), *Traffic Safety and the Driver,* New York: Van Nostrand Reinhold.

FINN, P., BRAGG, B. W. E. (1986), "Perception of the Risk of an Accident by Young and Older Drivers", *Accident Analysis and Prevention*, 18, 289-298.

FULLER, R. (1988), "Psychological Aspects of Learning to Drive", in ROTHENGATTER, J. A., DE BRUIN, R. A. (Eds), *Road User Behaviour: Theory and Research,* Assen: Van Gorcum, 527-537.

FULLER, R. (1991), "The Modification of Individual Road User Behaviour", in KOONSTRA, M. J., CHRISTENSEN, J. (Eds), *Enforcement and Rewarding: Strategies and Effects*, Leidschendam: SWOV Institute for Road Safety Research.

FULLER, R. (1992), "Learned Riskiness", *The Irish Journal of Psychology*, 13, 2, 250-257.

GREGERSEN, N. P. (1996), "Young Drivers' Overestimation of Their Own Skill. An Experiment on the Relation Between Training Strategy and Skill", *Accident Analysis and Prevention*, 28 (2), 243-250.

GREGERSEN, N. P., BREHMER, B., MOREN, B. (1996), "Road Safety Improvement in Large Companies: An Experimental Comparison of Different Measures", *Accident Analysis and Prevention*, 28 (3), 297-306.

GROEGER, J. A., CLEGG, B. A. (1994), "Why Isn't Driver Training Contributing more to Road Safety?", in GRAYSON, G. B. (Ed.), *Behavioural Research in Road Safety IV,* Crowthorne: Transport Research Laboratory.

HALL, J., WEST, R. (1994), "The Role of Formal Instruction and Informal Practice in Learning to Drive", paper presented at the *British Psychological Society Conference*, London.

HATAKKA, M., KESKINEN, E., KATILA, A., LAAPOTTI, S. (1997), "Self-Reported Driving Habits are Valid Predictors of Violations and Accidents", in ROTHENGATTER, T., CARBONELL VAYA, E., *Traffic and Transport Psychology: Theory and Application*, Amsterdam: Elsevier Science.

HAWKINS, F. H. (1987), *Human Factors in Flight,* Avebury: Gower.

HODGSON, J. D., BRAGG, B. W. E., FINN, P. (1981), "Young Driver Risk-Taking Research: The State of the Art", *National Highway Traffic Safety Administration*, Washington, D. C.: Contract No. DTNH 22-80-R-07360.

HOLDING, D. H. (1989), *Human Skills* (2nd edit.), Chichester: Wiley.

HOSKOVEC, J., STIKAR, J. (1992), "Training of Driving Skills with or without Simulators?", *Studia Psychologia*, 34 (2), 183-187.

JOHNSTON, A. N., REGIAN, J. W., SHEBILSKE, W. (1995), "Observational Learning and Training of Complex Skills in Laboratory and Applied Settings", in JOHNSTON, N., FULLER,

R., MCDONALD, N. (Eds), *Aviation Psychology: Training and Selection,* Aldershot: Avebury Aviation.

JOHNSTON, A. N. (1995), *Simulation and Training: Perspectives on Theory and Practice*, Working paper of the Aerospace Psychology Research Group, Dublin, Trinity College.

JONAH, B. A. (1986), "Accident Risk and Risk-Taking Behaviour among Young Drivers", *Accident Analysis and Prevention*, 18, 255-271.

LEWIN, I. (1982), "Driver Training: A Perceptual-Motor Skill Approach", *Ergonomics*, 25 (10), 917-924.

LUND, A. K., WILLIAMS, A. F. (1985), "A Review of the Literature Evaluating the Defensive Driving Course", *Accident Analysis and Prevention*, 17 (6), 449-460.

LYNAM, D., TWISK, D. (1995), *Car Driver Training and Licensing Systems in Europe,* TRL Report 147, Crowthorne: Transport Research Laboratory.

MATTHEWS, M.-L., MORAN, A. R. (1986), "Age Differences in Male Drivers' Perception of Accident Risk: The Role of Perceived Driving Ability", *Accident Analysis and Prevention*, 18, 299-313.

MICHIELS, W. O., SCHNEIDER, P. A. (1984), "Traffic Offences: Another Approach to Description and Prediction", *Accident Analysis and Prevention*, 16, 223-238.

NEWELL, A., ROSENBLOOM, P. S. (1981), "Mechanisms of Skill Acquisition and the Law of Practice", in ANDERSON, J. R. (Ed.), *Cognitive Skills and Their Acquisition,* Hillsdale, NJ: Lawrence Erlbaum.

QUIMBY, A. R., WATTS, G. R. (1981), *Human Factors and Driving Performance, Laboratory Report 1004,* Crowthorne: Transport and Road Research Laboratory.

REDDING, R. E., SEAMSTER, T. L. (1994), "Cognitive Task Analysis in Air Traffic Controller and Aviation Crew Training", in JOHNSTON, N., MCDONALD, N., FULLER, R. (Eds), *Aviation Psychology in Practice,* Aldershot: Avebury Technical.

RESNICK, L. B., FORD, W. W. (1978), "The Analysis of Tasks for Instruction: An Information-Processing Approach", in CATANIA, A. C., BRIGHAM, T. A. (Eds), *Handbook of Applied Behaviour Analysis: Social and Instructional Processes,* New York: Irvington.

TWISK, D. (1995), "Factors Contributing to the High Accident Liability of Novice Drivers and the Role of Driver Training", in *European Approaches to Reducing Young Driver Casualties,* Brussels: ETSC.

Chapter 7

ATTITUDES

Lars Åberg,
Dalarna University, Sweden

1. INTRODUCTION

The human factor is a major factor behind traffic accidents and knowledge about modification of driver behavior is of great importance for traffic safety research. Also, in the general debate the necessity of changing road user attitudes is frequently mentioned as a means of changing road user behavior and improving traffic safety. To what extent human behavior is determined by attitudes has been a topic in psychological research for a many years. In spite of this fact there is far from a general agreement among researchers concerning what is meant by attitudes, in what way and to what extent they are related to behavior and how they might be modified. The confusion among researchers is also shown in the discussions of unsafe driver behavior and attitudes.

The purpose of the present chapter is to present a short summary of current praxis for attitude research in the traffic psychology area. The chapter begins with a declaration of what is meant by the construct of attitude in the present context. Also, two models that explain the relationship between attitudes and behavior and that has been extensively used in general applied psychology as well as in traffic psychology will be presented. Road user attitudes, methods of attitude investigations and presentation of results from attitude research will be discussed. The present text is limited to traffic safety issues but the models for attitudes could easily be applied on behaviors like travel mode choice, environmental issues, purchase of vehicles, etc.

2. THE CONCEPT OF ATTITUDE

As there is no universally accepted definition of attitude, according to Manstead (1996) most social psychologists would accept a definition of attitude suggested by Eagly and Chaiken (1993), stating that attitudes are "tendencies to evaluate an entity with some degree of favor or disfavor, ordinarily expressed in cognitive, affective and behavioral responses." Entity refers to the object of an attitude and includes various things like individuals, inanimate objects, concepts, social groups, behaviors, and so on. In traffic psychology it would include different road-users like drivers, pedestrians, and cyclists as well as vehicles, safety equipment, traffic rules, enforcement, driver behavior, etc.

3. MODELS OF ATTITUDE-BEHAVIOUR RELATIONS

In general psychology several theories have aimed at explaining the relationship between attitudes and behavior (for a review see Forward, 1994). Examples of such attitude models are the theory of reasoned action suggested by Fishbein and Ajzen (1975), the model of attitude-behavior relations by Triandis (1980), the health belief model (e.g., Janz and Becker, 1984), and a theory of action (Huguenin, 1988). The different theories overlap a great deal although they stress different aspects of the problem (e.g., Connor and Norman, 1994; Forward, 1994). The Fishbein and Ajzen theory of reasoned action, has had the greatest impact on the applied field of psychology. This is probably due to theoretical merits but also because it is relatively easy to apply. The discussion of attitudes in the present chapter will focus on the theory of reasoned action (TRA) and on the theory of planned behavior (TPB); (Ajzen, 1985), the latter model is an extended version of the former.

The TRA suggested by Fishbein and Ajzen (e.g., Fishbein and Ajzen, 1975; Ajzen and Fishbein, 1980) has proved to be useful in traffic psychology, explaining drivers' and other road users' behavior. The theory, presented in Figure 1, is based on a principle of compatibility between attitudes and behavior. This means that attitudes and behavior should be measured on the same level for each of four elements (action, target, context and time). Compatibility is crucial for possibilities to achieve substantial correlations between attitudes and behavior. In order to predict a specific action of an individual toward a certain object in a special context and at a certain time, ideally one should measure the person's attitude toward performing that action toward that object in that context at that time. For example, when predicting use of safety belts among taxi passengers on Friday nights, one should ask about attitudes concerning that specific situation and not about opinions about safety belts in general.

In the theory the immediate determinant of behavior is the individual's intentions to act in a certain way. This means that the theory concerns behaviors that are the results of conscious decisions to perform or not perform the behaviors. Therefore, intentions represent a person's motivations to perform certain behaviors. Intentions are determined by the individual's attitude to the behavior and by the subjective norm of the individual. Attitude to a behavior is defined as person's evaluation of this behavior as good or bad. A taxi passenger who believes that seat belt wearing is a good thing is more likely to be belted than a passenger that does not like being belted. The subjective norm is an individual's beliefs about the opinion of important other persons concerning her or his performance of the behavior. A person may put on a belt in a taxi because he or she knows that members of his or her family are in favor of seat belt wearing. These two factors contribute to a person's intentions and the relative weight of each factor depends on the behavior of interest. For example, a man who wants to buy a new car decides to buy a car he has been dreaming about, a red sports car without much consideration about other persons' opinions. If the same man was married and had children he may still have a longing for the red sports car but he might believe that his wife would rather have a station wagon and that may ultimately determine his buying behavior.

Both attitude and social norms can be measured directly, as described above, as well as indirectly. Attitudes are, according to the model, determined by behavioral beliefs, that is, beliefs about the consequences of performing the behavior and of outcome evaluations, that is, evaluations of each consequence. For example, a driver may decide to avoid driving in winter because he or she experiences a risk of losing control over the vehicle and the loss of control might end up in a crash. Each behavioral belief is multiplied with the corresponding outcome evaluation and the sum of all such products becomes an index of the attitude with regard to the behavior. This index is an indirect measure of attitude. In a similar way the determinants of the subjective norm are normative beliefs, that is, beliefs that significant other persons expect an individual to act in a certain way and motivations to comply, i.e., the individual's willingness to comply with the views of other important persons. The normative beliefs are multiplied with motivations to comply and a sum of the products is an indirect measure of subjective norm.

The TRA was originally developed to deal with volitional behaviors and does not explain habitual behaviors or behaviors that do not involve conscious decision-making. Autonomous decision-making is a significant aspect of the task of driving. It is often assumed that an experienced driver performs many driving acts, like everyday driving from home to work, without awareness of specific acts. Many drivers can report experiences when they intend to drive to a certain place, partly along the a road they normally drive, and then they "wake up" to find themselves on the workplace parking lot where they had no intention at all to go to.

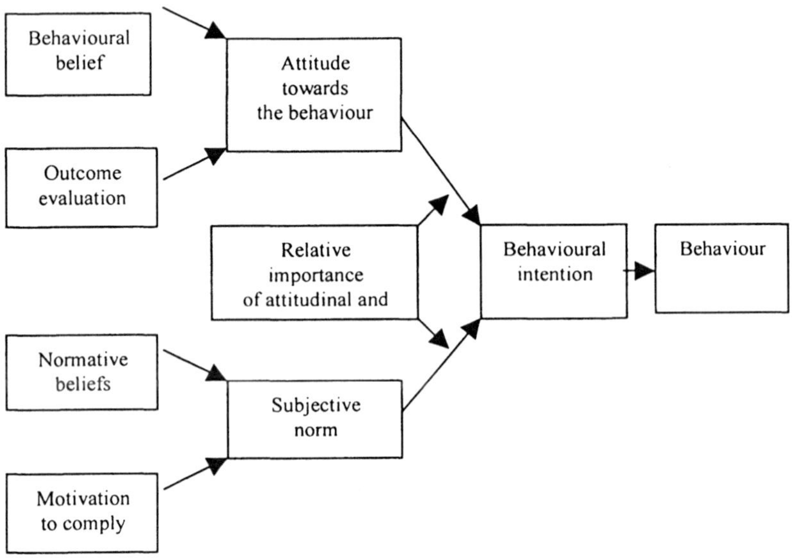

Figure 1. The theory of reasoned action (after Fishbein and Ajzen, 1975)

Partly to expand the theory to cover also habitual involuntary behaviors, Ajzen (1985) added a new factor to the theory of reasoned action, namely perceived behavioral control. The revised theory is called the theory of planned behavior (TPB) and it is shown in Figure 2. Perceived behavior control is believed to determine behavior indirectly via behavioral intentions but a direct path to behavior is also suggested. According to the theory, intentions are determined by attitudes together with subjective norms and perceived behavioral control. Behavior is determined by intentions and by perceived behavioral control. Perceived behavior control is (in a way similar to attitudes and subjective norms) determined by a sum of products of control beliefs and perceived power.

4. OPERATIONALISATION OF MODEL

CONCEPTS

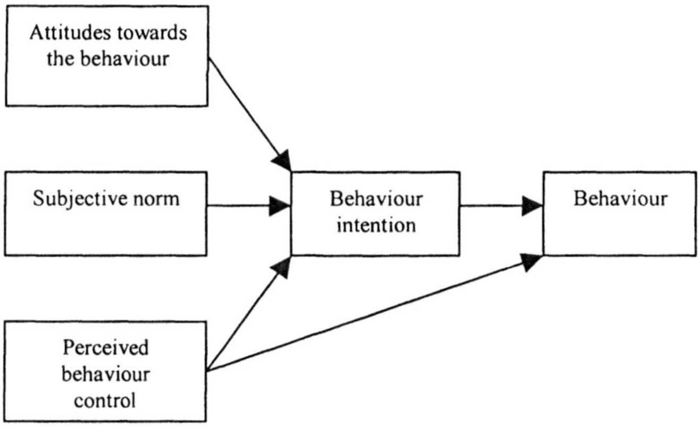

Figure 2. The theory of planned behaviour (after Ajzen, 1988)

In accordance with the principle of compatibility between attitude, norms, etc. and behavior it is important to carefully define the behavior to be predicted. All model concepts should be formulated on the same level of behavioral specificity, that is, action, target, context and time frame should be specified. For example, to investigate bike riding (action) wearing a helmet (target) when traveling to work (context) coming month (time). One advantage of the TRA and TPB compared to other theories is that there are clear guidelines how to formulate questions and response scales concerning the model concepts (e.g., Ajzen and Fishbein, 1980; Ajzen, 1985). In Table 1. a number of examples of such questions and response scales in a fictitious study of seat belt attitudes are presented. Examples of items cover some, of the TPB variables like self-reported behavior, behavioral intention, a direct attitude measure and behavioral belief and outcome evaluation used for an indirect attitude estimate. Other examples concern subjective norm, normative belief, motivation to comply, perceived behavior control, control belief and perceived power.

Table 1. Examples of item wordings and response formats for the different constructs of the theory of planned behavior

Behavior:

I wore a helmet when I went by bike to work during the past four weeks

Always 1 2 3 4 5 6 7 Never

Behavioral intention:

I intend to wear a helmet next time I go by bike to work

Likely 1 2 3 4 5 6 7 Unlikely

Attitude:

Me wearing a helmet next time I go by bike to work would be

Good 1 2 3 4 5 6 7 Bad

Behavior belief:

Wearing a helmet next time when I go by bike to work makes me safer:

Unlikely 1 2 3 4 5 6 7 Likely

Outcome evaluation:

Being a safe cyclist would be

God 1 2 3 4 5 6 7 Bad

Subjective norm:

Most people who are important to me think I:

Should 1 2 3 4 5 6 7 Should not

wear a helmet next time I go by bike to work

Normative belief:

My friends think I

Should 1 2 3 4 5 6 7 Should not

wear a helmet next time I go by bike to work

Motivation to comply:

With regard to bicycle helmet how much do you want to do what your friends think you should

Not at all 1 2 3 4 5 6 7 Very much

Perceived behavioral control:

For me to wear a helmet when going to work by bike is

Difficult 1 2 3 4 5 6 7 Easy

Control belief:

There is very little room for a helmet in the place that I work

Likely 1 2 3 4 5 6 7 Unlikely

Perceived power:

The storage problem at work makes my wearing a helmet

Less likely 1 2 3 4 5 6 7 More likely

5. ATTITUDES DRIVER BEHAVIOUR AND TRAFFIC SAFETY

The theory of reasoned action, the theory of planned behavior and also the health belief model has been applied in many different domains of human behavior including behavior in traffic. Factors believed to affect road user behavior like attitudes and subjective norms have been the subject of a number of studies where the theories have been used with success in understanding road user behavior, for example, the use of safety equipment like safety belts (Wittenbraker, Gibbs and Kahle, 1983; Budd and Spencer, 1986; Nelson and Moffitt, 1988; Sutton and Hallett, 1989; Stasson and Fishbein, 1990), the use of child restraints (Gielen, Eriksen, Dulbray and Rist, 1984; Webb, Sanson-Fischer and Bauman, 1988), the wearing of cyclist helmets (Arnold and Quine, 1994), avoiding motorcycle accidents (Rutter, Quine, and Chesman, 1993), and acceptance of automatic violation-registration systems (Laan, 1998).

One important task of traffic psychology is to improve traffic safety by modifying dangerous road user behavior. It is therefore important to identify risky behaviors that are frequent and under the road user control. Reason, Manstead, Stradling, Baxter and Campbell (1990), and Åberg and Rimmö (1998), investigated drivers' self-reported aberrant behavior and found that volitional behaviors such as violations of traffic rules are more frequent than non-volitional errors or mistakes. It has also been found that an important predictor of accidents is deliberately made violations of traffic rules (e.g., Parker, West, Stradling and Manstead, 1995). There are several studies reported in the literature where traffic psychologists and others have studied road users' beliefs and evaluations, but relatively few investigations have been based on theoretical models. However, the relations between attitudes and violations based on the TRA or TPB frameworks are being investigated in a growing number of studies. For example, frequent and dangerous violations such as speeding on highways as well as the less frequent but equally dangerous behaviors of drunk driving and aggressive driving have been investigated (e.g., Åberg, 1993; Carbonell, Tejero and Canet, 1997; Jonah and Dawson, 1982; Parker, Lajunen, and Stradling, 1998; Parker, Manstead, Stradling, Reason, and Baxter, 1992; Vogel and Rothengatter, 1984).

6. ATTITUDE MODELS AND DRIVER BEHAVIOUR

With a growing number of investigations, in which TRA or TPB have been used, there is an increasing body of knowledge about the appropriateness of these models for traffic psychology phenomena. Although the use of theoretical models for explanations of road user behavior has proved to be useful for researchers, there are some problematic issues that need to be discussed.

The TRA and TPB stipulate that the independent variables attitude, norm and perceived behavior control actual behavior through behavioral intention. However, most studies have relied on self-reports of actual behavior. It is well known that subjective reports may be biased towards socially desirable responses and especially so if the behavior concern violations of laws.

There are few studies that report a relationship between self reported and observed violations in traffic and all of them concern speeding behavior. The correlations reported vary between 0.12 and 0.58 (Åberg, Larsen, Glad and Beilinson, 1997; Fildes, Rumbold and Leening, 1991; Haglund and Åberg, 2000; Vogel and Rothengatter, 1984) with the highest coefficients, between 0.52 to 0.58, reported for driving too fast on highways. As a comparison, in a meta-analysis of 88 studies Kraus (1995) reported a mean correlation of .38 between attitudes and later behavior from different areas. Thus there seems to be a relatively good agreement between self-reports and actual behavior, especially considering the fact that self-reports like other subjective measures contain measurement errors. In Haglund and Åberg (2000) a test-retest reliability estimate was made for self-reports of the speed normally held and a coefficient of 0.71 was obtained. Haglund (2000) reported a correlation of 0.75 for free flowing vehicles on highways. A correction for attenuation of the relations between self-reported and point estimates of speed ($r = 0.55$) resulted in an estimated relation between the behavioral outcomes measured within the contexts of the TRA and TPB and normal speed held on the road of 0.75.

Another issue that has been debated concerns the measures of attitude. In the TRA and TPB a measure of attitude can be obtained directly through Likert scales or indirectly through a sum of evaluations x belief products. Both direct and indirect measures of attitude have been used in different studies and according to Rothengatter and Manstead (1997) they appear to measure the same concept. These authors also discussed the relative importance of attitudes and social norms. Regression analyses reported in different studies of speeding, drinking and driving have shown different relations between attitudes and social norms. In Dutch studies attitude explained a large part of the variance in intention (beta weights: 0.52-0.78) and social norm explained less (betas between 0.08 and 0.29), whereas in British studies the social norm has explained more (betas between: 0.26 and 0.30) and attitude less (beta about 0.08-0.13) of the variance of intention. The disparity between studies might be effects of differences in methodology used although effects of cultural differences in traffic law obedience between countries cannot be ruled out. It should also be mentioned that substantial correlations around 0.50 have often been reported between attitude and social norm (Rothengatter and Manstead, 1997). In the

TPB a third major predictor is introduced, namely perceived behavior control. Although there are fewer studies of TPB on road user behavior reported perceived behavioral control have been found to add significantly to explanations of intentions and behavior in traffic (e.g., Parker and Manstead, 1996).

7. EXPANDING THE TPB THROUGH INCLUSION OF NEW CONSTRUCTS

The TPB model was developed as an attempt to improve the TRA as the new construct, perceived behavior control, was added to attitudes and subjective norm. The new concept was introduced as a response to criticism towards the TRA. One argument was that TRA could only handle volitional behavior while much daily behavior, especially driving, are more or less autonomous and habitual. Habit has also been suggested as a new variable in the TPB by Verplanken, Aarts, van Knippenberg, and Moonen (1998), in a study of car choice. Also other attempts to include new constructs have been made. Beck and Ajzen (1991) suggested that expanding the model by including a personal norm would improve predictions of socially unacceptable behaviors like crimes. Parker, Manstead and Stradling (1995), in an attempt to investigate the personal norm studied effects of two new constructs – personal norm and anticipated regret – and found highly significant improvements of the model's power to predict self-reported behaviors like "driving through traffic lights that have turned red" and "overtaking on inside." Thus, new variables have been found to add to the predictability of TPB but at the same time it they might be weakening the conceptual simplicity and power of the original theory (Rothengatter, 1993). In Table 2. examples of some of the new scales are presented.

Table 2. Examples of item wordings and response formats for concepts of an extended theory of planned behavior

Moral norm:

It would be quite wrong for me to ride a bicycle to work without wearing a helmet

Agree 1 2 3 4 5 6 7 Disagree

Anticipated regret:

If I were not wearing a helmet next time I go by bike to work I would feel bad

Likely 1 2 3 4 5 6 7 Unlikely

Habit:

Putting on a helmet before cycling is something I do without thinking
Agree 1 2 3 4 5 6 7 Disagree

---

8. METHODOLOGICAL SHORTCOMINGS

The TRA, and later, the TRB have been widely and successfully used in applied social psychology (Manstead, 1996) including traffic psychology. The reason for the success in generating research is probably that the theory is easy to apply. The model provides a simple and clearly defined set of pathways between the constructs of the model and behavior and there are detailed instructions for operationalization of different constructs. However, there are discrepancies between results from different studies (some of them have been discussed in this chapter) using the same theoretical framework that is difficult to explain. According to Rothengatter and Manstead (1997), the causes of the discrepancies might be differences in methodology, specification of target behavior or cultural differences. The authors suggest cross-cultural studies to solve the dilemma with: a) a common method for measuring model components; b) a common definition of target behavior; c) a common method for sampling of subjects; and d) a common framework of analysis. It is very important for the trustworthiness of results based on the TRA and the TPB that different studies yield similar results especially as the results may be used as guidelines for practical applications. The last suggestion, a common framework of analysis, is highly important from a theoretical point of view and therefore testing of models will be discussed in the following point.

9. TESTING THE ATTITUDE MODELS

The TRA and TPB are causal models containing causal paths from attitude via intended behavior to behavior indicating an indirect causal link between attitude and behavior. The implication is that a change in attitude might result in a change in behavior. To test the causal aspects of such models it is necessary to perform experimentally controlled investigations with manipulation of the predictors or independent variables in the model. However, most of the relations discussed in the literature are correlational in nature and from such relationships it is difficult to draw conclusions about causality (Manstead, 1996). Experimental studies where road users attitudes have been manipulated with subsequent measurements of effects on actual behavior are not known to the present author. The published reports concern studies in which questions about model constructs and self-reported behavior are asked concurrently in questionnaires mailed or distributed in other ways to samples of road users. In such cross-sectional studies relations between constructs will be of a correlational nature and it will not be possible to draw conclusions about causality. Even through cross-sectional studies, as such, cannot answer questions about causality, they offer valuable information about the way different variables are interrelated. There are also methods for statistical analyses (path analysis and structural equation modeling) available that makes it possible to compare data obtained in a study with a pattern of relations stated by a causal model. These methods of analysis do not give information about causality but it is possible to rule out causal models that are not in agreement with the data. Although path analysis very well might be applied on TRA or TPB data, the current practice is to use conventional analysis of regression. Some of the confusion concerning results from different studies that have been observed might be an effect of the analyses performed. The results from an analysis of regression are easy to interpret if the independent variables are non-correlated but when, as often is the case with attitudes and social norms, the variables are interrelated the situation is more complex. Therefore, if the aim is to predict behavior, regression analysis may work very well but for explanation of interrelations path analysis is more suitable. This difference will be exemplified below on road user data from a questionnaire based on the TPB.

10. DIFFERENT METHODS OF DATA ANALYSIS

To investigate why drivers' commit traffic law violations a questionnaire based on the TPB was sent by mail to a random sample of 2 700 Swedish license holders with a return rate of 68.4% (1 847 returned questionnaires). Some results from the study have been presented earlier (Åberg, 1999). Questions were asked about 13 different traffic violations, two of which (speeding 10 km/h over limit on roads with 30 or 50 km/h speed limits) will be used in the present context. The model constructs used are direct measures of attitude (Att), subjective norm (SN), perceived behavior control (PBC), behavior intentions (Int), and normal – self-reported – behavior (Beh). Three kinds of analyses will be exemplified: Analysis of regression; path analysis; and structural equation modeling (LISREL). Only the final results of each analysis are presented.

The results from the three different regression analyses will first be presented in a traditional way. In the first analysis the behavioral intention was regressed on the three model constructs. In the second self-reported behavior is regressed on the same three variables and in the third analysis behavior was regressed on behavioral intention and the other three model constructs. The results are presented in Table 3.

Table 3. Results of three analyses of regression where intention and behavior are regressed on TPB variables. Each variable is a sum of answers about speeding on 30km/h and 50km/h roads

A. Behavior intention regressed on attitude, social norm, and perceived control			
Construct	R^2	R^2-change	β
Att	.24	.24	.31
PBC	.34	.10	.32
SN	.37	.03	.19
B. Self-reported behavior regressed on attitude, social norm, and perceived behavior control			
Construct	R^2	R^2-change	β
Att	.19	.19	.26
PBC	.32	.13	.36
SN	.34	.02	.16
C. Self-reported behavior regressed on intention, attitude, social norm, and perceived behavior control			
Construct	R^2	R^2-change	β
Int	.42	.42	.47
Att	.44	.02	.11
PBC	.48	.04	.21
SN	.48	.00	.07

The results of the three analyses presented in Table 3. give information about direct effects of the different model constructs for prediction of intention and behavior, respectively. An alternative way of presenting the results from Table 3. (section A and C) is to rearrange them into a path analysis according to Figure 3. with the beta-weights as path coefficients. In this way both the direct and indirect effect (that are stipulated in the TPB model) as well as inter-correlations among independent variables can be shown. The arrangement of results according to the path analysis procedure offers a possibility to view the results arranged in accordance with the original model. Path analyses also give directions for improvements of the model for a better fit with data. The figures presented in Figure 3. show that attitude may be more important than social norm. However, if it could be assumed on theoretical grounds that social norm should have a causal effect on attitude, then social norm would become more influential on behavior than attitude, if both direct and indirect effects were considered. Another result from the present analysis is, quite in line with the TPB, that perceived behavior control has the strongest direct effect on behavior beside behavior intention.

Structural equation modeling is a path analytic approach that is more statistically advanced. Although this method has been applied in attitude research in various fields, there are only few attempts within traffic psychology (Åberg, 1993; Carbonell, Tejero and Canet 1997). In LISREL (Jöreskog and Sörbom, 1993) a path analysis is performed with the purpose of modeling the latent structure of data, free from errors of measurement. In Figure 4. the results from a LISREL-analysis performed on the same data as in the regression analyses are presented. In this analysis the two sets of answers for 30 km/h and 50 km/h roads are treated separately. A comparison between the results presented in Figure 3. and Figure 4. shows a quite similar patterns of path coefficients. As the variables included in the LISREL analyses are corrected for deficient reliability in the answers, the path coefficients become higher in Figure 4. One important advantage of the LISREL approach is the possibility to test a fit between a theoretical model and data. In Figure 4. this is represented by RMSEA = 0.06 (the figure indicates a fair but not good (< .05) fit between the model and the questionnaire data). The information can be used for a decision about the ability of the theoretical model to provide an acceptable explanation of the empirical data. Furthermore, the LISREL program gives suggestions about how to change the theoretical model to improve the fit.

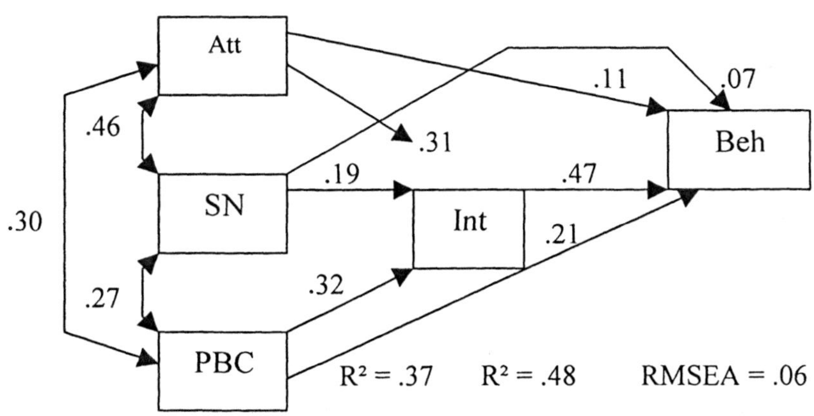

Figure 2. Result of a path analysis based on figures in Table 1. The constructs denote Att: attitude; SN: social norm; PBC: perceived behaviour control; Int: intention; Beh: self-reported behaviour. Each construct is based on a the sum of two items 30 km/h and 50 km/h respectively.

The examples in Table 3. are merely an illustration of current praxis for presenting results from studies of the TRA and TPB with the main focus on prediction of behavior. If the purpose is to explain rather than predict behaviors, alternative methods of analysis, such as LISREL, are available that might improve the understanding of results obtained and provide tests of the fit between theoretical models and data. However, it should be kept in mind that, even if the advanced methods offer better possibilities for data handling, the theoretical underpinnings of the models are highly important. In structural modeling on correlational data it is never possible to obtain proof for causal relations in a model. Successful structural equation modeling must rely on a good theoretical model that motivates the causal paths within the model.

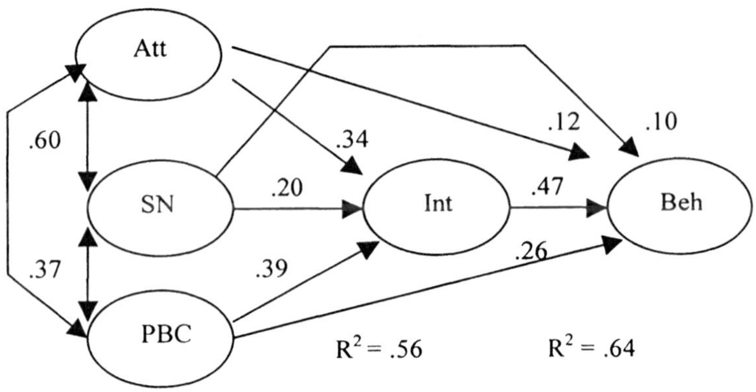

Figure 3. Result of a LISREL-analysis. The constructs denote Att: attitude; SN: social norm; PBC: perceived behaviour control; Int: intention; Beh: self reported behaviour. Each construct is based on independent contributions from two items 30 km/h and 50 km/h respectively

11. CONCLUDING REMARKS

The purpose of the present text is to give a short report of the state of art in attitude research within the field of traffic psychology. In general it is shown that interest about effect of attitudes on road user behavior is growing in the field of traffic psychology. Development of theoretical models like the TRA and TPB have stimulated a lot of researchers and the models have been applied to explain behaviors such as various traffic violations, use of safety restraints, choice of transport means, etc. The models popularity among researchers might be explained by the simplicity in model structure and in instructions for operationalization of model constructs. However, discrepancies in results reported from different studies points to a need for more standardized methods of measurement and for more standardized procedures and presentations of data analysis. Structural equation modeling provides a method of analyze that might take care of some of the problems. This method of analysis focus on the understanding of relationships between models constructs rather than prediction of behavior only. Furthermore, the method makes it possible to test the fit between model and data and in that way making comparisons between different studies more reliable and meaningful.

12. REFERENCES

ÅBERG, L. (1993), "Drinking and Driving: Intentions, Attitudes, and Social Norms of Swedish Male Drivers", *Accident Analysis and Prevention,* 25, 289-296

ÅBERG, L., (1999), "The Role of Attitudes in Decisions to Violate Traffic Regulations", paper presented at the *European Conference of Transport Psychology* in Angers, France, June 1999.

ÅBERG, L., LARSEN, L., GLAD, A., BEILINSON, L. (1997), "Observed Vehicle Speed and Drivers' Perceived Speed of Others", *Applied Psychology: An International Review*, 46, 287-303.

ÅBERG, L., RIMMÖ, P.-A. (1998), "Dimensions of Aberrant Driver Behaviour", *Ergonomics,* 41, 39-56.

AJZEN, I., FISHBEIN, M. (1980), *Understanding Attitudes and Predicting Social Behaviour,* Englewood cliffs, NJ: Prentice Hall.

AJZEN, I. (1985), "From Intentions to Actions: A Theory of Planning Behaviour", in KUHL, J., BECKMAN, J. (Eds.), *Action, Control: From Cognition to Behaviour*, Berlin: Springer Verlag.

AJZEN, I. (1988), *Attitudes, Personality and Behaviour*, Milton Keynes: Open University Press.

ARNOLD, QUINE, L. (1994), "Predicting Helmet Use among Schoolboy Cyclists: An Application of the Health Belief Model", in RUTTER, D. R., QUINE, L. (Eds.), *Social Psychology and Health: European Perspectives,* Avebury, Aldershot.

BECK, L., AJZEN, I. (1991), "Predicting Dishonest Actions Using the Theory of Planned Behaviour", *Journal of Research in Personality*, 25, 285-301.

BUDD, R. J., SPENCER, C. P. (1986), "Lay Theories of Behavioural Intention: A Source of Response Bias in the Theory of Reasoned Action?", *British Journal of Social Psychology*, 25, 109-117.

CARBONELL, E. J., TEJERO, P., CANET, F. (1997), "An Alternative Methodological Approach to Assessing the Theory of Reasoned Action when Applied to Drinking-Driving Intention", in ROTHENGATTER, J. A., CARBONELL, E. J. (Eds.), *Traffic and Transport Psychology, Theory and Application,* Amsterdam: Pergamon.

CONNOR, M., NORMAN, P. (1990), "Comparing the Health Belief Model and the Theory of Planned Behaviour in Health Screening", in RUTTER, D. R., QUINE, L. (Eds.), *Social Psychology and Health: European Perspectives,* Avebury, Aldershot.

EAGLY, A. H., CHAIKEN, S. (1993), *The Psychology of Attitudes,* Fort Worth, TX: Harcourt Brace Jovanovich.

FISHBEIN, M., AJZEN, I. (1975), *Belief, Attitude, Intention, and Behaviour*, New York: Wiley.

FORWARD, S. (1994), "Theoretical Models of Attitudes and the Prediction of Drivers' Behaviour", *Uppsala psychological reports*, No 434.

GIELEN, A. C., ERICKSEN, M. P., DALTOY, L. H., ROST, K. (1984), "Factors Associated with the Use of Child Restraint Devices", *Health Education Research*, 6, 297-306.

HAGLUND, M. (2000), *Stability in Drivers' Speed Choice*, Sweden: Department of Psychology, Uppsala University.

HAGLUND, M., ÅBERG, L. (2000), "Speed Choice in Relation to Speed Limit and Influences from Other Drivers", *Transportation Research*, 3F, 39-51.

HUGUENIN, R. D. (1988), "The Concept of Risk and Behaviour Models in Traffic Psychology", *Ergonomics*, 31, 557-569.

JANZ, N. K., BECKER, M. H. (1984), "The Health Belief Model: A Decade Later", *Health Education Quarterly*, 11, 1-47.

JONAH, B., DAWSON, N. E. (1982), "Predicting Reported Seat Belt Use from Attitudinal and Normative Factors", *Accident Analysis and Prevention*, 14, 305-310.

JÖRESKOG, K. G., SÖRBOM, D. (1993), *LISREL8: Structural Equation Modeling with the SIMPLIS Command Language,* Chicago, IL: Scientific Software International.

KRAUS, S. J. (1995), "Attitudes and the Prediction of Behaviuor: A Meta-Analysis of the Empirical Literature", *Personality and Social Psychology Bulletin*, 21, 58-75.

LAAN, VAN DER, J. D. (1998), *Acceptance of Automatic Violation-Registration Systems,* Groningen: H. Van Brug.

MANSTEAD, A. S. R. (1996), "Attitudes and Behaviour", in SEMIN, G. R., FIEDLER, K. (Eds.), *Applied Social Psychology*, London: Sage Publications.

NELSON, G. D., MOFFIT, P. B. (1988), "Safety Belt Promotion: Theory and Practice", *Accident Analysis and Prevention*, 20, 27-38.

PARKER, D., MANSTEAD, A. S. R. (1996), "The Social Psychology of Driver Behaviour", in SEMIN, G. R., FIEDLER, K. (Eds.), *Applied Social Psychology*, London: Sage Publications.

PARKER, D., MANSTEAD, A. S. R., STRADLING, S. (1995), "Extending the Theory of Planned Behaviour: The Role of Personal Norm", *British Journal of Social Psychology*, 34, 127-137.

PARKER, D., WEST, R., STRADLING, S., MANSTEAD, A. S. R. (1995), "Behavioural Characteristics and Involvement in Different Types of Traffic Accident", *Accident Analysis and Prevention*, 27, 571-581.

PARKER, D., LAJUNEN, T., STRADLING, S. (1998), "Attitudinal Predictors of Interpersonally Aggressive Violations on the Road", *Transportation Research*, 1F, 11-24.

PARKER, D., MANSTEAD, A. S. R., STRADLING, S. G., REASON, J. T., BAXTER, J. S. (1992), "Intentions to Commit Driving Violations: An Application of the Theory of Planned Behavior", *Journal of Applied Psychology*, 77, 94-101.

REASON, J. T., MANSTEAD, A. S. R., STRADLING, S. G., BAXTER, J. S., CAMPBELL, K. (1990), "Errors and Violations: A Real Distinction", *Ergonomics*, 33, 1315-1332.

ROTHENGATTER, J. A. (1993), "Road User Attitudes and Behaviour", in GRAYSON, G. B. (Ed.), *Behavioural Research in Road Safety III*, Crowthorne: Transport Research Laboratory.

ROTHENGATTER, J. A., MANSTEAD, A. S. R. (1997), "The Role of Subjective Norm in Predicting the Intention to Commit Traffic Violations", in ROTHENGATTER, J. A., CARBONELL (Eds), *Traffic and Transport Psychology, Theory and Application*, Amsterdam: Pergamon.

RUTTER, D. R., QUINE, L., CHESMAN, D. J. (1990), "Do Motorcyclists' Beliefs Predict Accident Liability?", in BENJAMIN, T. (Ed.), *Driving Behaviour in a Social Context*, Caen: Paradigme, 172-180.

STASSON, M., FISHBEIN, M. (1990), "The Relation between Perceived Risk and Preventive Action: A Within-Subject Analysis of Perceived Driving Risk and Intentions to Wear Seat-Belts", *Journal of Applied Social Psychology*, 20, 261-273.

SUTTON, S., HALLETT, R. (1989), "Understanding Seat-Belt Intentions and Behavior: A Decision-Making Approach", *Journal of Applied Social Psychology*, 19, 1310-1325.

TRIANDIS , H. C. (1980), "Values, Attitudes, and Interpersonal Behavior", in HOWE, H. E. Jr., PAGE, M. M. (Eds.), *Nebraska Symposium on Motivation*, 1979, Lincoln, NE: University of Nebraska Press.

WEBB, G. R., SANSON-FISHER, R. W., BOWMAN, J. A. (1988), "Psychosocial Factors Related to Parental Restraint of Pre-School Children in Motor Vehicles", *Accident, Analysis, and Prevention*, 20, 97-94.

WITTENBRAKER, J., GIBBS, B., KAHLE, L. R. (1983), "Seat Belt Attitudes, Habits, and Behaviors: An Adaptove Amendment to the Fishbein Model", *Journal of the Applied Psychology*, 13, 406-421.

VOGEL, R., ROTHENGATTER, J. A. (1985), *Motieven van snelheidsgedrag op autosnelwegen: een attitude onderzoek*, Report VK 84-09, Traffic Research Centre, University of Groningen, Haren, The Netherlands.

Chapter 8

MOTIVATIONAL AND EMOTIONAL ASPECTS INVOLVED IN DRIVING

Rosa Bañuls, Luis Montoro,
University of Valencia, Traffic Research Institute, Spain

1. THE ROLE OF THE EMOTIONS IN DRIVING

It is evident that individuals display different behaviors when driving and especially when involved in accidents. These differences seem to be related in various ways to physical, mental and socio-environmental characteristics of the driver. It is also known that a wide range of situational variables that are all too common in our daily lives, such as fatigue, use of drugs, alcohol or certain medication; sleepiness and transitory behavioral alterations such as stress or anxiety, have a negative influence on driving.

In this chapter we will address both the positive and negative roles that the emotions play in driving, taking into account that individual differences are important when dealing with emotional processes in general. The individual when driving not only makes use of his perceptive, sensory and motor abilities, but also depends on a large number of emotional and motivational variables, that in turn can influence the level of safety of a decision or act at any given moment.

The concept of emotion has evolved from the mentalist's position that was adhered to the discipline of philosophy to the presently held view that is defended by scientific psychology. The shift from the philosophical to the psychological perspective is probably due to Darwin's work (1872) *The Expression of Emotions in Man and Animals*. The thesis that Darwin defends in this book places special emphasis on the biological importance of the emotions, pointing out that emotional behavior is necessary for the survival of both mankind and animals. The emotions evolved along with the species, they exhibit a degree of inter-cultural universality that is evident and they are important in mankind's developmental maturity, a counterpoint to learning from the environment.

However, we presently know that not all emotions are adaptive for man. Among these negative emotions are excessive social anxiety or social phobia, which can make some individuals avoid a large number of social situations – with considerable negative consequences – and lack of control over reactions due to anger which can lead to difficulties in the workplace, trouble with the law and, of course, serious problems when driving.

Emotional reactions comprise a set of responses that include primary reactions, such as anger and fear, universal expressions, such as laughter and crying, automatic bodily reactions like blushing and piloerection, and mood states, such as happiness and sadness. When an emotional reaction takes place, changes are observed in three different types of canals of emotional expression: (a) at a subjective-experiential level (feelings of happiness, sadness, anger, etc.); (b) at a corporal or physiological-organic level (changes in heart or respiratory rate, increase in sweating, changes in muscular tension, etc.); and (c) at an observational-motor level (smiling, crying, facial expressions of fear or anger, etc.). In summary, when we talk about emotional reactions we are referring to three different ways of expressing the same phenomenon, which Lang (1968) christened with the names: cognitive, physiological and motor responses.

If these three types of responses elicited by emotional reactions are transferred to the context of traffic, we can reflect upon the positive or negative roles that emotions can play in driving.

The subjective experiential component of emotional reactions is strongly related to the achievement of higher order cognitive activities, such as memory and thinking (Cano-Vindel, 1989). A person who experiences a powerful emotional state of anger, anxiety or euphoria manifests serious attentional deficits, which interfere with the performance of a complex task such as driving. For example, the fear or anxiety that new drivers experience during the first few hundred kilometers of driving, often leads them into risky situations due to lack of decisiveness, insecurity, etc. Another example of this negative relationship is depression: a depressed person that drives can easily put their life and the lives of others in danger. Pathological sadness, irritability, totally irrational cognitive evaluations of one's past, present and future, low self-esteem, etc., which accompany a depressed driver, can turn into a dark veil that does not permit one to see the road clearly and interferes with the tasks that are necessary for safe driving.

Emotional experiences are usually classified according to three basic bipolar dimensions: hedonic tone (pleasure-displeasure), perceived intensity (high-low) and the degree of control perceived (high-low). It seems logical to conclude that the emotional states that are located in the extremes of any of these three dimensions will be dangerous states for driving. A person who feels very affected by emotional incidents, or a person who perceives that they have little control over their behavior when they find themselves enveloped in an emotional state, should reconsider their ability to drive.

The physiological component, or physiological responses, of an emotional reaction determine, or are directly related to, the intensity of the emotional behavior. Activation or arousal has been considered a bipolar dimension, with two extremes – minimum and maximum activation. However, psychophysiological studies show that activation is multidimensional. There can be a great degree of discrepancy between, for example, cortical, sympathetic, parasympathetic, and motor activation, which makes the concept of general activation seem inadequate for explaining behavior. Consequently, when we simply talk about activation and use only one physiological response as an indicator of this "activation," the intensity of this emotional response cannot be generalized to represent the individual's level of activation because many other physiological responses measured at the same moment would manifest different degrees of "activation."

In spite of these observations, the concept of activation has been widely used within the context of psychology of the emotions since Cannon's Model in 1915. The Yerkes-Dodson's Law regarding the relationship between activation and achievement has been with us since the beginning of the century. According to this law, optimum achievement in complex tasks is attained with intermediate levels of activation and achievement diminishes when there are very high or very low levels of activation. Consequently, the relationship between activation and achievement can be described as having the form of an inverted "U." Considering that driving is certainly a complex task, one can conclude that optimum performance would be achieved in this task with intermediate levels of activation. However, there are still some problems to resolve regarding which physiological responses would be the most representative of this "activation."

Also, it is often difficult to differentiate if the physiological activation at a given moment is due to emotional stimuli, the motivational system, or other factors. The studies concerning psychophysiological specificity of the emotions are scarce and their interpretation is ripe with controversy. In any case, strong psychophysiological specificity of the emotions does not seem to exist, which makes it difficult to assign a source or a cause to specific physiological activity. Nevertheless, taking into account the above mentioned difficulties, in general, low levels of physiological activation are associated with loss of attention, somnolence, prolonged response time, etc.; on the other hand, high levels of activation are associated with disorganized thinking, lack of concentration, impulsiveness and the tendency to act without thinking, etc.

Lastly, the motor responses or observable components of the emotional reactions serve to express and communicate our emotional state, intentions, etc., to others. Expressions of anger are frequent in driving, often escalating into explosions of violent conduct which can end in aggression with unforeseeable consequences. It seems that driving fosters the expression of negative emotions, especially anger. Driving is a complex task, which includes a diversity of situations many of which imply physical risk or threat. When a driver cognitively evaluates a traffic situation, taking into account that it can have negative consequences for him, it triggers a negative emotional reaction. After evaluating the situation, if the subject concludes that it is dangerous for him, the reaction that is produced is one of stress and includes a mixture of negative emotions, such as anxiety and anger. The verbal or non-verbal expression of anger typically triggers a second reaction of increased anger or rage, which in turn leads to conflicts between individuals. The result of this process is an increment in negative emotions and a loss of control over the cognitive system and behavior. Specifically, these consequences include negative changes in risk perception, attitudes toward traffic violations, risky decision making, reckless driving which translate into behaviors such as not maintaining safe headway, competing with other road users, making quick and unjustified changes in speed, accelerating when another driver attempts to overtake, etc.

Once again individual differences are important mediators in this general emotional process, producing different results in function of a variety of variables such as: beliefs and values, attribution style, level of subjective risk, problem solving skills, social skills, personality variables – for example, expression of anger – etc. The subject's previous state (organism, experience with the situation, predisposition, will, his thoughts, training with regard to involuntary physiological responses, etc.) can greatly modulate the emotional reaction, to the point of diminishing or increasing its intensity, suppressing it altogether, or modifying the type of emotion that is experienced.

Nowadays, according to statistics, the context of traffic has become one of the social situations in which there is the greatest number of violent offences (Barthelmess, 1996). According to a study conducted in Spain by the publication entitled *Tráfico* (*Tráfico*, n° 27) in which 23.000 readers responded, the feelings that are most frequently associated with driving are, first, anxiety, second, aggressiveness, while other more positive emotions, such as sense of control and pleasure are experienced with much less frequency. In a similar study Reig and Soler (1986) point out that one out of every ten drivers describes themselves as aggressive at the wheel.

For Whitlock (1971), violence at the wheel is just another expression of social violence and the vehicle is "private territory on the road," almost equivalent to one's own home. According to this perspective, the more aggressive conduct of male drivers is explained in terms of the traditional role of the male as protector of the personal territory. The greater aggressiveness that is found among young drivers is explained by Whitlock alluding to the fact that the car may be the very only possession for individuals in this age group and for this reason their "territorial instincts" are especially concentrated in their car. Western society tends to consider the vehicle as something absolutely private and personal which the driver is prepared to defend (Lofland, 1973). In this way, the automobile has become an important mechanism in ameliorating the negative effects of high population density, something which offers protection from others (Mehrabian, 1975).

In any event, the driver's aggressive response is mediated by a series of factors which are fundamentally attributional, among which the following are worth mentioning (Tortosa, Montoro and Soler, 1987; Montoro, Soler and Tortosa, 1988):

The driver's interpretation regarding the cause of the aggressor's behavior. Any driving maneuver, for example, to suddenly brake, will presumably not produce a hostile reaction if "it is interpreted" as justified. However, the same sudden braking, when perceived as an unnecessary capricious act by another driver, will probably insight an aggressive reaction in the drivers that are circulating around him.

The aggressor's conduct after he commits his act. If there is some kind of an apology after an incorrect maneuver – for example, to start to overtake another vehicle when a third car has already begun the same maneuver – there will be a more positive attribution and it is less probable that aggressive reactions will appear in others. If, on the contrary, there is an indifferent attitude or the guilty driver displays hostile gestures, then the response in others is more apt to be violent.

The magnitude of the damage or annoyance. There is usually a direct relationship between the damage received or that one thinks that they could have received, and the intensity of the aggressive response.

The characteristics of the aggressor. It seems that people react with greater aggressiveness towards drivers who are members of certain ethnic minorities, or because of their sex, age or personal appearance. The negative attributions that are made concerning some of these groups and social stereotypes play an important role here.

The relationship, knowledge and identification one has with respect to the aggressor. For example, the same maneuver made by a friend or by a stranger will not produce the same effects because it will be perceived differently. This difference is due to the fact that the attribution of responsibility and intention can vary markedly depending on the relationship one has with the person who is theoretically the "aggressor."

The characteristics of the aggressor's vehicle. The aggressive response will also vary according to a whole constellation of characteristics which are influenced by prejudices and attributions, regarding such characteristics as the country or province of the vehicle's license plate, the make, model and year of the vehicle, the presence of ostentatious accessories or extravagant ornaments, etc.

Characteristics of the situation. For example, the presence of passengers in any of the vehicles involved, the type of relationship or kinship the passengers have with the driver, traffic density, traffic jams, etc. mediate in our perception and, as a result, influence the kind of response that we will display in front of others.

The psychophysiological state of the aggrieved driver. Psychophysiological states such as anxiety, stress, relaxation, fatigue, etc. are situational variables that also modulate the judgments that we are going to make concerning other road users.

Contrary to the popular belief that road rage is directly caused by the words, gestures and actions of others, it must be pointed out that on many occasions this response can be provoked or strengthened by factors that have little to do with the constant process of social interaction. The physical environment in which the driver finds himself immersed (the noise, heat, traffic jams, etc.) can greatly influence his psychophysiological state, the quality of his driving performance, his judgments and desires, and, in short, his way of interacting with others when driving.

In addition, agents outside the driving context can produce an elevated level of activation, which, in turn, can lead to the manifestation of aggressive conduct at the wheel. In our society today, as Pinillos (1985) points out, the overcrowding of our cities, the excessive number of interpersonal relationships that are imposed upon us, and the over stimulation produced by the urban environment elevate the level of physiological activation and irritability of the people who live and work there, causing an increment in the number of aggressive acts that appear while driving. Even an individual's personal and family history has a clear influence upon driving styles.

2. THE SPECIFIC PROBLEM OF ANXIETY IN THE CONTEXT OF DRIVING

It is undeniable that a person is not always in optimum psychophysiological condition when he is behind the wheel of a car. Driving is simply another activity and manifestation of man's daily life, and, as such, it is not immune to certain alterations, such as stress or anxiety. Anxiety, whether provoked by the traffic environment or by individual circumstances, can alter the physical or psychological conditions that are necessary in order to accomplish the complex task of driving safely, which makes it a variable that sheds light on the driving behaviors that are effective, efficient and safe.

The fact that a large number of subjects who have been involved in serious traffic accidents display high levels of anxiety and insecurity, that anxiety is one of the emotional reactions most frequently experienced during driving, and that it is a variable which is present, implicitly or explicitly in a large number of models developed to explain driver behavior (although none of them analyze this topic in depth) has made it apparent that we need to direct our attention to this variable. Moreover, within the field of Psychology and Traffic Safety the research on this variable is considerably behind compared to other fields.

In spite of the important role that this variable plays as a cause of accidents, the term anxiety or stress only appears occasionally in the extensive literature that exists concerning Psychology and Traffic Safety, and when it does appear it is with respect to traffic accidents.

In general, the research on anxiety and driving can be grouped into two broad categories: (a) The studies that investigate how the stress and anxiety which originates outside the driving situation affects the way an individual drives, and, as a consequence, becomes a direct or indirect causal agent of a large percentage of traffic accidents; and (b) The studies that center on how the driving situation itself produces anxiety in the population exposed to traffic or increases the levels of stress that already exits due to our modern lifestyle.

2.1. How Anxiety Produced Outside the Driving Situation Affects the Driver

The line of research on anxiety and traffic that has been most fruitful has been the one, which takes into account the effects of stressful life events on the driver and their relationship with traffic accidents.

The initial work of Holmes and Rahe (1967), carried out at the end of the sixties, concerning the appearance of stress as a consequence of the adjustments to life changes, prompted other studies that had as their objective to determine if changes in the life of a subject and the degree of subsequent adjustment could be significantly related to traffic accidents. These studies consider stress as an event or set of circumstances that supposedly requires a person to elicit a response that is unfamiliar, and they point out that the effects of these life events generally reduce the individual's capacity to respond to the outside world, leading to the appearance of physiological and behavioral (irritability, impulsive conduct, difficulties concentrating) maladjustments, which make it difficult or altogether impossible to perform the driving task correctly.

Selzer *et al.* (1968), demonstrated that social stress – associated with personal and economic conflicts – was significantly greater in a group of drivers that had been involved in serious accidents, compared with another group of drivers without accidents. One year later, Brenner and Selzer

(1969) concluded that the risk of suffering a mortal accident is five times greater in subjects that have experienced severe social stress compared with those who have not had this experience. Based on these findings, Selzer and Vinokur (1974, 1975) develop an instrument to predict traffic accidents as a function of the stress produced by the subjects' life events. This work was based on a multidimensional conception of stress and focused on what are called life changes and the subjective response to these changes, in the experiences of negative stress in various life contexts (work, matrimony, etc.) and in the physiological and behavioral manifestations of the stress. In 1975, this instrument was used to establish the relationship between life events, subjective stress and traffic accidents. In addition, this study showed that the accumulation of such events is significantly correlated with tension and cognitive and behavioral indicators of anxiety, although only in the case of life events which the subject does not desire (Vinokur and Selzer, 1975). These findings suggest that the desirability of these events from the individual's perspective is a crucial determinant of the way in which stress will be associated with accidents.

Similar results are offered by Fernández-Ríos *et al.*, (1987) who confirm the existence of statistically significant differences between drivers who were involved in accidents and those who were not, in the number of disturbing circumstances they experienced as well as in the degree of importance they gave to these events. The authors conclude that the stress that is generated by negative economic events (a cut in salary, denial of a loan, etc.) and negative health related events (physical illness, physical incapacity, lack of adequate medical assistance, etc.) are the type of stress that provides the best classification of drivers with and without accidents.

In an attempt to investigate further the relationship between life events and traffic accidents, Holt (1979) classified accident victims into two categories depending on their guilt and responsibility for the accident, i.e., guilty or not guilty. These authors found that the guilty subjects had experienced a greater number of unfortunate and problematic circumstances during the previous twelve months compared with the drivers who were not guilty.

In addition, McMurray (1970) observed that in the days, and even months after a highly stressful event, the possibility of suffering of an accident is substantially greater. Partial support is provided by the study conducted by Sobel and Underhill (1976) who explored the relationship between lifestyles and family, and driving record, pointing out that family problems and social stress in males (not in females) serve as a predictor of the probability of suffering a traffic accident.

Lastly, Finch and Smith (1970) discovered that eighty percent of a sample of subjects who died in traffic accidents had experienced a considerably stressful event during the twenty-four hours prior to the accident. However, although the authors attributed the cause of the accident to the effects of stress, they did not take into account that some people use the automobile as an instrument of suicide (McDonald, 1964; Crancer and Quiring, 1970; Signori and Bowman, 1974; McGuire, 1976), which makes us consider it premature, in light of the research presented, to be able to establish an absolute relationship between these two variables.

In general, the thesis that underlies these studies is the relevance of the variable *stress derived from life events* in explaining and predicting traffic accidents. It should be noted that this line of research was developed during the sixties and that presently it has been pretty much forgotten, or, at best, is the object of tough methodological criticism fundamentally directed at the lack of scientific rigor used in the evaluation of stress (Thoits, 1983). Today, the research regarding stressful life events has led the way to the consideration of cognitive variables that take into account the subjective impact that life events have on an individual.

There is no doubt that life events have a great influence over driving behavior, incrementing accident risk, especially if the individual turns to alcohol or drugs to solve the problematic situation (Wagenaar, 1983). When a subject finds himself in a state of anxiety, he usually does not possess adequate coping strategies, thus the availability of drugs and an automobile offer alternative means of reducing psychological tension. The use/abuse of certain medications and alcohol in order to reduce or manage the anxiety can make the driver more accident prone (Campbell and Singer, 1987; Cartwright *et al.*, 1996). Nevertheless, more rigorous research needs to be conducted before cause-effect relationships can be established, especially because these studies, in general, ignored the interaction between the driver and the traffic environment and the way in which the latter affects the individual's behavior. This factor is precisely what will be discussed in the following section.

2.2. How Anxiety Produced by the Traffic Environment Affects the Driver

2.2.1 The Specific Case of Professional Drivers

Traffic environment as a source of anxiety has been the object of research in recent years. Generally, these studies concentrate on specific groups of professional drivers: especially drivers who work in public transportation. The work that these drivers do is ranked among the most stressful and unhealthy modern-day occupations (Evans and Carrère, 1991). Gardell *et al.* (1982) found that bus drivers characterized their work as producing a high level of tension. Evans *et al.* (1987) informed that these professionals show significantly high levels of psychophysiological stress while they are working.

The data concerning accidents indicate that this occupational group has a substantially higher level of mortality and morbidity when compared with other professions, and that these professionals have higher percentages of absenteeism attributable to the stress that is produced while carrying out this activity (Backman, 1983; Dijkstra, 1983; Gardell *et al.*, 1982; Morris *et al.*, 1966; Netterstrom and Laursen, 1983; Rissler and Aronson, 1983; Syme *et al.*, 1983; Nijhuis and Bullinga, 1991). However, it is important to point out that very little analysis has been carried out regarding the specific aspects of the working environment, which are stressful.

The long-term impact of the demands of driving on the health and well-being of the drivers is evident. These drivers are more prone to suffer cardiovascular disease, in particular hypertension, stomach and intestinal problems, neurovegitative disorders, back problems, headaches and problems falling asleep (Mulders *et al.*, 1988; Costa *et al.*, 1988; Winkleby *et al.*, 1988).

In addition, as mentioned above, urban bus drivers, compared to individuals who have other occupations, also have higher percentages of absenteeism, attributable to the stress, which is produced in the traffic environment (Long and Perry, 1985; Kompier and Di Martino, 1995; Mulders *et al.*, 1988). Moreover, due to their poor health, the majority of drivers must retire before they are sixty years old (Bundesanstatlt and Arbeitsschutz, 1985). The simple routine that they are obligated to carry out during long runs has a cumulative impact on their personal health, their psychological adjustment, the way they act on the job and their family relationships. There are indices in the scientific literature that show that the mere act of traveling acts as a stressor, interfering with one's life and one's work, by reducing the amount of free time one has at the same time that it is associated with higher absenteeism, higher rates of accidents on the job and lower productivity (Costa *et al.*, 1988).

Some recent studies on this topic suggest that certain elements of the public transportation drivers' psychosocial environment contribute to high levels of stress related to their job. The characteristics that are most frequently mentioned as leading to stress in bus drivers are time pressures, frustration for being unable to attend entirely to the passengers, restriction from movement, constant vigilance, problems with the family and other social activities because of the rotational working hours, lack of control over working conditions (Finkelman *et al.*, 1977; Petry, 1980; Garbe, 1981; Jonah *et al.*, 1981; Gardell *et al.*, 1982; Feickert and Forrester, 1983; Melfort *et al.*, 1983; Patterson *et al.*, 1986; Reig, 1987; Mulders *et al.*, 1988; Bartone, 1989; Duffy and McGoldrick, 1990; Evans and Carrère, 1991; Carrère *et al.*, 1991; Bailer and Trakle, 1994; Evans, 1994; Evans and Eschwing, 1985; Kompier and Di Martino, 1995*)*.

Concerning this point, Raggatt (1991) found that among long distance bus drivers, driving during excessively long periods without a break, was the best simple predictor of maladaptive behavior, such as the use of stimulants and problems sleeping. Conversely, maladaptive behavior consistently predicts symptoms of stress. From this perspective, Bailer and Trakle (1994), after verifying the serious health problems experienced by public bus drivers produced by their high levels of stress, proposed mixed job strategies that prevent monotony and reduce the time spent at the wheel.

The problems that affect bus drivers are similar to those that *long distance truck drivers* suffer. In a study carried out by Vivoli *et al.* (1993) concerning long distance truck drivers, they found important individual differences in the ratio of catecolamines secretion, associated with variations in driving conditions. The ratio of epinephrine secretion was particularly high when the weather and the traffic conditions were more stressful (fog, traffic congestion, etc.). At the same time, the norepinephrine secretion in urine generally increased at the end of the work shift and while driving in fog, an increment that showed a correlation with state-anxiety scores. These findings led Hentschel *et al.* (1993) to conclude that, in addition to technical advances in the vehicles, it is necessary to consider psychological and situational variables such as anxiety in order to prevent accidents within this group of drivers.

2.2.2 Stressors that Arise within the Traffic Context

Not only professional driving can cause anxiety in the driver. The traffic situation itself, the environment through which the driver has to drive, the interaction with other drivers, etc., can provoke reactions of anxiety in the driver which can lead to greater levels of hostility and competitive behavior – triggering displays of direct aggression toward other drivers – as well as increasing the predisposition toward less cautious and more reckless driving, with a tendency toward making decisions which are more dangerous than usual or involve a reduced evaluation and perception of risk. In some of the stages of the exhaustion phase of stress, the anxiety makes even concentrating difficult, reduces attention and sensory thresholds and hinders the ability to integrated information. The result of many of these states is logically an accident (Fernández-Ríos *et al.*, 1987).

Also, there are times when the overwhelming number of signs on the road, which provide information or prohibit certain actions, make it impossible for the driver to process in there entirety and lead to what is known as perceptive stress, directly related to certain kinds of accidents. For this reason, in specific areas there has been a massive elimination of traffic signs.

In addition, pollution, noise, heat and traffic congestion can lead to reactions of anxiety (Finkelman *et al.*, 1977; Jonah *et al.*, 1981; Reig, 1987; Bartone, 1989; Evans and Carrère, 1991; Carrère *et al.*, 1991; Bañuls, 1993; Bañuls *et al.*, 1985). As a result, it is evident that the task of driving a vehicle includes the almost chronic presence of stressors of varying intensities.

Mackie *et al.* (1974), investigated the effects of heat and humidity on drivers. These authors found a significant elevation of body temperature and blood pressure, and an increment and variations of drivers' heart rate while driving in conditions of high temperature and humidity. These conditions are also associated with a reduction in one's subjective assessment of alertness, the activation of the central nervous system and the general level of activity of the driver.

Also, Neumann *et al.* (1978), conducted research investigating driver's reactions to high levels of noise and heat. The results revealed significantly higher levels of anxiety under conditions of high temperature and noise. Specifically, subjects showed a reduction in their responses in both visual and auditory tasks, an increments in heart rate, galvanic skin response, blood pressure and adrenocortisol secretion.

In a similar line of research, Wyon *et al.* (1996) warned about the stressful effects that heat can have on the driver. In their study, which compared the performance of two groups of drivers who drove under two different temperature conditions, they found that driving under high temperatures increases excitability and can alter the driver's level of vigilance and reaction times. Specifically, reaction times become longer and a portion of the traffic signals on the road are ignored.

Another factor that drivers are exposed to while they are driving is *atmospheric pollution*. Taking into account that the greatest concentrations of carbon monoxide is produced during rush hours and at the busiest intersections (Chaney, 1978), several laboratory experiments have demonstrated that levels of carbon monoxide equivalent to those found during rush hour can reduce visual activity in signal detection tasks (Halperin *et al.*, 1959; Horvath *et al.*, 1971).

But perhaps the stressor that has generated the greatest number of studies has been *traffic congestion*, fundamentally during the "rush hours." These situations are potentially stressful because of the delays they cause and the hostility that they sometimes provoke (Turner *et al.*, 1975). Evans and Carrère, 1991; Gulian *et al.* (1989a and b); and Bañuls *et al.* (1995), consider these types of situations particularly stressful because they reduce the driver's perceived control over his environment. Among the effects on the individual who is driving, these authors point out elevated blood pressure, cardiovascular changes, increments in electro dermal activity, skeletal muscle tone, and respiratory rate, reductions in motor skill and in general, an elevated index of somatic symptoms indicative of physiological anxiety.

A study conducted by Novaco *et al.* (1979) presents evidence that people who travel by automobile along congested routes show higher blood pressure, reduced motor skills, as well as an elevated index of somatic symptoms compared with those who drive along routes with little traffic. Schaeffer *et al.*, (1988) also found that drivers exposed to a lot of traffic congestion showed elevated blood pressure and deficits in responding to stimuli, specifically, driving to work along congested routes is associated with significant increments in systolic and diastolic blood pressure, which, for these authors, is indicative of physiological anxiety.

In a study by Evans and Carrère, (1991) in which they analyze the relationship between traffic congestion and psychophysiological anxiety among drivers who work in public transportation, they found that the catecolamines secretion in urine increased greatly as a function of being exposed to rush hour traffic congestion.

Other investigations designed to analyze the consequences of exposure to real or simulated traffic situations informed of considerable increments in heart rate (Simonson *et al.*, 1968; Shiomi, 1974), skin conductance (Heimstra, 1970), catecolamine secretion (Bellet *et al.*, 1969; Mulders *et al.*, 1988) and changes in respiratory rhythm (Shiomi, 1974).

More specifically, the highest heart rate is registered in highly complex traffic situations, such as driving in fog, in traffic jams, etc. (Hunt and May, 1968; Simonson *et al.*, 1968; Taggart *et al.,* 1969; Rutley and Mace, 1970; Aronow *et al.,* 1972; Littler *et al.*, 1973; Zeier, 1979; Vivoli *et al.,* 1993). It has also been demonstrated that sinus arrhythmia (normal variation in heart rate) is increased in boring or monotonous conditions, apparently reflecting a reduction in activation of the Central Nervous System and, possibly, a drop in alertness to respond to unexpected events (O'Hanlon and Kelley, 1977). At the same time, the levels of *catecolamines* in urine have been proven to be valid and reliable indicators of occupational stress during a workday (Evans *et al.*, 1987; Mulders *et al.*, 1982), as well as of psychological stress related in laboratory conditions.

From these investigations it can be deduced that exposure to certain conditions in the traffic environment are related to considerable increments in the activity of the Sympathetic Nervous System. Although it is true that it is not possible to use any physiological response indistinctly as an indicator of activation, a certain constancy has been found across subjects with respect to several physiological measures. The skin conductance response and heart rate appear to be the best indicators of a subject's state of anxiety, although some limitations have to be taken into account. Nevertheless, even though no measure of physiological responses taken in isolation seems to predict anxiety under various traffic conditions, a strong relationship exists between the cognitive evaluation that the driver makes of traffic situations and the physiological responses that he manifests (Shiomi, 1974; Jonah *et al.*, 1981; Hoyos and Kastner, 1986). Thus, Robertson (1978) using the *Driver Behavior Inventory* (Gulian *et al.*, 1988; 1989a y b) – a questionnaire which evaluates driver's stress – together with physiological measures, finds that two factors in the inventory (alertness while driving, and frustration and tension on being unable to overtake another vehicle) and an acceleration of the heart rate in car drivers occur concurrently. Similarly, Vivoli *et al.* (1993) found a relationship between levels of epinephrine in urine and scores on a scale measuring state anxiety.

2.2.3 The Transactional Perspective

Despite the fact that the nervous reactions that are often produced while driving are relatively minor, various studies point out that different environmental factors that are potentially anxiety producing can lead to deficits in the driver's abilities and reduce the overall level of driving safety (Altman, 1975; Stokols, 1976; Rule and Nesdale 1976; Matthews *et al.*, 1996b). However, interactionist theories regarding anxiety suggest that subjects' reactions to traffic events are not uniform, instead they are mediated by personal and social variables (Appley and Trumbull, 1967; Glass and Singer, 1972; Lazarus and Launier, 1978).

Based on these assumptions, the research group that developed the *Driver Behavior Inventory*, (Gulian, 1987; Gulian *et al.*, 1988, 1989a and b, 1990; Matthews *et al.*, 1991; Matthews *et al.*, 1996a and b), focus on "driver's stress" in an attempt to discover the dimensions that define it and the factors that contribute to its appearance and duration. In these studies the term "driver's stress" is used as a synonym for the subjective perception of stress reported by drivers solely in reference to the task of driving and does not include the possible effects of factors outside the driving context. Nevertheless, these authors consider "driver's stress" to be a function of factors which are intrinsic to driving (traffic conditions) and factors which are extrinsic to driving (one's personal life). On the one hand, family or work related problems can contribute substantially to driver's stress if they interfere with the driver's ability to attend to the demands of the traffic environment. At the same time, stress produced by traffic events may not only be experienced while driving, but also later on during other activities. The emotional demands of driving not only can reduce the driver's capacities on the road, in addition they can give rise to emotional and behavioral problems on arriving home from work (Glass and Singer, 1972; Sherrod, 1974).

The authors point out that satisfactory living conditions (health, family, work, etc.) can lessen the driver's level of stress, whereas a problem in any one of these areas can influence the driver's response to specific traffic incidents and driving in general, incrementing the driver's stress at all levels.

The perspective that these authors use in studying driver's stress is based on Lazarus' Transactional model of psychological stress. According to this model, stress appears when the subject's perception of the demands of the task, in this case driving, exceed the driver's abilities to cope with the situation (Lazarus and Folkman, 1984). Driving is assumed to be a stressful task because it involves a set of circumstances, which are outside the realm of the driver, which at any given moment may require extraordinary or unusual responses. These authors point out that marked individual differences exist with regard to susceptibility to stress while driving. This perspective emphasizes that the cognitive evaluation of the events is an important determinant of the level of the stress response. Driver's stress is defined as the set of responses associated with the perception and evaluation of the driving task. Implicit in this transactional perspective is the idea that driver's stress is probably a complex syndrome determined by the interaction of situational and personal factors. The stress response can be an emotional reaction (e.g., anxiety), a physiological response (e.g., an increase in heart rate) and/or a behavioral response (e.g., adopting an aggressive driving style). Consequently, anxiety is considered to be a stress reaction, a subjective experience that is directly related to the seriousness and the duration of the threat. At the same time, these authors point out that physiological and behavioral responses to traffic events are affected by the driver's evaluation of the situation (Hoyos and Kastner, 1986; Gulian, 1987).

These authors think that driver's stress can be experienced on two different but related levels: (1) According to the classification of stress producing factors established by Lazarus and Cohen (1977), at an initial level, stress could be induced by traffic events that occur only on certain occasions over which the driver has only limited control and consequently become situations likely to produce anxiety in drivers. The highest level of anxiety in these situations would debilitate the driver's ability to process information rapidly and satisfactorily, impeding the use of adequate response patterns. (2) On a second level, driver's stress can be the result of continued exposure to traffic situations, which exceed the driver's ability to function in a safe and satisfactory manner. Once again according to Lazarus and Cohen (1977), stress would be the result of small stressors, which occur daily, the effects of which are cumulative. Here, driver's stress is the consequence of emotional, cognitive and physiological responses that accumulate in traffic situations, such as long trips and daily journeys (Stokols and Novako, 1981).

2.2.4 Research on Personality Traits

An entirely different point of view is offered by Sholomo Giora Shoham and collaborators (Shoham, 1975; Shoham *et al.*, 1974; 1976; 1977; 1984), who address the problem of anxiety in the context of driving by conducting research on personality traits, adopting a position based on *trait theory.* For these authors, anxiety is understood to be a personality trait, a predisposition to respond which is relatively independent of the situation. Their research is based on the hypothesis that subjects who are accident prone and/or inclined to commit traffic violations do not all possess the same personality traits, but rather their behavior expresses one of two different personalities, that is: the anxious or the reckless driver. The anxious driver, due to structural defects in his innate personality, tends to enter into a state of anxiety when he is confronted with risky situations in traffic. In other words, the driver's level of anxiety (state) at a given moment is affected by the interaction of two components: the characteristics of his personality (trait) and features of the stimuli in the traffic situation.

The anxious driver is defined as a subject with a high level of trait-anxiety, who possesses a great ability to learn, is easily conditioned and who is not impulsive. The extremely anxious driver accepts the normative system of traffic laws as legitimate and is a citizen who accepts and abides by the law in all contexts, including traffic situations. Because the traffic regulations are highly internalized, the desire to take risks is reduced. This type of driver is not a thrill seeker and does not take unnecessary risks while driving. However, when this type of driver is involved in an ambiguous or conflictive situation, he tends to take greater risks in order to cope with the situation and release tension. This driver mixes his emotions with his decisions, and tends to develop a state of anxiety in situations in which one must decide rapidly and almost instinctively. Traffic situations which create pressure or are ambiguous will produce an approach-avoidance conflict (similar to that described by Fuller in his Approach-Avoidance model) in the driver, which leads to confusion, loss of self-control and control over the vehicle, which will probably lead to committing a traffic violation or being involved in an accident.

Also based on the theory of trait-anxiety, is the research carried out by K. Tiwari and P. B. Behere in 1983. These authors tried to prove that accidents are not a matter of chance, that personality variables play an important role in causing accidents. They focused their attention on exploring how anxiety as a personality trait can increase the probability of having an accident, finding that subjects who were involved in accidents possessed higher levels of trait-anxiety in comparison to accident-free subjects. However, in this study state-anxiety was unable to differentiate between these two groups of drivers.

2.3. The Evaluation of Anxiety in the Context of Driving

The measurement of anxiety is one of the topics which psychological evaluation has dedicated the most effort (Miguel-Tobal and Cano-Vindel, 1989). The fact that distinct theoretical positions regarding anxiety have lead to different ways of comprehending this phenomenon has brought with it the development of a great number of instruments and techniques of evaluation, which, in many cases, share few common elements.

Nevertheless, in spite of the plethora of viewpoints that have existed and still exist concerning the concept and evaluation of anxiety, in recent years this confusion is giving way to more clarity, although not precisely in the field of Traffic Safety.

As mentioned previously, traditionally, anxiety produced in traffic situations has been investigated by means of the evaluation of the driver's physiological responses to traffic events (Robertson, 1987; Gulian, 1987). Several studies designed to analyze the consequences of exposure to real or simulated traffic situations considered the increment in neuroendocrine responses while driving a sign of the driver's effort to cope with high pressure situations in order to maintain a satisfactory level of activity. These studies inform of considerable increments in heart rate (Simonson *et al.*, 1968; Hunt and May, 1968; Taggart *et al.,* 1969; Rutley and Mace, 1970; Aronow *et al.,* 1972; Littler *et al.*, 1973; Shiomi, 1974; Mackie *et al.*, 1974; O'Hanlon and Kelley, 1977; Helander 1978; Neumann *et al.*, 1978; Zeier, 1979; Vivoli *et al.,* 1993), blood pressure (Mackie *et al.*, 1974; Neumann *et al.*, 1978; Novaco *et al.,* 1979; Evans *et al.*, 1987; Schaeffer *et al.*, 1988), skin conductance (Hulbert, 1957; Heimstra, 1970; Helander 1978; Zeier, 1979), catecolamines secretion (Bellet *et al.*, 1969; Mulders *et al.*, 1982; Evans *et al.*, 1987; Mulders *et al.*, 1988; Evans *et al.*, 1987; Evans and Carrère, 1991; Vivoli *et al.,* 1993), and changes in respiratory rate (Shiomi, 1974).

In general, the physiological measures are relatively free from voluntary influences of the subject, and consequently free from bias and, when measured using objective methods, they are highly reliable. However, these measurements require personnel who are experts in these specific techniques and the use of these apparatus, which, in many cases, makes it difficult to employ these methods. Also, these types of measures are usually taken in laboratory conditions and, as a result, they share the problems inherent in measures of motor behavior and self-ratings (Miguel-Tobal, 1990).

In comparison to this type of measurements, the most commonly used instruments to evaluate anxiety, from any theoretical perspective, have been questionnaires, inventories and scales. However, many of them do not include information that is pertinent to traffic safety because they are based on trait models, which do not take into account the situational component.

In the previous section we repeatedly alluded to the scarcity of studies on anxiety and driving. This paucity is also evident in the number of self-evaluations designed specifically to evaluate anxiety and/or stress produced in traffic situations. Specifically, in a review of the literature only two were found: the *Driver Behavior Inventory* (Gulian *et al.*, 1988, 1989a and b) and the *Anxiety Questionnaire* (Shoham *et al.*, 1974).

The *Driver Behavior Inventory* (D.B.I.) is based, as said before, on Lazarus's Transactional model of psychological stress. For the authors the principal objective of the development of this instrument was to determine the dimensions of driver's stress and to identify the variables outside the driving situation capable of predicting this stress. This questionnaire originally consisted of 97 items that included the following areas: demography, accidents and attitudes toward accidents, state of health and its consequences for driving, work and personal relations, job related, domestic, and personal problems, emotions and attitudes toward driving, toward traffic situations and toward other road users, and coping strategies in general situations and specifically in traffic situations. This instrument measures the frequency with which one experiences specific behaviors (always-never) and categorizes behaviors and opinions on a 4-point scale (a lot-not at all).

The analysis of the D.B.I. reveals the existence of a general driving factor (General driver's stress), as well as five factors that define the general evaluation that the subject makes of driving: 1) dislikes driving and informs of anxiety; 2) irritation on overtaking other vehicles and on being overtaken by other drivers; 3) aggressive driving; 4) increased concentration and vigilance; 5) frustration and tension on being unable to overtake another vehicle. The appearance of driver's stress is related to increased aggression, dislike for driving, frustration and irritation provoked by the interaction with other drivers, primarily in relation to overtaking, and increased alertness which was only a positive factor in complex traffic situations.

After several attempts to refine the D.B.I., and because the authors considered that this instrument did not provide a comprehensive measure of the affective reactions experienced while driving, Matthews *et al.*, (1996a; 1996b) present a new version of the questionnaire called the Driver Stress Inventory (DSI), in which they include measures of fatigue and thrill seeking, and thus, according to the authors, incrementing the capacity of the questionnaire to identify reckless drivers in order to develop specific programs of intervention for these drivers.

In contrast, the *Anxiety Questionnaire* (Shoham *et al.*, 1974) is based on Taylor's (1953) *Manifest Anxiety Scale* and the *Sarason and Mandler's Anxiety Questionnaires for Drivers.* This scale consists of 8 items that require an affirmative or negative response. The items represent the physiological component of anxiety (before a long trip I feel my heart beat rapidly, my hands sweat while I am driving), as well as the cognitive responses of anxiety (driving when there is a lot of traffic makes me nervous, not feeling safe leads to more cautious driving and fewer accidents, when I am in a hurry I drive worse than usual), although the situational spectra that is included is minimal and the responses are only treated as manifestations of trait-anxiety.

The remaining investigations used the *Manifest Anxiety Scale* (Taylor, 1953), the *State-Trait Anxiety Inventory: Trait Scale* (Spielberger *et al.*, 1970) or direct physiological measures of anxiety. No studies were found that employed direct motor responses of anxiety, probably due to the difficulties entailed in controlling the situational context in which these behaviors occur.

In Spain, a general inventory, called I.S.A.T. (Inventory of Anxiety Producing Traffic Situations), has been developed to evaluate the anxiety, which is elicited in a representative sample of driving situations which are potentially anxiety producing. The I.S.A.T. provides a global score, as well as scores pertaining to four specific situation areas: situations that involve evaluation (self assessment or external evaluation), criticism and aggression, external impediments and delays, and situations that involve evaluation by an authority figure (Bañuls, 1993; Bañuls *et al.*, 1993; 1994a, b and c; 1995; 1996). Because situations that involve evaluations have been classically considered as anxiety provoking – as evidenced by the importance of this factor in self-rating scales, such as the *Inventory of anxiety evoking situations and anxious responses* (Miguel-Tobal and Cano-Vindel, 1988), *Inventory of attitudes toward general situations* (Bermúdez, 1983), the *S-R Inventory of General Trait Anxiousness* (Endler, 1978; Flood and Endler, 1980), and the *Endler Multidimensional Anxiety Scales-Perception* (Revised) (King and Endler, 1989) – the component "Situations Involving Evaluation" is represented strongly in two of the factors in this questionnaire, the first of which is the most heavily weighted, explaining more than half of the total variance.

3 REFERENCES

ALTMAN, I. (1975), *The Environment and Social Behavior: Privacy, Personal Space, Territory and Crowding,* Monterey, Calif.: Brooks-Cole.

APPLEY, M. H., TRUMBULL, R. (1967), *Psychological Stress*, New York: Appleton-Century-Crofts.

ARONOW, W. S., HARRIS, C. N., ISBELL, M. W., ROKAW, M. D., IMPARATO, B. (1972), "Effect of Freeway Travel on Angina Pectoris", *Annals of Internal Medicine*, 77, 669-676.

BACKMAN, A. L. (1983), "Work Conditions of Bus Drivers in Finland", paper presented at the *Conference on the Working Environment in Urban public Transport*, Stockholm.

BAILER, H., TRANKLE, U. (1994), "Mischarbeit mit dem Ziel der Verringerung von Belastungen und Beanspruchungen: Uberlegungen am Beispiel Fahrtatigkeiten im offentlichen Personennahverkehr", *Zeitschrift-fur-Arbeits-und-Organisationspsychologie,* 38 (3), 126-130.

BAÑULS EGEDA, R. (1993), *Aspectos emocionales y actividad de conducción: El Inventario de Situaciones Ansiógenas en el Tráfico (I.S.A.T.)*, Doctoral Thesis, Valencia.

BAÑULS EGEDA, R., CARBONELL VAYA, E., MIGUEL TOBAL, J. J. (1993), "Inventario de Situaciones Ansiógenas en el Tráfico", in FORNS., M., ANGUERA, M. T. (Eds.), *Contribuciones recientes a la evaluación psicológica,* Barcelona: P.P.U.

BAÑULS EGEDA, R., CARBONELL, E., MIGUEL TOBAL, J. J. (1994a), "Anxiety and Traffic Safety", in *23rd International Congress of Applied Psychology*, Madrid.

BAÑULS EGEDA, R., CASANOVES, M., CHISVERT, M. (1996), "Emotional Anxiety in Traffic Situations as a Function of Driving Experience", in *International Conference on Traffic and Transport Psychology*, Valencia.

BAÑULS EGEDA, R., LÓPEZ-LATORRE, M. J., TEJERO, P. (1994c), "Anxiety and Recklessness in Driving", in *23rd International Congress of Applied Psychology*, Madrid.

BAÑULS EGEDA, R., SANMARTÍN, J., CANO-VINDEL, A. (1994b), "Procedure Followed in the Elaboration of an Instrument for Evaluating Anxiety when Driving", in *23rd International Congress of Applied Psychology*, Madrid.

BAÑULS, R., CANO, A., CARBONELL, E., MIGUEL TOBAL, J. J. (1995), "Reacciones Emocionales, Diferencias Individuales y Tráfico", in MONTORO, L., CARBONELL, E., SANMARTÍN, J., TORTOSA, F. (Eds), *Seguridad Vial: Del Factor Humano a las Nuevas Tecnologías*, Editorial Síntesis.

BARTHELMESS, W. (1996), "Agressiviness in Road Traffic", in *International Conference on Traffic and Transport Psychology*, Valencia.

BARTONE (1989), "Predictores of Stress-Related Illness in City Bus Drivers", *Journal of Ocupational Medicine*, 31 (8), 657-663.

BELLET, S., ROMAN, L., KOSTIS, J. (1969), "The Effect of Automobile Driving on Catecholamine and Adrenocortical Excretion", *The American Journal of Cardiology*, 24, 265-368.

BERMUDEZ, J. (1983), "Modelo interactivo de ansiedad: implicaciones y contrastación empírica", *Revista de Psicología General y Aplicada*, 38, (6), 1003-1030.

BRENNER, B., SELZER, M. L. (1969), "Risk of Causing a Fatal Accident Associated with Alcoholism, Psychopathology and Stress: Further Analysis of Previous Data", *Behavioral Science*, 14, 490-495.

CAMPBELL, F., SINGER, G. (1987), *Estrés, drogas y salud*, Barcelona: Ariel.

CANO-VINDEL, A. (1989), *Cognición, emoción y personalidad. Un estudio centrado en la ansiedad*, Madrid: Publicaciones de la Universidad Complutense.

CARRÈRE, S., Evans, G. W., PALSANE, M. N., RIVAS, M. (1991), "Job Strain and Occupational Stress among Urban Public Transit Operators", *Journal of Occupational Psychology*, 64, 305-316.

CARTWRIGHT, S., COOPER, C. L., BARRON, A. (1996), "The Company Car Driver: Occupational Stress as a Predictor of Motor Vehicle Accident Involvement", *Human relations*, 49 (2), 195-208.

CHANEY, L.W. (1978), "Carbon Monoxide Automobile Emissions Measured from the Interior of a Traveling Automobile", *Science*, 199, 1203-1204.

COSTA, G., PICKUP, L., DIMARTINO, V. (1988), "Commuting – A Further Stress Factor for Working People: Evidence from the Eureopean Community, II: An Empirical Study", *International Archives of Occupational and Environmental Health*, 60, 377-385.

CRANCER, A., QUIRING, D. L. (1970), "Driving Records of Persons Hospitalized with Suicidal Gestures", *Behavioral Research in Highway Safety*, 1, 33-42.

DIJKSTRA, A. (1983), "Personal Functioning, Health and Absenteeism in a Public Transport Company", in *Conference on the Working Environment in Urban public Transport*, Stockholm.

DUFFY, C. A., MCGOLDRICK, A. E. (1990), "Stress and the Bus Driver in the UK Transport Industry", *Work and Stress,* 4 (1), 17-27.

ENDLER, N. S. (1978), "The Interaction Model of Anxiety. Some Possible Implications", in LANDERS, O. M., CHRISTINA, R. W. (Eds), *Psychology of motor behaviors and sport,* Human Kinetics, Champaing.

EVANS, G. W., CARRÈRE, S. (1991), "Traffic Congestion, Perceived Control, and Psychophysiological Stress among Urban Bus Drivers", *Journal of Applied Psychology*, 76 (5), 658-663.

EVANS, G. W., PALSANE, M. N., CARRÈRE, S. (1987), "Type A Behaviour and Occupational Stress: A Cross-Cultural Study of Blue-Collar Workers", *Journal of Personality and Social Psychology*, 52 (5), 1002-1007.

EVANS, L. (1984), "Human Behavior Feedback and Traffic Safety", *Human Factors*, 27, 55-76.

EVANS, L., ESCHWING, R. C. (1985), *Human Behavior and Traffic Safety,* New York/London: Plenum Press.

FEICKERT, D., FORRESTER, K. (1983), "Stress Factors in Urban Public Transport Work", in *Conference on the Working Environment in Urban Public Transport,* Stockholm.

FERNANDEZ-RIOS, M., MORENO-JIMENEZ, B., GIMENEZ-GARCIA, L. (1987), "El estrés en los accidentes de tráfico", *Mapfre Seguridad*, 26, 13-20.

FINCH, J. R., SMITH, J. P. (1970), *Psychiatric and Legal Aspects of Automobile Fatalities,* Springfield, III Thomas.

FINKELMAN, J. M., ZEITLIN, L. R., FILIPPI, J. A., FRIEND, M. A. (1977), "Noise and Driver Performance", *Journal of Applied Psychology*, 62 (6), 713-718.

FLOOD, M., ENDLER, N. S. (1980), "The Interaction Model of Anxiety: An Empirical Test in an Athletic Competition Situation", *Journal of Research in Personality*, 14, 329-339.

GARBE, C. (1981), *Gesundheitszustand und gesundheitliche Risiken von Linienbusfabrern in Berlin (West)*, Berlin: Dietrich Reimer.

GARDELL, B., ARONSSON, G., BARKLÖF, K. (1982), "The Working Environment for Local Pulic Transport Personnel", in *Report from the Swedish Work Environment Fund,* Stockholm.

GLASS, D. C., SINGER, J. (1972), *Urban Stress: Experiments on Noise and Social Stressors,* New York: Academic Press.

GULIAN, E. (1987), "Driver Stress: A Literature Review", in *CPERU Report 1,* Aston Business School, Aston University, Birmingham, U. K.

GULIAN, E., DEBNEY, L. M., GLENDON, A. I., DAVIES, D. R., MATTHEWS, G. (1989b), "Coping with Driver Stress", in MCGUIGAN, F. J., SIME, W. E., WALLACE, J. M. (Eds.), *Stress and Tension Control* (vol. 3), New York: Plenum Press.

GULIAN, E., GLENDON, A. I., MATTHEWS, G., DAVIES, D. R., DEBNEY, L. M. (1990), "The Stress of Driving: A Diary Study", *Work-and-Stress*, 4 (1), 7-16.

GULIAN, E., GLENDON, A. I., MATTHEWS, G., DAVIES, D. R., DEBNEY, L. M. (1988), "Exploration of Driver Stress Using Self-Reported Data", in ROTHENGATTER, J. A., DE

BRUIN, R. A. (Eds.), *Road User Behaviour: Theory and Research*, Assen/Maastricht, The Netherlands: Vam Gorcum.

GULIAN, E., MATTHEWS, G., GLENDON, A. I., DAVIES, D. R., DEBNEY, L. M. (1989a), "Dimensions of Driver Stress", *Ergonomics*, 32 (6), 585-602.

HALPERIN, M. H., MCFARLAND, R. A., NIVEN J. L., ROUGHTON, F. W. (1959), "The Time Course of the Effects of Carbon Monoxide on Visual Thresholds", *Journal of Physiology*, 146, 583-593.

HEIMSTRA, N. W. (1970), "The Effects of Stress-Fatigue on Performance in a Simulated Driving Situation", *Ergonomics*, 13 (2), 209-218.

HELANDER, M. (1978), "Applicability of Drivers' Electrodermal Responses to the Design of the Traffic Environment", *Journal of Applied Psychology, American Psychologial Association*, 63, (4), 481-488.

HENTSCHEL, U., BIJLEVELD, C., KIESSLING, M., HOSEMANN, A. (1993), "Stress-Related Psychophysiological Reactions of Truck Drivers in Relation to Anxiety, Defense, and Situational Factors", *Accident Analysis and Prevention*, 25 (2) 115-121.

HOLMES, D. S., RAHE, R. H. (1967), "The Social Readjustment Rating Scales", *Journal of Psychosomatic Research*, 11, 213-218.

HOLT, L. (1979), "Stressful Life Events Preceding Road Traffic Accidents", *The British Journal of Accident Surgery*, 13, 111-115.

HORVATH, S. M., DAHMS, T. E., O'HANLON, J. F. (1971), "Carbon Monoxide and Human Vigilance: A Deleterious Effect of Present Urban Concentrations", *Archives of Environmental Health*, 23, 343-347.

HOYOS, C. G., KASTNER, M. (1986), *Belastung und Beanspruchung von Kraftfahrern*, Bergisch Gladbach: Bundesanstalt.

HULBERT, S. F. (1957), "Drivers's G.S.R.s in Traffic", *Perceptual and Motor Skills*, 7, 305-315.

HUNT, T. J., MAY, P. I. (1968), *A Preliminary Investigation into a Psychological Assessment of Driving Stress*, London: Metropolitan Police, Accident Research Unit.

JONAH, B. A., BRADLEY, J. S., DAWSON, N. E. (1981), "Predicting Individual Subjective Responses to Traffic Noise", *Journal of Applied Psychology*, 66 (4), 490-501.

KING, P. R., ENDLER, N. S. (1989), "Improving the Assessment of Situation Perception with Respect to Anxiety", *Personality and Individual Differences*, 10, 1063-1069.

KOMPIER, M., DI MARTINO, V. (1995), "Review of Bus Drivers' Occupational Stress and Stress Prevention", *Stress Medicine*, 11(4), 253-262.

LANG, P. J. (1968), "Fear Reduction and Fear Behaviour: Problems in Treating a Construct", in SHLEIEN, J. M. (Ed.), *Research in Psychotherapy, III*, American Psychological Association, Washington.

LAZARUS, R. S., COHEN, J. B. (1977), "Environmental Stress", in ALTMAN, I, WOHLWILL, J. (Eds.), *Human Behaviour and Environment*, New York: Plenum Press.

LAZARUS, R. S., FOLKMAN, S. (1984), *Stress, Appraisal and Coping*, New York: Spinger Publishing Company.

LAZARUS, R. S., LAUNIER, R. (1978), "Stress-Related Transactions between Person and Environment", in PERVIN, L. A., LEWIS, M. (Eds), *Perspectives in Interactional Psychology*, New York: Plenum Press.

LITTLER, W. A., HONOUR, A. J., SLEIGHT, P. (1973), "Direct Arterial Pressure and Electrocardiogram during Motor Car Driving", *British Medical Journal*, 2, 273-277.

LOFLAND, L. H. (1973), *A World of Strangers: Order and Action in Urban Public Space,* New York: Basic Books.

LONG, L., PERRY, J., (1985), "Economic and Occupational Causes of Transit Operator Absenteeism: A Review of Research", *Transport Reviews, 5, 247-267.*

MACKIE, R. R., O'HANLON, J. F., MCCAULEY, M. (1984), "A Study of Heat, Noise, and Vibration in Relation to Driver Performance and Physiological Status", in Washington, D.C.: *Report *DOT 11S -801-215*, National Highway Traffic Safity Administration, Dec. 1974.

MATTHEWS, G., DESMOND, P., JOYNER, L., CARCARY, B. (1996b), "A Comprehensive Questionnaire Measure of Driver Stress and Affect", in *International Conference on Traffic and Transport Psychology*, Valencia.

MATTHEWS, G., DORN, L., GLENDON, A. (1991), "Personality Correlates of Driver Stress", *Personality and Individual Differences*, 12 (6), 535-549.

MATTHEWS, G., SPARKES, T., BYGRAVE, H. (1996a), "Attentional Overload, Stress, and Simulated Driving Performance", *Human Performance,* 9 (1) 77-101.

MCDONALD, J. M. (1964), "Suicide and Homicide by Automobile", *American Journal of Psychiatry*, 121, 366-370.

MCGUIRE, F. L. (1976), "Personality Factors in Highway Accident", *Human Factors*, 18 (5), 433-442.

MCMURRAY, L. (1970), "Emotional Stress and Driving Performance: The Effect of Divorce", *Behavioral Research in Highway Safety*, 1, 100-114.

MEHRABIAN, A. (1975), *Public Spaces and Private Spaces,* New York: Basic Books.

MELFORT, J., REINERS, H., SCHUH, J. (1983), *Arbeitsbedingungen von Linienbus- und StraBenbahnfabrern der Dortmunder stadtwerke aktiengesellshaft,* Wilhelmshaven: Wirtschaftsverlag.

MIGUEL-TOBAL, J. J., CANO-VINDEL, A. (1988), *Inventario de situaciones y respuestas de ansiedad (ISRA)*, (2° edición) TEA, Madrid.

MIGUEL-TOBAL, J. J., CANO-VINDEL, A. (1989), "Avances en la evaluación de la ansiedad y sus implicaciones en la elección del tratamiento", *Terapia del Comportamiento*, n° 22-23, 125-128.

MIGUEL-TOBAL, J. J. (1990), "La ansiedad", in MAYOR, J., PINILLOS, J. L. (Eds), *Tratado de Psicología General, 8: Motivación y emoción*, Madrid: Alhambra.

MONTORO, L., SOLER, J., TORTOSA, F. (1988), "La atribución: Lo que esperamos de los demás conductores", *Revista Tráfico*, *33,* 48-49.

MORRIS, J. N., KAGAN, A., PATTISON, D. C., GARDNER, M. J. (1966), "Incidence and Prediction of Ischaemic Heart-Disease in London Busmen", *Lancet*, 2, 7463, 553-559.

MULDERS, H., MEIJMAN, T., MULDERS, B., KOMPIER, M., BROERSEN, S., WESTERINK, B., O'HANLON, J. (1988), "Ocupational Stress in City Bus Drivers", in ROTHENGATTER, J. A., DE BRUIN, R. A. (Eds.), *Road User Behaviour: Theory and Research,* Assen/Maastricht, The Netherlands: Vam Gorcum.

MULDERS, H. P., MEIJMAN, T. F., O'HANLON, J. F., MULDER, G. (1982), "Differential Psychophysiological Reactivity of City Bus Drivers", *Ergonomics*, 25, 1003-1011.

NETTERSTROM, B., LAURSEN, P. (1983), "Incidence and Prevalence of Ischaemic Heart-Disease among Bus Drivers in Copenhagen", in *Conference on the Working Environment in Urban public Transport*, Stockholm.

NEUMANN, E. S., ROMANSKY, M. L., PLUMMER R. W. (1978), "Passenger Car Comfort and Travel Decisions", *Journal of Transport Economics and Policy*, 12, 231-243.

NIJHUIS, F. J., BULLINGA, R. (1991), "Preventie: werken aan werkplek en organisatie", Health problems of bus drivers (1990, Nijmegen, Netherlands), *Gedrag and Gezondheid Tijdschrift voor Psychologie and Gezondheid*, 19 (1), 36-39.

NOVACO, R. W., STOKOLS, D., CAMPBELL, J., STOKOLS, J. (1979), "Transportation, Stress, and Community Psychology", *American Journal of Community Psychology*, 7 (4), 361-380.

O'HANLON, J. F., KELLEY, G. R. (1977), "Comparison of Perfomance and Physiological Changes between Drivers who Perform Well and Poorly during Prolonged Vehicular Operation", in MACKIE, R. R. (Ed.), *Vigilance, Theory, Operational Perfomance and Physiological Correlates*, New York: Plenum Press.

PATTERSON, P. K., EUBANKS, T., RAMSEYER, R. (1986), "1985 Survey of Back Injury and Other Health Factors among TRI-MET Bus Drivers", Unpublished paper, Oregon Health Science University.

PETRY, H. (1980), "Arbeits – und sozialmedizinische Probleme beim Fahrpersonal offentlicher Verkehrsbetriebe", *Verkehr und Technic*, 33, 16-18 and 53-56.

PINILLOS, J. L. (1985), "Personalidad y conducción", in *I Reunión Internacional de Psicología de Tráfico y Seguridad Vial*, 397-403, Madrid: Dirección General de Tráfico.

RAGGATT, P. T. F (1991), "Work Stress among Long-Distance Coach Drivers: A Survey and Correlational Study", *Journal of Organizational Behaviour*, 12, 565-579.

REIG, A. (1987), "Estrés y conducción", in SOLER, J., TORTOSA, F. (Eds.), *Psicología y Tráfico*, Valencia: Nau Llibres.

REIG, A., SOLER, J. (1986), *Accidentes de tráfico y comportamiento: un estudio empírico en Alicante*, Instituto Juan Gil Albert/Diputación de Alicante.

RISSLER, A., ARONSON, G. (1983), "Stressors, Psychophysiological Reactions and Health Complaints among Urban Bus Drivers", in *Conference on the Working Environment in Urban public Transport*, Stockholm.

ROBERTSON, S. A. (1978), "Stress Related to Driving: An Exploratory Study", *Transport Studies Unit Report 420*, Oxford: Oxford University.

ROBERTSON, S. A. (1987), "Literature Review on Driver Stress", *Transport Studies Unit Report 364*, Oxford: Oxford University.

RULE, B. G., NESDALE, A.R. (1976), "Environmental Stressors, Emotional Arousal, and Aggression", in SARASON, I. G., SPIELBERGER, C. D. (Eds.), *Stress and Anxiety*, New York: Halsted Press.

RUTLEY, K. S., MACE, D. G. W. (1970), "Heart Rate as a Measure in Rod Layout Design", London: *Research Report No. RRL-LR 347*, Road Research Laboratory, British Ministry of Transportation.

SCHAEFFER, M. H., STREET, S. W., SINGER, J. E., BAUM, A. (1988), "Effects of Control on the Stress Reactions of Commuters", *Journal of Applied Social Psychology*, 18, 944-957.

SELZER, M. L., VINOKUR, A. (1974), "Life Events, Subjective Stress, and Traffic Accidents", *American Journal of Psychiatry*, 131 (8), 903-906.

SELZER, M. L., VINOKUR, A. (1975), "Role of Life Avents in Accident Causation", *Mental Health and Society*, 2, 36-54.

SELZER, M. L., ROGERS, J. E., KERN, S. (1968), "Fatal Accidents: The Role of Psychopathology, Social Stress and Acure Disturbance", *American Journal of Psychiatry*, 124, 1028-1036.

SHERROD, D. R. (1974), "Crowding, Perceived Control, and Behavioural Aftereffects", *Journal of Applied Social Psychology*, 4, 171-186.

SHIOMI, K. (1974), "The Roles of Anticipation and Defense Mechanism Experimentally Induced Anxiety", *Psychologia*, 17 (3), 150-158.

SHOHAM, S. G. (1975), *Traffic Offenders' Behaviour, Judgement and Punishment,* Report to the Ministry of Transportation: Tel Aviv University.

SHOHAM, S. G., GEVA, N., MARKOVSKI, R., KAPLINSKY, N. (1974), *Relationship between Traffic Offences, Anxiety, Risk-Taking and Internalization of Traffic Norms,* Report to the Ministry of Transportation: Tel Aviv University.

SHOHAM, S. G., GEVA, N., MARKOVSKI, R., KAPLINSKY, N. (1976), "Internalization of Norms, Risk-Perception and Anxiety as Related to Driving Offences", *The British Journal of Criminology*, 16, 142-155.

SHOHAM, S. G., RAHAV, G., BLAU, J., KAPLINSKY, N., MARKOVSKI, R., SNAKED, Y., STEIN, Y., WOLF, B. (1977), *The Reckless and Anxious Drivers: Some Initial Parameters,* Report to the Ministry of Transportation: Tel Aviv University.

SHOHAM, S. G., RAHAV, G., MARKOVSKI, R., CHARD, F., BARUCH, I. (1984), "Anxious and Reckless Drivers", *Deviant Behavior*, 5, 181-191.

SIGNORI, E. I., BOWMAN, R. G. (1974), "On the Study of Personality Factors in Research on Driving Behavior", *Perceptual and Motor Skills*, 38, 1067-1076.

SIMONSON, E., BAKER, C., BURNS, N., KEIPER, C., SCHMITT, O. H., STACKHOUSE, S. (1968), "Cardiovascular Stress (Electrocardiographic Changes) Produced by Driving an Automobile", *American Heart Journal*, 75, 125-135.

SOBEL, R., UNDERHILL, R. (1976), "Family Disorganization and Teenage Autoaccidents", *Journal of Safety Research*, 8, 8-18.

SPIELBERGER, C. D., GORSUCH, R. L., LUSHENE, R. E. (1970), *Manual for the State-Trait Anxiety Inventory*, Consulting Psychologist Press, Palo Alto, California.

STOKOLS, D., NOVACO, R. W. (1981), "Transportation and Well-Being: An Ecological Perspective", *Human Behavior and Environment Advances in Theory and Research*, 5, 85-130.

STOKOLS, D. (1976), "The Experience of Crowding in Primary and Secondary Environments", *Environment and Behaviour,* 8, 49-86.

SYME, S. L., RAGLAND, D., HOLMAN, B. L., MORSE, L., FISHER, J. M. (1983) "Hypertension in San Francisco Bus Drivers: Preliminary Considerations with Emphasis on Psychosocial Factors", in *Conference on the Working Environment in Urban public Transport*, Stockholm.

TAGGART, P., GIBBONS, D., SOMERVILLE, W. (1969), "Some Effects of Motor-Car Driving on the Normal and Abnormal Heart", *British Medical Journal,* 4, 130-134.

TAYLOR, J. A. (1953), "A Personality Scale of Manifest Anxiety", *Journal of Abnormal and Social Psychology*, 48, 285-290.

THOITS, P. A. (1983), "Dimensions of Life Events that Influence Psychological Distress: An Evaluation and Synthesis of the Literature", in KAPLAN, H. B. (Ed.), *Psychosocial Stress, Trends in Theory and Research,* Accademic Press.

TIWARI, K., BEHERE, P. B. (1983), "Anxiety Dimensions and Motor Vehicle Accidents", *Indian Journal of Clinical Psychology*, 10 (2), 469-471.

TORTOSA, F., MONTORO, L., SOLER, J. (1987), "La conducta agresiva", *Revista Tráfico (Dirección General de Tráfico)*, 27, 40-41.

TURNER, C. W., LAYTON, J. F., SIMONS, L. S. (1975), "Naturalistic Studies of Aggressive Behavior: Aggressive Stimuli, Victim Visibility and Horn Honking", *Journal of Personality and Social Psychology*, 31, 1098-1107.

VINOKUR, A., SELZER, M. L. (1975), "Desirable Versus Undesirable Life Events: Their Relationship to Stress and Mental Distress", *Journal of Personality and Social Psychology*, 32 (2), 329-337.

VIVOLI, G., BERGOMI, M., ROVESTI, S., CARROZZI, G., VEZZOSI, A. (1993), "Biochemical and Haemodynamic Indicators of Stress in Truck Drivers", *Ergonomics,* 1993, 36 (9), 1089-1097.

WAGENAAR, A. C. (1983), *Alcohol, Young Drivers, and Traffic Accidents,* Lexington: Lexington Books.

WHITLOCK, F. A. (1971), *Death on the Road: A Study in Social Violence,* London: Tavistock Publications.

WINKLEBY, M., RAGLAND, D., FISHER, J., SYME, S. L. (1988), "Excess Risk of Sickness and Disease in Bus Drivers: A Review and Synthesis of Epidemiological Studies", *International Journal of Epidemiology,* 17, 255-262.

WYON, D. P., WYON, I., NORIN, F. (1996), "Effects of Moderate Heat Stress on Driver Vigilance in a Moving Vehicle", *Ergonomics*, 39 (1), 61-75.

ZEIER, H. (1979), "Concurrent Physiological Activity of Driver and Passenger when Driving with and without Automatic Transmision in Heavy City Traffic", *Ergonomics*, 22, (7), 799-810.

PART THREE
CHANGING DRIVING COGNITIONS, ATTITUDES AND BEHAVIOR

Chapter 9

DRIVER IMPROVEMENT

Günter Kroj,
BASt, Federal Highway Research Institute, Germany
Elisabeth Dienes,
Hungarian Coordination Council for Work Psychology

1. INITIAL SITUATION AND ASPECTS OF INFLUENCING ROAD USERS

The increasing scarcity in public and private funds on the one hand and the improved scientific basis of the road safety work on the other led to the development of a kind of risk management of safety work: based on accident and behavioral studies, especially relating to certain high-risk groups – for example, novice drivers, drinking drivers (DWI), multiple offenders and high-risk types of road users, such as cyclists, moped riders – there was an increasing emphasis on target group programs. Evaluation research findings helped to optimize program effectiveness in the past years.

The evaluation studies on driver improvement programs provided a series of new information reaching beyond the scope of this safety measure. Participants' realistic descriptions of the causes and consequences of undetected drinking and driving offenses were particularly revealing – throwing important light on how to approach the problem of undetected cases. Furthermore, important approaches to the diagnosis of driver ability, future publicity actions and safety campaigns on this sector should also be derived from the detailed explanations of the conflicts leading to a drinking and driving offense and of the personal consequences of a drinking and driving conviction.

Present efforts towards a more systematic linkage between road safety education, driver training and examination, safety training and driver improvement programs and their integration into the existing legal, administrative and organizational framework appear to be more promising road safety efforts in the medium term than demands for radical changes in the legislation relating to the punishment of drivers or the search for alternatives to license suspension or revocation (Winkler *et al.*, 1988 and 1991; Spoerer and Ruby, 1996). Driver improvement programs can already be considered as making an important and irreplaceable contribution to the road safety work in a lot of countries. They cannot be, however, a substitute for legal sanctions and measures.

Driver improvement programs will be effective only as a functioning part of an overall system of measures providing information, support and feedback. Efficient police control here plays as eminently an important role as does an efficient central system to register offensive and inoffensive modes of road user behavior. From the viewpoint of social pressures and process of life-long learning, driver improvement programs can effect very little compared with the lessons learned daily from the consequences of offending against the road traffic rules. For that reason the optimization of a central system to register and evaluate inoffensive and offensive road user behavior is seen as an important future task (Winkler, 1985).

Another task of future driver improvement programs also lies in the area of sociopolitical decisions: the rate of drinking and driving offenses evidently increases as a nation's alcohol consumption rates increase (Norström, 1983). Driver improvement is essentially a concept addressing the individual. However, consideration should also be given to the establishment of a special driver improvement program system attempting to influence the general situations and conditions of offensive behavior parallel to the efforts directed at the high-risk driver as an individual (Winkler, 1985). This is a point increasingly emphasized in U.S. programs to prevent drinking and driving offenses (Mosher, 1983).

Driver improvement programs concerned with adult education, further training of adults, and the social reintegration of adults with previous convictions require that the experience acquired in the field of adult education be considered to a greater extent than hitherto in the prior education and further training of program leaders and the optimization of program models. What driver improvement programs basically offer is a chance of adult socialization in a society where more appropriate means to achieve the same at an earlier period had not been available (Hebenstreit *et al.*, 1982; Winkler *et al.*, 1988).

In conclusion let us consider once more the effectiveness of police controls, legislative and educational measures. Deliberations as to the learning theory and research findings from studies on general prevention revealed that, when in doubt, police actions are rated as more effective than legal measures. Legal measures generally come too late to affect motivation directly in psychological terms. The law is generally on a still remoter plane to immediately affect behavior. Previous experience has shown that short-term effects, such as disconcertion, or temporary restraint can be assumed to result from legislative measures. Of importance, however, is law implementation once a law is introduced, that is its application and execution. And that is where one is referred back again to the control measures of the police and traffic authorities.

For about 30 years now the rehabilitation of traffic offenders has been in the range of preventive traffic safety measures. There are different types of courses for the treatment of specific target groups. Some of them have shown their efficiency in an impressive manner (Utzelmann *et al.*, 1985; Stephan, 1988; Winkler *et al.*, 1991; Dienes, 1994; Jacobshagen, 1997).

By one way or the other all courses were established to improve the driver's ability and motivation to drive safely, especially in those cases in which the driver's ability to drive is assumed to show deficiencies, either by legal interpretation or as a result of a medical and psychological assessment.

The experiences in this field, found primarily by traffic psychologists since 1968, have been published in many articles and at various conferences (Zuzan, 1979; Huguenin and Hess, 1982; Bundesanstalt für Straßenwesen, 1985; Zuzan and Michalke, 1989; Siegrist, 1994). However, if someone is interested in being informed thoroughly on all main aspects regarding the rehabilitation of drivers, must have been unsuccessful in searching a description, which covers the whole field. But a good attempt to close this gap was published last year (Spoerer and Ruby, 1996).

Examples of evaluation research and the development of countermeasures follow in section 2.

2. MAIN RESULTS AND PRACTICAL APPLICATION

A short overview on different kinds of DI courses and on the amount of psychologists' actual professional activities in this field in some European countries (A, CH, D, H) are given in Tables 1 and 2.

These figures demonstrate that more than 750 especially trained and supervised traffic psychologists have conducted, since the last decade, more than 5,000 DI-courses per year for more than 50,000 participants in Austria, Germany, Hungary and Switzerland. The long term rehabilitation activities for multiple offenders and above all for DWI are not included in this statistic. But from one of the biggest German DI agencies (AFN) the recent figure for long term rehabilitation participants was published: In 1996, two percent of all participants needed long term rehabilitation courses (nine months or more), other DI-agencies offer individual treatment in such cases. In this section descriptions and overviews only on the group-based DI-measures are provided.

Table 1. Participants of different DI-courses in Austria, Germany and Hungary in the year 1996 conducted by psychologists (sources: KfV, 1977; TÜV, 1997; MOT, 1997)

DI-courses / Country	FOR DWI	FOR MULTIPLE OFFENDERS	FOR YOUNGSTERS AND NOVICES IN THE PROBATIONARY LICENSE PERIOD	A
AUSTRIA	1.785	–	3.846[1)] 3.043[2)]	8.674
GERMANY	26.243	2.233	8.839[1)]	37.315
HUNGARY	5.500	2.500 [2)]	–	8.000
A	33.528	4.733	15.728	53.989

[1)] for DWI, [2)] for offenders without alcohol

Table 2. Number of course leaders

DI Courses / Country	Nr. of Course Leaders: Psychologists	Others
Austria	100	–
Germany	500	4.000
Switzerland	30	30
Hungary	120	60

2.1. Programs for Young and Inexperienced Drivers

On average, the mileage per accident of young drivers aged eighteen to twenty-four has been found to be half the mileage of drivers from other age groups. Their accidents are more frequently caused by speeding and they are more often involved in "single car accidents" than is the case with other groups of drivers. The car leaving the roadway in a curve at high speed is a particularly frequent type of accident with young drivers. The accidents of young and inexperienced drivers are generally caused by the fatal combination of youthful risk-taking, lack of driving experience (OECD, 1975; Böcher *et al.*, 1977), and special kinds of lifestyles (Berg, 1991, 1993; Moe, 1993; Schulze, 1991, 1996). The driver improvement programs for young and inexperienced drivers, developed on the basis of this knowledge, aim at improving attitudes and skills by means of practical exercises and group dynamics. The first phase of these programs, carried out on a trial basis in four German Bundesländer was completed in 1978. Driving instructors directed the practical exercises and the exchange of information on the physical conditions of driving, and professional psychologists moderated the group dynamic sessions. The second phase began in September 1979 in North Rhine-Westphalia and Baden-Württemberg, in co-operation with a great number of driving instructors from the two states. The program aimed particularly at producing: Better social attitudes to traffic, a decrease in risk-taking, an improved knowledge of driving physics, and better control over a car in critical situations. Participation in the improvement program was limited to eight to twelve persons. A total of seven lessons were offered within a period of three weeks (v. Hebenstreit *et al.*, 1982).

The revision of the German Highway code and the law governing driving instructors (effective as of May 13, 1986) led to the introduction of the probationary driving license on November 1, 1986. The main objective of this measure was to reduce the high risk especially of young novice drivers by means of a probationary period of two years and by a mandatory improvement program for novices committing serious traffic offences during the probationary period. Based on present data, about 50,000 of the one million novice drivers per year will have to submit to this improvement program. It was that the general preventive effect of this measure will contribute to greater circumspection and a higher degree of risk awareness of young novices in the late eighties and the nineties proved by a long termed evaluation study by the Federal Highway Research Institute (Hansjosten and Schade, 1997).

Almost 90 % of novice drivers in class three are young adults under the age of twenty-five years, roughly two thirds of whom obtain their driving license by the age of eighteen. Every seventh novice driver (fourteen percent) does not manage to pass the current two-year trial period without being recorded in the central index of traffic offenders, this category being filled on average by three and a half times more men than women. This ratio becomes even more detrimental to the men when driving fines, alcohol and speed offenses are taken into consideration. Forty percent of the offenses committed during the trial period are coupled with an accident, way above the average for motorists. With novice drivers it was proven that there is a particularly close relationship between road discipline and road safety. Twenty-nine percent of novice drivers in class three recorded in the central index of traffic offenders during the trial period have their names entered again within two years. Amongst men beginning their trial period between the ages of nineteen and twenty-four, the recurrence quota is as high as 45 %.

The information gained from this investigation leads to suggestions for using the database for further analyses, for statistical methods of presenting legal probation, for classification of target groups, for the development of individual measures to affect behavior which are aimed at specific problems in the area of speed and priority regulations, for extending the trial period for young novice drivers and for introducing stipulations (Hansjosten and Schade, 1997).

Since 1986, obligatory courses for DWI youngsters and novices in the probationary period are conducted in Germany by specially trained and supervised psychologists. In a recent study it was proven that participants readily accepted these coursed and above all had 54 percent fewer instances of recidivism in comparison to non participants (Jacobshagen, 1997).

2.2. Programs for Multiple Offenders

At present, multiple offenders in Germany are characterized by more than one entry in the Central Traffic Registry (point demerit system). In view of the demerit points associated with the entries, offenders can be divided into two risk groups: Persons with nine to under fourteen points, persons with fourteen to under eighteen points.

There is no point system in Hungary. The system of Driver Improvement started in 1992. It is for the drivers, whose driving license has been withdrawn. Participation in one type of course is one of the conditions for getting it back. There is no type of program for drivers who want to volunteer to improve their driving habits.

Faulty driver behavior is often found in people who are not sufficiently familiar with the dangers involved in motorized road usage and therefore are not able to cope with them. For twenty years, the German Federation of Driving Instructors' Association has offered driving seminars which attempt is to show drivers the various aspects of traffic in the light of inherent risks.

Target groups are drivers with less than fourteen demerit points at the Central Traffic Registry. Participation is voluntary; "experiences," "insights," and "attitudes" relating to safe road usage are discussed in small groups of eight to twelve participants, with the guidance of specially-trained driving instructors.

The seminars (four sessions of 35 minutes each) are designed to identify and remove theoretical and practical deficiencies. The participation is recommended to drivers upon receipt of an official notice from the Registry. Drivers are also advised that participation may be followed by receiving "plus" points at the Registry. In addition, there is a driving demonstration on the road for each participant (45 minutes). A standard proforma is used to record the observations during these practical demonstration on the road.

The German program for multiple offenders offered by the Medico-Psychological Institutes of TÜV, Rhineland and TÜV, Renish-Westphalian has been considered for two decades as a measure designed to change the behavior of multiple offenders having a high (non-alcohol) accident probability. The target groups are drivers with fourteen or more demerit points at the Central Traffic Registry and, in particular, those having had two or more accidents. The improvement program consists of six sessions, the first being an individual talk. The five sessions thereafter are held as group discussions. Groups comprise eight to twelve participants. The program is designed for drivers with almost habitual risky behavioral modes. Gaining an insight into the dangerous consequences of faulty behavioral modes is an important objective of this program. Risk evaluation and risk compensation in traffic are also part of it. In the individual sessions, a behavioral analysis is made to prepare drivers for changing their self-image and becoming more self-critical. The program is directed by professional psychologists with additional specialized training.

Both above described programs have been proven to affect the safety-relevant attitudes of participants in a positive manner. A lasting success concerning the observation of traffic regulations could not be resorted (Utzelmann and Haas, 1985).

The Hungarian system has seven different programs. Two of them are closely related. One course is for drivers who are not familiar enough with traffic laws; the other is for drivers who have some technical problems with driving (poor habits).

The next two courses are for drivers who have a "conspicuous" (sensational) behavior on the road, i.e., for risky drivers who do not obey the traffic rules. But, they are not effected by alcohol.

The next three are for drivers who were driving under different level of alcohol effect.

2.3. Programs for DWI (Driving while Intoxicated)

During the seventies, a series of measures was developed in Austria, Germany, and Hungary, and in Switzerland during the early nineties, aiming at influencing road users. In 1978, a project team was set up at the Federal Highway Research Institute (BASt) to survey programs and evaluate their success. On the basis of their results, courts and local road traffic authorities in Germany have been, since the early eighties, increasingly interested in the application of improvement and rehabilitation courses for alcohol impaired drivers (v. Hebenstreit *et al.*, 1982).

In Hungary, till the end of 1997, participation was compulsory in one of the DI programs if someone's driver's license was withdrawn. Since 1998, the decision ordering participation lies with the courts or local road traffic authorities.

For *novice driver* committing drinking and driving offenses in the period of probation 1986 a special alcohol safety program has been designed in Germany. It deals exclusively with the drinking and driving problem, striving to help participants comprehend what is involved in acquiring modes of behavior that avoid drinking and driving in the future. Instructors are specially trained psychologists.

Driver improvement programs for *first-time drinking and driving offenders* were introduced in Germany in 1978. Their objective is to supplement the general preventive effect of fines and license revocation. The programs aim at changing drinking and driving. In addition, the participants are motivated to prove that they have succeeded in overcoming their uncontrolled drinking because the certificate awarded after the program can have positive legal consequences for the participant; in particular a reduction in the period of license revocation. The programs are conducted by traffic psychologists specially trained for this task, and subject to continuous training. Judges, according to German law, are expressly obliged to protect the public from drunk drivers, which is best ensured by license revocation. In this sense, a driver improvement program can only be regarded as an additional safety measure. However, the data, which are available from evaluation studies, reveal a clear reduction in the relapse frequency of program participants compared with the general relapse probability of first-time offenders within the 36 months observation period. Previous experience has shown that the quality of programs depends heavily on the quality of program leader training and on how correctly the suitability of offenders for program participation is assessed (Stephan, 1988).

Participation in an improvement program *for drivers with several previous drinking and driving offenses* is restricted by assignment criteria and as decreed by the highest federal state authorities. In general a driver, who after two previous drunken driving convictions and expiration of the revocation period has provided a medical and psychological report attesting his shortcomings, may be admitted to a program believing that these shortcomings could be corrected in a specific driver improvement program.

In the meantime, the decrees of the various German federal states also name the institutions which are officially approved driver improvement centers, including admission criteria and assignment procedures, the program models implemented in a particular state, form and contents of the certificate testifying program participation, and their legal status as far as the traffic authorities dealing with these cases are concerned (Nickel and Schell, 1993).

In addition to the incorporation of the programs into the existing administrative system and their positive effects on license renewal, the success of improvement measures is mainly due to their content, organization and supervision. Due to intensive teamwork in small groups, all three German program models for drivers with previous drinking and driving offenses (individual approach, behaviorist approach, LEER model) have proven successful (Winkler *et al.*, 1991). They all succeeded in redirecting driver motives for participation, which are often superficial at the initial stage, and helping them accept the chance to solve their personal and drinking problems. These changes include drinking habits, control of alcohol consumption, avoidance of occasions when alcohol will be consumed and critical self-assessment of one's driving ability.

In addition, these driver improvement programs also produce a clear reduction in the relapse frequency, confirmed by a monitoring period of 5 years following the program participation. The program participants who had received unfavorable medical and psychological reports clearly displayed a lower relapse rate (thirteen percent) than the drivers in the control group (eighteen percent) with better medical and psychological rating (Winkler *et al.*, 1991).

2.4. Elderly road users

In view of the anticipated growing percentage of elderly road users and drivers, measures reducing risks and at the same time ensuring a high degree of mobility, particularly behind the wheel, are of special importance. Gerontological studies revealed that the rate of decrease in adjustment of the elderly to internal and external stresses of all kinds largely depends on the individual. Favorable social and physical environmental conditions should be created for elderly drivers and specific information campaigns used to point out ways of reducing their risks in traffic. Maintaining an acceptable level of ability and skill, for example, by means of refresher courses, and especially retaining are of importance in this connection. But drivers in other age groups also have to be made aware of the conditions of elderly drivers.

In order to gain an insight into the possible effects on road safety and the necessary countermeasures to be derived, a representative survey of 4,364 persons from the age of eighteen years and an intensive survey questioning 328 male and female drivers aged fifty-five years and over were undertaken in 1986 by BASt (Hartenstein, *et al.*, 1990). The most important results are:

- Driver performance decreases only slightly as people grow older;
- Elderly male and female drivers feel unable to get along without a car and rate their physical and mental condition higher than non-motorized elderly people do;

- Compared with those not having a car at their disposal, the life-long activities of male and female drivers hardly undergo any changes as they grow older, even at an advanced age;
- Only 50 % of those over the age of 80 years give up driving their own car;
- The self-image, especially that of elderly male drivers, contradicts the rating of the remaining car driving population, a fact likely to result in conflicts.

Based on the results of the investigation, special measures for elderly male and female drivers do not seem to be required, especially sincethese measures would also not be well accepted by the elderly. Information about structural changes should address all drivers. Improvements on the sector of automotive engineering and the design of transport systems would be to the benefit of elderly male and female drivers, but not only to them (Hartenstein *et al.*, 1990).

Additionally a field study and laboratory tests were conducted by BASt (Ellinghaus *et al.*, 1990). A group of elderly drivers aged 60 years and over (N = 80) was compared with a group of experienced drivers aged between 40 and 50 years (N = 30) by means of the following tests:

- Driving tests on the traffic scene;
- Psychophysical and psychological tests; and
- Questionnaires.

The study was undertaken with the objective of clarifying questions, such as whether and how the psychophysical performance parameters change with age, whether the driving behavior of elderly drivers differs from that of younger drivers and if so, what the differences are and, finally, whether and how the age-related decline in the performance of elderly drivers is compensated.

The authors have arrived at the result that elderly drivers are able to cope just as well as younger drivers do with driving tasks involving average requirements. However, very difficult traffic situations, above all those requiring quick action, may result in conflicts. This explains why elderly drivers tend to avoid difficult situations and those involving stress and also, on the whole, drive less than younger drivers do. The strategy of the elderly of avoiding high-risk situations, but also their general tendency of abstaining from risk taking, seems to be the principal reason for the relatively low accident experience of this group (Ellinghaus *et al.*, 1990).

During the early nineties, a project group of the German Traffic Safety Council (DVR) integrated the results of these studies in a long-term concept of countermeasures for this target group, which is being redrafted at the moment.

3. DISCUSSION

Why are countermeasures often ineffective in the medium term?

Alcohol as a drug has been known in the Western World for thousands of years and intensively consumed for centuries, and till quite recently hardly any society has succeeded in integrating alcohol consumption into innocuous patterns of behavior thus neutralizing the risks associated with its consumption.

Social customs as part of the culture as a whole are slow in changing – often it takes centuries before a change is accomplished. Recent efforts in the field of legal sociology revealed that an adequate comprehension of the effect of laws can be expected only if they are understood as part of a socio-cultural tradition rooted in the structure of economy and society (Kerner, 1985). In this connection the pressures of being caught and discrimination against behaviors on moral grounds play an important role. But the laws affecting administrative rules and regulations have to be distinguished from those directly and centrally affecting criminal acts. Behavioral changes on the former level are presumably easier and faster to accomplish than on the latter. The established way of thinking in a lot of countries is still far from a process of reorientation. This is borne out by the fact that drinking and driving offenses at BAC values over 50 mg/100 ml constitute a violation of administrative regulations and not a criminal act.

The motor vehicle has been in use for about one hundred years, and on a widespread basis only about 40 years. In most European countries it has seriously affected the life-styles of not more than two generations. The process of reorientation to change dominant social attitudes (drinking customs, desire to assert oneself in traffic) and to practice and slowly accomplish new behavioral modes will presumably be long and difficult. This discrepancy between attitude and behavior also has to be considered, in particular as regards road users. To have an insight into the risks involved in one's actions is the necessary condition of reorientation, but for the motorist, it is a hardly ever sufficient reason to motivate changes in behavior.

In view of the complicated structure of social control in its entirety, a solution to the problem may be sought by concentrating on primary prevention. Primary prevention, however, can hardly be achieved in the short term by means of formal judicial methods of social control (Kerner, 1985; Snortum, 1988; Ross, 1988). Driving under the influence of alcohol is still generally regarded as normal behavior and a petty or pardonable offense. There is, however, a tendency that is being more negatively assessed in the past years. Most drivers (95-98 %) today probably remain below the legal BAC limit of 50 mg/100 ml respective 80 mg/100 ml (Paschke and Pfafferott, 1994). The percentage of drunk drivers caught and sentenced is very low however – according to recent surveys, not higher than 2 % (Kerner, 1985). A denser network of police controls and much stiffer sentences would be of comparatively little marginal benefits. This still would presumably not result in catching and sentencing considerably more than 2 % of all drunk drivers. In addition, the studies on drinking patterns and motivation analyses point to the presence of deeply rooted personality factors in the population of drinking drivers which can only be successfully dealt with on an influencing level preceding that of sanctions, i.e., customs, habits, and attitudes with respect to alcohol consumption.

4. PERSPECTIVES

From the foregoing the following results: Control and sanctioning have to remain part of, or integrated still more effectively into, a broad concept of safety campaigns which are repeated on a continuous basis with education measures combating the disastrous combination of drinking and driving.

A considerable percentage of probably legitimate sanctions cannot be imposed owing to the failure rates of breath and blood tests, which are still relatively high. Improved equipment and a more stringent standardization of at-the-scene application might increase the rate of convictions noticeably (penalty notice, order of summary punishment or conviction and sentence). With respect to the preventive effect on the individual it should, however, be pointed out that, from the criminological standpoint, the fact of being caught by the police alone suffices to achieve the essential sanctioning effect intended by the learning theory, regardless of the nature of further consequences.

The number of drinking and driving offenses among young drivers has been particularly high and adds considerably to the risk factors of little driving experience and the taste of thrill, which is typical of the young and novice drivers. With due regard to the limited possibilities of really influencing this development, the following combination of measures appears to be appropriate: Driver instruction accompanied by relatives, probationary driving license, practice of "avoidance strategies" with respect to alcohol consumption as part of driver education, publicity actions and safety campaigns in schools and companies on a continuous basis.

The discussion concerning instruction, driver improvement or selection can give rise to the impression that the road safety problems of one or several groups of road users can be completely overcome by the application of such instruments. However, one must point out that the effectiveness of these measures is limited.

Traffic psychology should therefore not only exert influence at the final intervention level by using existing instruments for behavioral modification.

Rather, one should intervene at a higher level to avoid the danger of taking token action – for instance at the point where traffic concepts or systems are created, in order to influence behavioral control earlier (Huguenin, 1995).

It is not always possible to pursue such a course, but it is one of the tasks of traffic psychology to demonstrate that it is feasible to intervene not only *on* but also *for* the human being in a systematic manner. To promote this view, particular efforts and methods are needed, for example, marketing, which can be used not only to influence the road user but also policy – and decision makers (OECD, 1993).

For twenty years BASt, bfu and KfV have conducted periodical workshops on DI to exchange information between legislators, administrators, practitioners and researchers to identify the legal and socio-cultural background for DI, and to clarify the scientific progress and initiate promising developments influencing road users and decision makers (BASt, 1997).

5. REFERENCES

BASt (1985), *3. Internationaler Workshop Driver Improvement,* Heft 50 Unfall- und Sicherheitsforschung Straßenverkehr, Bergisch Gladbach: Bundesanstalt für Straßenwesen.

BASt (1997), *6. Internationaler Workshop Driver Improvement*, 20-22 Oktober 1997, Berlin, Proceedings in preparation, Bergisch Gladbach: Bundesanstalt für Straßenwesen.

BERG, H. Y. (1996), "Lifestyle, Traffic and Young Drivers - An Interview Study", in *Heft M 52 Berichte der BASt*, Bergisch Gladbach: Bundesanstalt für Straßenwesen, 370-378.

BÖCHER, W., KROJ, G., PFAFFEROTT, I., SOGEMEYER, H., SPOERER, E., WINKLER, W. (1977), *Typische Fehlverhaltensweisen von Fahranfängern und Möglichkeiten gezielter Nachschulung,* Heft 8 Unfall- und Sicherheitsforschung Straßenverkehr, Köln: Bundesanstalt für Straßenwesen.

ELLINGHAUS, D., SCHLAG, B., STEINBRECHER, J. (1990), *Leistungsfähigkeit und Fahrverhalten älterer Kraftfahrer*, Heft 80 Unfall- und Sicherheitsforschung Straßenverkehr, Bergisch Gladbach: Bundesanstalt für Straßenwesen.

HARTENSTEIN, W., SCHULZ-HEISING, J., BERGMANN-GRIES, J., KRAUß, F., RUDAT, R., SMID, M. (1990), *Lebenssituation, Einstellung und Verhalten älterer Autofahrer und Autofahrerinnen,* Heft 79 Unfall- und Sicherheitsforschung Straßenverkehr, Bergisch Gladbach: Bundesanstalt für Straßenwesen.

HEBENSTREIT, v. B., HUNDHAUSEN, G., KLEBE, W., KROJ, G., SPOERER, E., SCHNEIDER, W., WALTER, R., WINKLER, W., WUHRER, H. (1982), *Kurse für auffällige Kraftfahrer*, Heft 12 Projektgruppenberichte der BASt, Köln: Bundesanstalt für Straßenwesen.

HUGUENIN, R. D., HESS, E. (1982), *Driver Improvement-Rahmenbedingungen und Methoden der Verhaltensbeeinflussung in der Ausbildung, Weiterausbildung und Nachschulung von Fahrzeuglenkern,* bfu-Report 5, Bern: bfu.

HUGUENIN, R. D. (1995), "Driver Instruction, Diagnostics and Improvement in Switzerland – New Perspectives", paper presented on *7th European Congress on Work and Organizational Psychology*, Györ, 19-22 April 1995.

HUGUENIN, R. D. (1995), "Zur Lage in der Schweiz", in, RISSER, R. (Ed.), *Die Einflüsse verkehrspsychologischer Arbeiten auf die Praxis von Planern, Politikern und Behörden,* Bericht zum 35, BDP-Kongress für Verkehrspsychologie (Baden bei Wien), Bonn: Deutscher Psychologen Verlag GmbH, 59-70.

KERNER, H. J. (1985), Untersuchungen zu "Alkohol und Fahren", *Gesetzgebung, polizeiliche Überwachung und Strafgerichtsbarkeit in der Bundesrepublik Deutschland,* Bergisch Gladbach: Forschungsberichte der Bundesanstalt für Straßenwesen, Band 11, Heft 115.

KfV (1997), *Verkehrspsychologischer Informationsdienst,* Folge 45 Kuratorium für Verkehrssicherheit, Wien, 17-18.

MOE, D. (1996), "Young Drivers Risk Taking Behaviour and Accidents – An Interview-Study", in *Heft M 52 Forschungsberichte der BASt*, Bergisch Gladbach: Bundesanstalt für Straßenwesen, 359-363.

MOSHER, J. F. (1983), "Severe Intervention: A New Approach for Preventing Drinking and Driving", *Accident Analyzes and Prevention,* 15, 483-498.

NICKEL, W., SCHELL, A. (1993), *Qualitätssicherung in der medizinisch-psychologischen Eignungsbeurteilung und bei der Rehabilitation auffälliger Kraftfahrer*, Zeitschrift für Verkehrssicherheit (39, Heft 4), Köln: Verlag TÜV-Rheinland GmbH, 156-160.

NORSTRÖM, T. (1983), "Law Enforcement and Alcohol Consumption Policy as Countermeasures against Drunken Driving; Possibilities and Limitations", *Accident Analyzes and Prevention,* 15, 513-522.

OECD (1975), *Young Driver Accidents,* RRG S8, Paris: Organization for Economic Co-operation and Development.

OECD (1993), *Marketing of traffic safety,* Paris: Organization for Economic Co-operation and Development.

PASCHKE, S., PFAFFEROTT, I. (1994), "Meinungen und Einstellungen deutscher Autofahrer/innen im europäischen Vergleich", *Deutscher Ergebnisbericht zur internationalen SARTRE – Studie,* Bonn: Deutscher Psychologen Verlag.

ROSS, A. L. (1988), "Deterrence-Based Policies in Britain, Canada and Australia", *in Social Control of the Drinking Driver*, Chicago, London: University of Chicago Press, 64-78.

SCHADE, F. D., HANSJOSTEN, E. (1997), *Legalbewährung von Fahranfängern,* M 71 Berichte der BASt, Bergisch Gladbach: Bundesanstalt für Straßenwesen.

SCHANZ-LORENZ, I. (1995), *Zur verkehrspsychologischen Nachschulung alkoholauffällig gewordener Kraftfahrer in der Europäischen Union*, Inaugural-Dissertation, Eberhard-Karls-Universität, Stuttgart.

SCHULZE, H. (1991), *Zur Ökologie jugendlichen Freizeit- und Verkehrsverhaltens*, Fortschritte der Verkehrspsychologie 90, Köln: Verlag TÜV Rheinland.

SCHULZE, H. (1996), *Lebensstil und Verkehrsverhalten junger Fahrer und Fahrerinnen*, M 56 Berichte der BASt, Bergisch Gladbach: Bundesanstalt für Straßenwesen.

SIEGRIST, S. (1994), *5. Internationaler Workshop Driver Improvement*, bfu-Report 23, Bern: bfu.

SNORTUM, J. R. (1988), "Deterrence of Alcohol-Impaired Driving: An Effect in Search of Course", in *Social Control of the Drinking Driver*, Chicago, London: University of Chicago Press, 189-226.

SPOERER, E., RUBY, M. M. (1996), *Zurück ans Steuer, Theorie und Praxis der Rehabilitation auffälliger Kraftfahrer,* Heft 39 Schriftenreihe Faktor Mensch im Verkehr, Braunschweig: Rot-Gelb-Grün.

SPOERER, E. (1976), *Rehabilitation von alkoholauffälligen Kraftfahrern in Nordamerika. Bericht zum Forschungsprojekt 7517.1 im Auftrag der Bundesanstalt für Straßenwesen,* Köln: BASt.

STEPHAN, E. (1988), *Wirksamkeit der Nachschulungskurse bei erstmals alkoholauffälligen Kraftfahrern, Bestandsaufnahme nach drei Jahren,* Heft 170 Forschungsberichte der Bundesanstalt für Straßenwesen, Bergisch Gladbach.

TÜV (1997), *Jahresbericht 1996 des FA-R,* Technischer Überwachungsverein Hannover/Sachsen-Anhalt e. V. Hannover.

UTZELMANN, H. D., HAAS, R. (1985), *Evaluation der Kurse für mehrfach alkoholauffällige Kraftfahrer*, Heft 53 Schriftenreihe Unfall- und Sicherheitsforschung Straßenverkehr, Bergisch Gladbach: Bundesanstalt für Straßenwesen.

WINKLER, W. (1985), *Driver improvement 1984 – Anspruch und Wirklichkeit,* Heft 50 Schriftenreihe Unfall- und Sicherheitsforschung Straßenverkehr: Bergisch Gladbach. Bundesanstalt für Straßenwesen.,15-20.

WINKLER, W., JACOBSHAGEN, W., NICKEL, W. R. (1988), *Wirksamkeit von Kursen für wiederholt alkoholauffällige Kraftfahrer,* Heft 64 Schriftenreihe Unfall- und Sicherheitsforschung Straßenverkehr, Bergisch Gladbach: Bundesanstalt für Straßenwesen.

WINKLER, W., JACOBSHAGEN, W., NICKEL, W. R. (1991), *Rückfälligkeit von Teilnehmern an Kursen für wiederholt alkoholauffällige Kraftfahrer nach fünf Jahren,* Heft 224 Forschungsberichte der Bundesanstalt für Straßenwesen, Bergisch Gladbach.

ZUZAN, W. D., MICHALKE, H. (1989), *4. Internationaler Workshop,* Wien: Kuratorium für Verkehrssicherheit.

ZUZAN, W. D. (1979), *Driver Improvement, Tagungsbericht über den Ersten internationalen Workshop "Psychologische Behandlungsmodelle für verkehrsauffällige Kraftfahrer",* Wien und Salzburg: Kuratorium für Verkehrssicherheit.

Chapter 10

THE EFFECTS OF SAFETY REGULATIONS AND LAW ENFORCEMENT

Stefan Siegrist,
Swiss Council for Accident Prevention bfu, Berne
Eva Roskova,
Comenius University, Slovakia, Bratislava

1. INTRODUCTION

There is no doubt about the relationship between breaking certain traffic laws and loss of health. Non-compliance with speed and alcohol limits is a major cause of road accidents at the individual as well as at group level (e.g., Evans, 1991). Experts estimate that reducing the average speed by 5 km/h causes fatal injuries to fall by 25%, and reducing driving with a blood-alcohol level (BAC) in excess of 0.5 parts per thousand results in a further reduction of 5 - 40% in fatalities (ETSC, 1997).

There are different strategies in influencing road-user behavior. Legislation and enforcement have proved to be efficient if certain strategies are followed. In order to understand how this works, how it can be improved and what supporting measures are necessary, we need to understand the psychological reasons for non-compliance. This chapter covers the main effects of regulations and enforcement on human behavior and discusses the explanations and improvement psychology can offer this system.

It is mainly three groups of individuals that are concerned with enforcement: decision-makers, the police and road users. Decision-makers and the police are experts. What they expect of a traffic system is safety and the likelihood of efficiency, social equality and protection of the environment. They need specific knowledge about injuries to set safety targets. While safety experts know that non-compliance with traffic regulations is a major cause of injuries, road users have learned that non-compliance tends not to be followed by an accident. For this reason safety is not of primary concern for road users. It is therefore necessary to discover the primary causes of their behavior on the road in order to improve safety.

In this chapter a definition of unsafe behavior is offered and possible psychological processes underlying such behavior are discussed. General aspects, the theoretical background and the main results concerning legislation in traffic are then presented. Future measures for which there are as yet no results are discussed. Then a close look is taken at speeding and drunk driving. The effects of the regulations, different enforcement and punishment strategies as well as of additional or alternative measures, such as campaigns, are discussed. In the following section studies are presented in more detail and discussed in relation to the nature of the offense in question. In the next part integrated measures aimed at increasing compliance with traffic regulations are presented.

It would go beyond the scope of this chapter to report on all the relevant studies and surveys because so much excellent work has been done in this field.

2. THE NATURE OF UNSAFE BEHAVIOUR AND THE REASONS FOR NON-COMPLIANCE WITH TRAFFIC REGULATIONS

2.1. The Organization of Driving Behavior

In order to plan our safety measures actively we need a definition of unsafe behavior and we need to know the psychological correspondents for unsafe behavior.

The Action Theory (e.g., von Cranach, 1982) provides an opportunity to describe planning, executing and steering an action, the function of these different levels and interaction among them. According to this theory, self-active systems, such as human beings, actively strive towards certain ends by means of behavior that is directed towards the achievement of future states. This system is not just activated by external stimuli and forces, but instantaneously activated by internal energy and steered by internal information. The information-processing system is differentiated according to function and hierarchy. Different subsystems serve different tasks (such as information input, steering the execution, etc.), and according to these subtasks they are either more central and normally superordinated or more peripheral and subordinated. There is mutual interaction among these levels.

According to this theory, the aim of a journey is the result of a comparison between the perception of the environment and one's existing state. This results in the selection of an objective and the choice of a program, which will be converted into behavior. Starting the engine and changing gear are subordinate tasks of the execution phase in which impulses of the inner system (such as driving motives) as well as of the external system (changing situation on the road as a function of social interaction) have a concrete influence on the objectives of the action as such.

2.2. The Nature and Categorization of Unsafe Behavior

It is not the occurrence of an accident or injury that qualifies the preceding action. The definition of unsafe behavior depends rather on the demands of a concrete task and situation. If these demands are not fulfilled to a certain extent, the accident risk increases and the behavior can be classified as unsafe. The moment of attainment of this state has to be defined for different actions, such as speeding and drunk driving. Analytic epidemiology results say something about unsafe behavior in physical terms (speed or BAC limit).

Psychologists attempt to classify different forms of errors. According to Reason (1994), there are three major levels – ability-based, rule-based and knowledge-based. On the lowest level mainly routine actions are executed. On the second level actions are chosen with respect to the demands of the current situation. On the third level new strategies are developed to overcome problems. There are different levels of error. Not paying attention is an error on the lowest level and application of the wrong rules is an error on the rule-based level.

In contrast to the workplace situation (which is the main field of application of Reason's approach), behavior in traffic is not mainly determined by the structure of the task. Road user behavior is also determined by values and motivation other than safety. Apart from errors that result from the incorrect execution of a task or subtask, it is necessary to consider that the main objective of the task in the view of road users differs from what system designers would expect. If the reason for behavior to correspond to a higher crash risk is, say, pleasure in driving rather than lack of ability, wrong application of rules or insufficient knowledge, this is known as a quasi-error.

2.3. Which Are the Main Types of Unsafe Behavior that Contribute to Road Accident Risk?

One can argue that making a distinction, for example, between errors and violations or between errors and quasi-errors (Table 1.) is not relevant, as both are deviations from goal-directed behavior (Rothengatter, 1997b). The point that is made here is that there may be different causes of the same kind of unsafe behavior, and different causes demand different solutions. Several traffic psychology papers have pointed out the central role of normative orientation, behavioral motives, the general social situation and the dynamics of the current social situation (e.g., Näätänen and Summala, 1976). So what we have referred to as quasi-errors are the main cause of injuries in road traffic rather than errors in the execution of the task. Offenses may be associated with errors but they are mainly the result of attitudes, norms and motivation. Furthermore, we have to distinguish between conscious decisions not to comply with a regulation (which we shall refer to as a violation) and not paying attention to the regulation (which is a quasi-error). Non-compliance with traffic regulations reflects the influence of attitudes and motives that are contradictory to safe behavior rather than a conscious decision to break the rules. Contravening the regulations does not necessarily mean that a driver has no respect for safety norms, nor that there is a lack of motivation to comply with the regulation in question. In some cases it may only mean that the desire to have pleasure (which is thought to be a consequence of speeding) is at the moment more dominant than other motivational factors, such as laws and enforcement work, among other reasons, because they remind the driver of existing norms and values. If all road users were violators who did not accept the regulations and had negative attitudes towards compliance, enforcement would probably have very little effect.

There is empirical evidence that higher-level errors (quasi-errors and partial violations) are the main reason for non-compliance:

- Men have slightly higher accident rates and more serious crashes than women, although they commit fewer errors than women when driving a car. But men tend to contravene the regulations more (Reason *et al.*, 1990);
- Young drivers are more likely to have an accident than older drivers, but there are differences within the young-driver group. The concept of lifestyle, which is a complex amalgam of values, attitudes and behavioral preferences, seems to offer an explanation for these differences (e.g., Gregersen and Berg, 1994);

This evidence supports the theoretical notion of the Action Theory, which maintains that higher-level processes have a more dominant effect on the objective and the main characteristics of the action.

What is the advantage of classifying the causes of unsafe behavior? It helps when attempting to match prevention strategies to the identified problems.

Table 1. Causes, types and prevention of unsafe behavior

Main cause of unsafe behavior	Type of unsafe behavior	Objective of prevention strategy
• ability • application of regulations • knowledge	**Error** Task is wrongly executed (e.g., application of wrong regulations, decisions based on wrong knowledge)	learning of basic skills (maneuvering), regulations, specific knowledge and action strategies; aptitude-oriented selection
• attitudes	**Violation** Conscious decision to contravene the regulations	changing attitudes
• norms, • attitudes • social influence	**Quasi-error** Normative, motivational or social tendencies are more dominant than safety	show relevance of a regulation in a specific situation; increase road users' awareness of risk, increase social pressure to behave safely
• aptitude	**Quasi-error** Lack of aptitude to decide whether to comply or not	identifying violators, analyzing reasons for non-compliance, offering psychological support

2.4. Compliance and Non-Compliance with Regulations – Theoretical Approaches

There are different theoretical approaches to explain compliance or non-compliance with traffic regulations and the role of police control.

In behaviouristic terms compliance is mainly a result of the fear of detection and the negative feedback that follows. In this sense the Deterrence Theory (Homel, 1988) attempts to describe or even explain road user behavior as a function of traffic regulation enforcement, which is a type of social control. It is assumed that individuals will be deterred from taking a particular action by the threat of punishment. The threat of being detected in the act of contravening a regulation may be either real or perceived. General deterrence is the mechanism that influences all road users through the threat of police control and the probability of being checked and punished. Specific deterrence is the impact of a concrete experience of detection and punishment. This experience may be personal or that of a friend or family member. It is important to note that, according to the Deterrence Theory, individuals will only be deterred from contravening the regulations if they believe that the risk of detection is high. Many results that show the need for a minimal level of enforcement seem to support this theory. However, it must be pointed out that this theory does not explain the psychological process that leads to a modification of behavior. A causal line from control intensity through fear of detection to behavioral change is hypothesized but not proven. This theory must leave open whether other motives (conformity) or cognitive processes (police control induces a memory effect) are more important than a negative emotional state (threat).

The possible cognitive operations that guide road user behavior as a function of detection probability and punishment are described in the Rational Choice Theory. This theory is parametric, the probabilities of different outcomes are external and not dependent on the decision chosen. According to his or her standards, a road user will comply with regulations if the expected benefit of compliance is greater than expected benefit of violation. For example, a driver's benefit from speeding is the sum of the following products: (Probability of detection by the police x Disadvantage of detection) + (Probability of non-detection by the police x Benefit of driving at the desired speed) (following Bjornskau and Elvik, 1992). This means that there is a static relationship between enforcement and road-user behavior – drivers are deterred from non-compliance by increasing the probability of detection or the disadvantage of detection (size of penalty). The Rational Choice Theory focuses mainly on risk (of accident or apprehension) and ignores the fact that road users are active agents who wish to achieve their own objectives, which also depend on motives not associated with the traffic system.

An interactive example of the normative influence of regulations and law enforcement might be provided by a theoretical game model. It was suggested by Bjornskau et al. (1992) that the use of a theoretical game model might be seen as an elaboration of the argument put forward by the various theories of behavioral adaptation (OECD, 1990) – to the effect that road users are not passive (parametric) players and recipients of safety measures. They take advantage of all available measures to achieve their objectives, which do not always include maximizing road safety. A theoretical game model takes account of the dependency of two or more players who jointly influence the outcomes as well as each other's payoff (for example, the degree of speeding and level of enforcement) and simultaneously considers actions chosen by the enforcer in response to road-user behavior. This means that car drivers will not comply with the regulations (such as speeding, drinking and driving, wearing a seat belt) if there is no police enforcement. This in turn leads to enforcement, which will cause car drivers to comply with the regulations. When car drivers observe regulations and norms, enforcers will stop enforcing, which will result in car drivers violating them again. According to the theoretical game model, changes in violation rates can only be expected if road users notice that the risk of detection has increased, so the increase in the objective risk of detection must be substantial for road users to notice it. Bjornskau and Elvik (1992) present empirical findings that support this relationship between the level of enforcement and the level of compliance. Of course, this theory represents an approach to a very specific question; it does not say anything about the main psychological causes of non-compliance.

The Theory of Reasoned Action (TRA) (Fishbein and Ajzen, 1975) and the Theory of Planned Behavior (TPB) (Ajzen, 1985, 1988) offer a way of explaining the independent influences of subjective norms and attitudes on specific behavior. The Theory of Reasoned Action – and the Theory of Planned Behavior – states that behavior and analysis of behavior should be based on intentional measures that are determined by attitudes and subjective norms. Attitudes are determined by beliefs and the evaluation of the outcomes of behavior; they reflect the personal tendency of an individual to perform this behavior. Subjective norms are determined by the individual's perception of social expectations to perform the behavior (normative beliefs), weighted by his or her motivation to comply with these perceived expectations (motivation to comply). So subjective norms reflect the subjectively perceived influence of the social environment on the subject's behavior. TPB (Ajzen, 1985, 1988) includes a third determinant of behavioral intention, perceived behavioral control (the ease of performing the activity or of avoiding it – volitional control). In terms of TPB, norms influence behavior in so far as each individual is motivated to comply with this information and to the extent that he or she is able to do so. According to TPB, drivers do not comply with traffic regulations and norms (such as speeding, wearing seat belts, drinking and driving) because:

a) they believe this non-compliance is acceptable, and are prepared to follow this belief,
b) they are unaware of their non-compliance with the regulations and laws,
c) the anticipated positive outcome of non-compliance outweighs any anticipated consequences of compliance (Ajzen, 1988, following Groeger and Chapman, 1997).

Parker *et al.* (1992) have demonstrated the ability of the Theory of Planned Behavior (TPB) to account for a driver's intentions to commit four specific driving violations: drinking and driving, speeding, tailgating and overtaking in risky circumstances. Empirical evidence showed that the addition of perceived behavioral control led to a significant increment in the amount of explained variance in intentions, thus supporting the theory. The relationship between subjective norms and behavioral intention was stronger than between attitudes toward behavior and behavioral intention.

Rothengatter (1988) shows that enforcement, which increases the objective risk of detection, can influence the level of compliance, although motivation and attitudes towards the prohibited behavior (in this case speeding) remain unchanged. Contrary to other theories, TPB offers a possible explanation for this finding: obviously police control represents a social influence on the subjective norms. This interpretation is supported by the fact that posting the percentage of drivers complying with the law has a considerable effect on the compliance level (Rothengatter, 1988; Van Houten and Nau, 1983). A questionnaire survey of observed drivers showed that "these results cannot be solely attributed to an implied threat of apprehension as Shinar and McKnight (1985) suggest" (Rothengatter, 1988). At least the choice of speed seems not to be exclusively a result of detection probability, it also depends on the motivation of at least a part of the driver population to behave similarly to the majority. This means that regulations are a factor influencing road user behavior, possibly independent of a road user's attitudes towards specific prohibited behavior (speeding, drunk driving); the norm seems to have a positive effect on behavior if the driver notices that compliant driving is common behavior. This fact also supports the need to use additional measures, such as information campaigns.

The Fishbein and Ajzen model seems to be valuable in explaining road-user behavior and psychological processes. It is able to explain changes in behavior that are not accompanied by changes in attitudes. However, questions still remain, or as Groeger states it: "Thus, while useful descriptively, one must conclude that TPB requires substantially more confirmatory evidence before we can accept it as an accurate explanation of behavior in traffic" (1995, p. 265).

3. LAWS AND REGULATIONS

The legal system is only one element of the traffic system. It provides the criteria to judge road-user behavior. The comparison between what is desirable and what is observed is more conclusive if traffic regulations represent a homogeneous set of prescriptions that are based on few values, mainly safety. If compliance with a regulation has positive effects on the traffic safety level, it will probably be easier to convince road users to comply with the regulation.

There are several conditions a regulation must fulfill in order to have a minimal chance of influencing road user behavior (e.g., Opp, 1971, for general conditions, Noordzij, 1976, for the traffic field). A regulation has to be:

- Known;

- Understood (content and intention of the regulation);
- Convertible;
- Easy to control;
- Accepted to a certain degree;
- In accordance with other regulations;
- In minimal accordance with situational prerogatives.

If a change in behavior is sought, road users have to know and understand what the regulation is about. It has to be formulated so as to contain a convertible request to the driver. In the case of speed limits this condition is easier to fulfill than in the case of driver state or tailgating. Conversion into behavior is also more likely when the number of regulations is small. The acceptance of traffic regulations depends on different aspects. A regulation is easier to accept if it does not contradict other regulations and characteristics of the traffic situation.

It is difficult to quantify the minimal acceptance of a regulation by road users that will produce positive effects. Without a minimal social acceptance the introduction of a regulation will probably not make sense. On the other hand, some regulations would no longer be necessary as acceptance has meanwhile reached nearly 100 percent. In the middle of the range of the 30 to 90 percent who observe the norm a regulation will probably have positive effects. In these cases regulations are both a cause and an effect of social standards. In a European Survey of drivers' opinions (SARTRE, 1994), it was found that attitudes towards regulations were closely related to the current legislation in the individual countries. After implementation the level of acceptance may rise, depending on the driver's experience, which itself depends on the enforcement process and supporting measures.

4. ENFORCEMENT, PUNISHMENT AND SUPPORTING MEASURES

A regulation on its own has no noticeable effect. It has to be communicated, explained, remembered and enforced. The focus of enforcement should be on regulations that have a strong link to safety and which have not yet been converted into behavioral standards. In some cases changed behavior is internalized by road users (wearing a motor cycle helmet, to some extent wearing a seat belt) and enforcement can be stopped or reduced.

Contrary to other regulations, such as those relating to robbery or homicide, traffic regulations are violated by a large majority of the (driver) population. Speeding, in particular, can hardly be attributed to a small group of road users, and is easy to identify. For this reason, enforcement must be understood and planned predominantly as a general preventive measure.

The most effective and efficient method of enforcement depends on the regulation in question. Laws and regulations place different demands on the driver. Either a specific action is completely forbidden (stop sign), an action that is in principle allowed is restricted in quantitative terms (alcohol, speed) or a specific action is explicitly requested (wearing a seat belt). Compliance with these demands can, in some cases, be observed and recorded without stopping the driver (wearing of a seat belt, speeding). In some cases the driver must be stopped in order to check compliance (drunk driving) (Table 2.). Later in this chapter we shall discuss drunk driving and speeding in more detail because the respective regulations are crucially important for road safety.

Table 2. Type of regulation and enforcement method

	An action is completely prohibited	An action is restricted	An action is explicitly required
drivers must be stopped	driving without a valid license	Drunk driving	
non-compliance can be observed without stopping drivers	stop sign	speeding	wearing of seat belt wearing helmet maintaining proper distance

Enforcement is easy to conduct in the case of regulations that explicitly require a specific action and which can be checked without stopping the driver. Speed differs from this group of regulations because it requires multiple decisions on the part of the driver and varies over time.

In general, we can say that enforcement is a necessary element of safety policy. Its effects on behavior have been documented (in this chapter we shall focus on the research into speeding and drunk driving). Above a minimal level of police control there is a relationship between the degree of enforcement and compliance: the more enforcement the more compliance. A measure of this minimal, critical level of control is the subjective likelihood of detection, which depends on the objective level of enforcement. This minimal level of enforcement must be defined with respect to the type of offense as well as in relation to the size and homogeneity of the group of offenders. However, it can hardly be reached without technical support, which means automatic speed checks and random as well as evidential breath testing. Although enforcement is effective in changing behavior, it only changes underlying psychological characteristics for a longer period if it is accompanied by information for drivers, by attitude-oriented campaigns and if enforcement itself is kept above the critical level for an extended period.

The effects of enforcement can be modified by taking additional, attitude-oriented measures, such as public information campaigns.

Increasing the consequences of detection has positive effects, but the severity is less important than the certainty and swiftness of punishment. Fines and license withdrawal have proven to be effective, whereas imprisonment does not seem to reduce recidivism or accident figures.

5. SPEEDING

Exceeding the speed limit is probably the most widespread violation today. High speeds have negative effects on the safety of road users, the efficiency of the traffic system and the environment (air pollution, noise). Non-compliance with speed limits is also one of the most extensively studied violations. Depending on the situation, type of road, country and level of law enforcement, between 20 % and 80 % of all drivers exceed the indicated limits.

Empirical evidence indicates that *speeding is the major cause of traffic accidents* and injuries. Speeding is the contributory cause of traffic accidents that is cited most often and is generally associated with accidents involving serious crashes and, in particular, fatalities. Speeding or non-adaptive speed is recorded as an accident-contributory "failure" in 46.6 % of the fatal accidents in Germany; 31.9 % in Switzerland, and 14.4 % in France (Rothengatter, 1993). With regard to speed distribution on various types of roads in the Netherlands, it has been found that more than 75 % of drivers exceed the 80 km/h limit on single carriageway roads and about two thirds of drivers exceed the 100 km/h limit on dual carriageway roads (Rothengatter, 1988). It has been estimated that speeding alone accounts for up to 30 % of serious injuries and fatal accidents (Salusjärvi, 1982). Several studies (cited in Waard and Rooijers, 1994) have demonstrated that reducing the average driving speed by 2 to 5 km/h can result in a reduction of up to 30% in injuries and fatal accidents.

There is also subjective evidence for the dangers of speeding. "Excessive speed as a factor in accidents" was an item included in a series of questions asking drivers what factors they thought "caused" road accidents (SARTRE, 1994). "Driving too fast" was widely thought to be a major cause of road accidents – second only to drinking and driving. There were considerable differences in the responses obtained in different countries – for example 87.5 % of drivers in the Netherlands and 88.0 % in the eastern part of Germany thought that "driving too fast" was the cause of an accident either "often," "very often" or "always"– compared to 60.6 %, 61.4 % and 67.2 % of drivers in the Czech Republic, France and Spain (SARTRE, 1994).

Why do drivers speed? What are the *psychological reasons for speeding*? Several models have been used to explain driver behavior in general and speeding in particular, but none has been generally accepted. There is a whole series of explanations as to why drivers exceed the speed limit, for example, (a) according to Rothengatter (1990), speed appears to be determined by four motivational factors, which can be termed: "The pleasure of driving," "traffic risk," "journey time," and "driving costs." Speeders and non-speeders differ on all four factors – drivers who exceed the speed limit believe this gives them more pleasure and do not accept that it will increase risk, and they value time more and costs less than those drivers who observe the speed limit; (b) when choosing driving speed the influence of the behavior of other drivers ("social comparison") probably also plays a significant role, (Connoly and Åberg, 1993, Van Houten *et al*, 1981, 1983); exceeding the speed limit implies a deviation from the speed choice of other drivers, a driver will speed because the other drivers are speeding (the results of SARTRE, 1994 documented that a large proportion of drivers (81.8 %) thought that the other drivers "break the speed limit" either "often," "very often," or "always"); (c) drivers believe speeding is acceptable and are prepared to act in accordance with this belief (attitudes, beliefs).

There have been numerous *attempts to influence speed behavior*. Speed checks to improve speed limit compliance have followed a range of approaches with varying degrees of success. Several studies (e.g., Harper, 1991; De Ward and Rooijers, 1994; Fildes, 1995; Rothengatter, 1997b) have described enforcement attempts aimed at managing speed, varying from attitudinal and behavioral enforcement messages to road design, sending warning letters to offending drivers, the on-view stopping of offenders, the mailing of fines, the use of speed cameras and varying policing strategies and then comparing the results obtained in empirical studies that have been carried out to measure the effectiveness of these approaches. There is no space to go into the detailed results of all these studies. A summary of several approaches to speeding behavior will be attempted, showing their effects on such behavior (police enforcement, probability of detection, intensity of enforcement, role of feedback, publicity campaigns and automatic surveillance systems). Finally, the main finding concerning improving the effectiveness of enforcement will be briefly discussed.

Traffic regulation enforcement, as an oriented-oriented approach, has demonstrable effects on speeding behavior. A number of papers have attempted to summaries the results of this research (e.g., Rothengatter, 1982; Shinar and McKnight, 1985). However, traffic regulation enforcement has a significant influence on the behavior of drivers in traffic, but the mechanisms, which result in impact on behavior, are still not fully understood. According to Hauer *et al.* (1982), mechanisms exist that are often labeled as "time halo" and "distance halo" which can explain the effects of enforcement work on a driver's choice of speed through the process of learning. The time halo effect is defined as the length of time during which the effect of enforcement is still present after police activity has been withdrawn. The distance halo effect is the number of kilometers from the enforcement site – downstream or upstream – in which the effect is maintained.

Obtrusive law enforcement has two effects: (a) preventive effect – passing drivers notice enforcement activities and most will be deterred from speeding; (b) repressive effect – on detected offenders, who either receive a warning or a ticket. Police can optimize these two effects. Use of a marked police car (obtrusive enforcement) will enhance deterrence, but will reduce detection. Groeger (1995) summarizes results of several studies concerning attempts to increase compliance by increasing the likelihood of detection. The results show that parked marked vehicles are in general considered to have a greater likelihood of reducing violations than moving ones, and the use of unmarked police vehicles tends to have very little overall effect. The use of *unobtrusive speed radar measurement* will increase detection levels but will be noticed less often by drivers. In order to be effective, Shinar and McKnight (1985) conclude that enforcement must increase the perceived risk of detection by maintaining a certain level of objective probability of detection. Obtrusive police enforcement that does not increase the objective probability of detection does not seem to affect speed choice or attitudes toward it. Rothengatter (1990) demonstrated that police enforcement that increases both the objective (by obtrusive policing) and subjective probability of detection (by radar checkpoints) does increase the compliance level but does not change motivations and attitudes or the perception of safety on the road. The enforcement of regulations can probably be looked on as an external variable that affects behavior without affecting the intention to display that behavior (oriented-oriented attempt).

Bjornskau and Elvik (1992) offer an overview of several studies regarding the relationship between *intensity of enforcement* and violation rate. The studies show that the effects of police enforcement are transient, the violation rate declines when enforcement is increased. In some studies only excessive speed was reduced, in others, the overall violation rate declines between 7 % and 35 %. If enforcement is encountered, violation rates drop momentarily, but return to their previous levels once the site of enforcement has been passed or as the time since the last enforcement was encountered increases. Intensifying enforcement without increasing the objective probability of apprehension does not influence speed choice.

Similar results have been obtained by changing the *consequences of being detected*. Shinar and McKnight (1985) conclude that any effect due to increased penalty severity lasts only as long as the increased enforcement and that "as enforcement declines to original levels, so does the effect, even though the more severe penalties remain in force" (Shinar and McKnight, 1985, p. 388). From the psychological point of view these results can be explained in the context of the Theory of Learning, which postulates that the likelihood of punishment rather than the severity of any punishment is the effective agent (Groeger, 1995).

Collective feedback plays an important part in the enforcement process. According to the findings of SARTRE (1994), a large proportion of drivers (81.8 %) thought that other drivers "break the speed limit" either "often," "very often" or "always." Although these results represent drivers' subjective assessments of other drivers, the role of the behavior of the other road users has a significant effect on driving behavior. It was suggested by Conolly and Åberg (1993) that drivers' adjustment of speed is affected by a comparison of their own speed with that of other drivers in a way that makes speeding a contagious offense. The effects of the normative influence of others' behavior has been demonstrated by Van Houten and colleagues (1981, 1983), who reported that the public posting of the percentage of complying drivers (collective feedback), i.e., those who had not violated the speed limit, had a positive effect on a number of speed violators. They demonstrated that increased compliance with the speed limit was not just due to the presence of text on the sign, but that the digits on the sign indicating the number of violators were essential to achieve a reduction in speed. This finding reflects a simple normative influence operating through "social comparison" (drivers speed when they perceive that other drivers break the speed limit; likewise, drivers reduce speed when public posting of the percentage of compliant drivers increases). From this viewpoint, increased compliance occurs because drivers are motivated to comply and the posting of reliable information about others' compliance makes it readily available. Van Houten and colleagues report that the effect of the signs has been shown to be greater than overt and covert enforcement, and lasts longer, with a lower incidence of speeding still apparent 20 months after the initial installation (positive effect even without police enforcement). In addition, there also seem to be fewer accidents. On the other hand, Groeger (1995) summarizes the results of attempts that were made to increase compliance with speed limits by presenting drivers with an indication of their current speed by means of a visual speed indicator ("Your speed is XXX") (following Dart, Hunter, 1976). Such information about individual behavior produced a positive effect and significantly reduced speed, especially where the visual speed indicator was combined with enforcement.

Individual feedback (see also automatic enforcement system), employment of in-vehicle feedback and in-vehicle warning systems has a similar effect on speeding behavior. A considerable reduction in speed can be achieved when verbal feedback is used and occurs immediately following deviation from the required speed. The frequency of speed violations reduces drastically when an in-vehicle warning system is used (following Rothengatter, 1997b). Rothengatter concluded that feedback, either applied individually in-vehicle or collectively by means of roadside signs, appears to be a powerful tool in reducing inappropriate speed.

Because feedback has proved to be a powerful way of increasing traffic compliance (see Van Houten and Nau, 1983; Rothengatter, 1997b), it should be considered an important ingredient in an *automatic policing system*. Such systems could be located either on-site (traffic lights) or in-vehicle, or both, and should fulfill several functions (Rothengatter, 1991): a) to monitor road-user behavior; b) to compare road-user behavior with normative, required behavior; c) to provide feedback to the driver if there is a discrepancy between the monitored behavior and required behavior, and that, if illegal behavior persists, to record the offense; d) to procure evidence that an offense has been committed. Information can by recorded on a "smart card," which can be read by the police and other interested parties (for example, insurance companies).

Modern technology can be used to solve the problem of poor speed adaptation and to improve road safety. Engineering solutions (such as speed limiters, intelligent vehicle and highway systems, AUTOPOLICE) could supplement police efforts and help to reduce the incidence of speeding. A speed limiter (SL) restricts vehicle speed to the legal limit. This system has been developed and tested in Sweden (Almquist *et al.*, 1991; Almquist, 1997) with the aim of studying how automatic speed limitation could become a key part of improved traffic safety in general, and to investigate the attitudes and driving behavior of test drivers using automatic speed limitation every day over an extended period. Whenever the speed limit changes, a transmitter emits pulses. Each vehicle is equipped with a special unit for receiving these pulses, which automatically limit the maximum speed of the vehicle. Such a system would be automatic and obligatory. Intelligent Vehicle and Highway Systems (IVHS) could involve a network of feedback or speed controls where the posted speed is read "off the road." In-vehicle recording of speed offenses (monitored from electronic road loops) is another option for enforcing speed compliance without police involvement (Fildes, 1995). The main objective of the AUTOPOLICE project is to design a computerized system to assist with the detection/deterrence and processing of traffic violations as they occur, to monitor the interaction between vehicle and environment and to provide information on speed, vehicle type, lane position and even mechanical condition of the vehicle. This system must be co-operative and adaptive (Harper, 1991). The fact that such systems do not yet exist poses problems for an ergonomic assessment of their impact.

Automatic enforcement would be effective in reducing accident occurrence, but the acceptance of such a system also has several contradictory aspects. Drivers tend to resist the idea that their freedom may be curbed and automatic enforcement might be a fairly unpopular measure and thus difficult to implement (Rothengatter, 1997a). The restriction in freedom of choice might lead to unwanted forms of behavior, compensatory behavior (more aggressive behavior at pedestrian crossings). Furthermore, the frustration resulting from a perceived loss of time could lead to even higher speeds during a low-speed maneuver (such as turning right or left) where the maximum speed is lower than that permitted by the SL system. Nevertheless, results of a Swedish study (Almquist, 1997) indicate very good acceptance of speed adaptation. Drivers reported that they had changed their driving behavior to the extent that driving generally had become much calmer. As drivers, they felt more comfortable and less stressed. Overall, SL is expected to have positive effects on road safety, fuel consumption, air pollution and noise emission.

Several problems associated with current enforcement programs (for example, the transient effects of enforcement attempts, the short-term halo effect, the long time between apprehension and punishment) provide powerful arguments for the development an automatic enforcement system.

Many authors, (among them Rooijers and De Bruin, 1990; Rooijers, 1988; Shinar and McKnight, 1985; Riedel *et al.*, 1988) support the view that media or *public information campaigns* can enhance the effects of enforcement but public information alone has no effect on violations (see Rothengatter, 1990). Riedel *et al.* (1988) investigated whether a publicity campaign with or without selective police surveillance can have an effect on speeding behavior on rural roads. They concluded that a publicity campaign is helpful in supporting police enforcement strategies by increasing the effectiveness of surveillance on speeding and therefore on their efficiency. Riedel *et al.* observed that the effects of a publicity campaign on speeding behavior accompanied by police enforcement are more effective than those of police enforcement alone. The combination of publicity and surveillance not only proved more effective than publicity information, it also proved more effective than intensified surveillance alone in comparable experiments. Public information can enhance the effects of police surveillance but it cannot replace police surveillance, as the effects of their own are very short-lived. However, police surveillance can only compel drivers to comply by means of threat rather than as the result of a change in attitude. Behavioral change resulting from a fundamental change in attitude towards speeding may be more effectively brought about by a publicity campaign.

To summarize, we can say that several studies indicate the necessity to improve compliance with traffic regulations with regard to those traffic regulation violations that appear to be accident-contributory factors (speeding). Drivers and police ought to be aware of the gravity of excessive speed. According to Rothengatter (1993), this can be brought about by providing the relevant information, but mainly by increasing the subjective probability of detection, because the high level of detection is most likely to change the belief system of drivers, either directly because the likelihood of negative consequences of their behavior would increase, or indirectly, as the subjective norm with regard to police would change. Results of the above studies indicate that the type of enforcement has significant effects on the choice of speed. As far as the review of different approaches is concerned, it can be concluded that feedback provision (collective and individual) and public information campaigns, both of these in combination with law enforcement, seem to represent a promising attempt to influence speeding behavior and bring speed under control.

Another important issue is that enforcement methods affect behavior but do not affect attitudes towards that behavior. Only when enforcement persists over a period of years and is coupled with massive media campaigns do attitudes adapt to the enforced behavior. This mechanism, which is crucial to the effective enforcement of regulations, is still not fully understood. Explanations of enforcement mechanisms are essential in order to understand how enforcement programs influence driver behavior both positively and negatively. They are thus necessary for developing effective practices, enforcement strategies and remedial programs, etc. Without this knowledge, enforcement procedures will tend to be ad hoc and will meet with varying degrees of success.

6. DRUNK DRIVING

In most countries, driving under the influence of alcohol is not forbidden but restricted in quantitative terms. The respective regulation indicates a critical *Blood Alcohol Concentration* (BAC-limit). One problem of BAC-limits is that they are difficult to translate into behavioral standards. High BAC-limits (0.8 or 1.0) are more difficult to convert because individual differences in the number of glasses allowed are bigger compared with a lower BAC-value.

Compared to other traffic offenses, drunk driving is quite rare, but dangerous. In European countries about 3 % of all journeys are associated with an illegal BAC (ETSC, 1995: overview of different studies), but about 30% of injured drivers were under the influence of alcohol. So alcohol is definitely one of the major causes of accidents and is also an aggravating factor.

Alcohol is also the major cause of accidents in the view of road users. Some 85% of European drivers say that alcohol is "often", "very often" or "always" the cause of accidents (SARTRE, 1994). While 93 % of Swedish drivers express this opinion, in western Germany the figure is about 20 % lower (72 %). This is a quite remarkable difference but the ratings concerning other accident causes differ even more.

The nature of non-compliance with BAC-limits is not similar to the nature of non-compliance with speed limits. The group of drunk drivers can be characterized mainly by their drinking behavior and their psychological reasons for drinking. Krüger (1995) demonstrated that only about 15 % of Germany's driver population drinks as much as is necessary to exceed the legal limit (BAC-limit in Germany, 1995: 0.8). Drivers who are over the limit of 0.5 or 0.8 very often have serious alcohol problems or even show symptoms of addiction. Compared to those drivers who keep drinking and driving separate, offenders were found to exhibit problematic actions (behavior patterns, attitudes) as well as problematic situation-related behavior (tendency to drink because of social control and situational characteristics, experience of negative emotions during drinking) (Siegrist, 1992). The degree and the significance of alcohol consumption are especially important in this group. Equally important are momentary feelings and how situational aspects are dealt with. So-called social drinkers quite seldom combine drinking and driving. Their risk of combining drinking and driving is determined more by situational factors. Although offenders have less negative attitudes towards the combination of drinking and driving, their non-compliance is not determined by a conscious decision to violate the regulation. Rather, it is a logical consequence of their usual behavior and social life.

In the view of the majority of road users, drunk driving is more common than it really is. In fact, many drivers think they have been drinking too much before driving. One explanation of this misunderstanding are high BAC-limits which do not allow drivers to determine whether their own behavior is legal or illegal. Because of this it is not surprising that the attitudes of non-offenders towards drunk driving are not as negative as might be expected. On the attitude level, conforming drivers tend to be identified as the extreme group whereas most of the non-offenders are quite similar to the offender-group (Kretschmer-Bäumel, 1989). This may be a reason why, in some countries, there is little social pressure to drink moderately before driving.

As a consequence of these specific causes of drunk driving and the perceived level of compliance, a preventive strategy can be developed. First the BAC-limit must be set at a level that makes it possible for drivers to be given clear advice regarding what they have to do. This is a necessary basis for safety campaigns that set out to explain the regulations in order to influence attitudes. Police control and punishment are no more than concomitant elements. In order to create social pressure aimed at keeping drinking and driving separate, non-offenders are also a target-group: the level of compliance must be communicated and offenders must be supported in trying to reduce their alcohol consumption. The main function of police control is to detect high-consumption groups and to demonstrate that the law is being enforced. Because car drivers have to be stopped in order to detect non-compliance, random and evidential breath testing are important elements, because we know that subjective control probability depends on the objective risk of detection.

The available empirical results on the effectiveness of regulations and enforcement strategies are difficult to generalize, because it is usually methodologically difficult to isolate the effect of a measure and because the results also reflect some aspect of the cultural background of a specific country. Nevertheless there are some tendencies. *Lowering the BAC-limit* from 0.8 or 1.0 to 0.5 would seem to be one element that has a positive effect on the offense rate as well as on injuries due to road crashes (Homel, 1994; Mathijssen and Noordzij, 1993). In a survey of European drivers a close relationship was shown to exist between drivers' attitudes towards drunk driving and the respective BAC-limit (SARTRE, 1994): "Drivers from countries with a legal BAC limit below 0.08% state more often that they:

- Never drive after drinking even a small amount of alcohol;
- Have been stopped and breathalysed by the police on at least one occasion;
- Expect to be stopped and breathalysed by the police on a typical journey" (p. 122).

Although drivers' opinions cannot be interpreted in terms of traffic safety it is possible to say that lowering BAC-limits is followed by an internalization of the new norm.

A high level of *subjective probability of control* has positive effects on compliance rates. As in the case of speeding, it depends on the objective probability of control. It can be enhanced by random breath testing, which allows a minimal level of control to be maintained. Random breath testing is one of the major elements in an effective strategy (Span and Stanislaw, 1995). Because road users tend to overestimate the offense rate dramatically, the impact of providing drivers with information on the percentage of drivers over the legal limit was tested (Nau *et al.*, 1993). In contrast to the case of speeding, no positive effects on behavior could be measured. One reason might be the nature of and the causes of the offense. It was noticed that many offenders do not have much control over their behavior, so this kind of weak social pressure (behave as others do) is not enough to change behavior. However, *providing drivers with information* is an important part of enforcement strategy. It has been shown that drivers must be informed about enforcement activities, and accompanying public education campaigns have positive effects (Söder, 1991) if they appeal to and target specific behavior rather than a general value. Automatic control systems that do not allow a drunken driver to start his car have not yet been evaluated.

Driver Improvement or remedial education programs were developed as an additional measure for high-consumption groups. Psychologists can contribute to the prevention and therapy of alcohol addiction, which has a positive effect on recidivism rates. In Europe this strategy has been followed, especially in the German-speaking countries. Evaluation studies support the hypothesis that it is a measure, which contributes to traffic safety (Winkler *et al.*, 1990).

There is serious evidence concerning the effective *consequences of detection*. Imprisonment and other severe consequences have no positive effects (Mann *et al.*, 1991). This result is confirmed by experience in Scandinavia that the retreat (Ross *et al.*, 1983) from long standard prison sentences does not seem to increase the extent of drunk driving. This is consistent with the Learning Theory, which maintains that the likelihood of punishment is more important than its severity. License withdrawal, on the other hand, is an effective measure (Mann *et al.* 1991; Tornrös, 1993). In most countries it is a swift and certain consequence and even reduces risk exposure.

7. INTEGRATING THE TRAFFIC – PSYCHOLOGICAL VIEW IN TRAFFIC LEGISLATION AND ENFORCEMENT

Legislation is a political process and is therefore quite different in nature from traffic psychology, which is a social science. In order to influence the traffic system, traffic psychology has to overcome this difference. The first step is to produce clear recommendations that can be integrated in prevention strategies at local level. A summary of existing knowledge produces the following points, which show how enforcement should be designed and what supporting measures enhance the effects.

- Each regulation should be analyzed with respect to the most important conditions a good regulation should fulfill;
- The whole set of regulations must be checked with respect to homogeneity, correspondence with other characteristics of the traffic system and the number of regulations;
- The different regulations require that road users groups be categorized with respect to their willingness to comply, their actual behavior and the reasons for persistent non-compliance;
- If compliance is not close to 100 %, a regulation must be enforced. Two strategies must be combined: the minimal level of highly visible enforcement and the detection of non-compliance. Enforcement must be kept at a minimal level in order to produce a minimal level of subjective probability of detection;
- In order to enhance the effects of enforcement, four supporting measures are necessary;
  - on a local level drivers must be informed about the police activity (frequency and detection rate) and the level of compliance,
  - attitude-oriented campaigns showing that safety is the main reason for the regulation in question,
  - communication of the level of traffic safety and the potential of enforcement,
  - driver improvement or remedial training for groups with a high likelihood of recidivism;
- Preference should be given to punishment that is immediate and certain, which influences the psychological reasons for non-compliance and also reduces exposure.

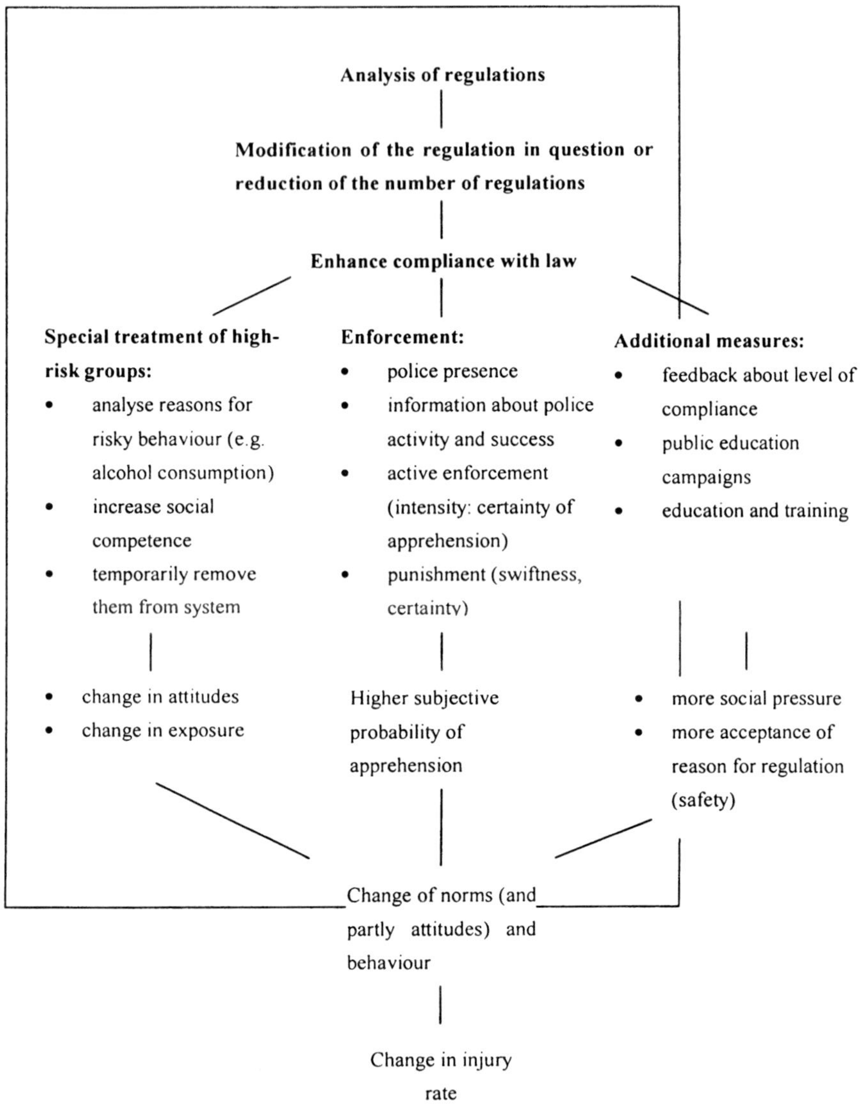

Figure 1. Illustrates the logical structure of this integrated approach to non-compliance with traffic regulations

8. REFERENCES

ÅBERG, L., LARSEN, L., GLAD, A., BEILINSSON, L. (1997), "Observed Vehicle Speed and Drivers' Perceived Speed of Others", *Applied Psychology: An International Review*, 46 (3), 287-302.

AJZEN, I. (1985), "From Intentions to Actions: A Theory of Planned Behaviour", in KHUL, J., BECKMAN, J. (Eds.), *Action Control: From Cognition to Behaviour* (pp.11-38), Berlin: Springer-Verlag.

AJZEN, I. (1988), *Attitudes, Personality and Behaviour,* Milton Keynes, England: Open University Press.

ALMQUIST, S. (1997), *Automatic Speed Adaptation in Built-up Areas, Driver Attitudes and Behaviour,* results from a field trial in real traffic in Eslov, Sweden, Lund Institute of Technology, Lund University.

ALMQUIST, S., HYDEN, C., RISSER, R. (1991), *Use of Speed Limiters in Cars for Increased Safety and a Better Environment,* Transportation Research Record 1318, Transportation Research Board, National Research Council, Washington, D.C.

BJORNSKAU, T., ELVIK, R. (1992), "Can Road Traffic Law Enforcement Permanently Reduce the Number of Accidents?", *Accident Analysis and Prevention,* 24 (5), 507-520.

CONNOLY, T., ÅBERG, L. (1993), "Some Contagion Models of Speeding", *Accident Analysis and Prevention,* 25, 57-66.

CRANACH, VON, M. *et al.* (1982), *Goal-Directed Actions,* London: Academic Press.

DART, O. K., HUNTER, W. W. (1976), "Evaluation of the Halo Effect in Speed Detection and Enforcement", *Highway Safety,* Traffic Records and Law Enforcement, Transportation Research Board, National Research Council, Washington, D.C.

DE WAARD, D., ROOIJERS, T. (1994), "An Experimental Study to Evaluate the Effectiveness of Different Methods and Intensities of Law Enforcement on Driving Speed on Motorways", *Accident Analysis and Prevention,* 25, 751-765.

EVANS, L. (1991), *Traffic Safety and the Driver,* New York: Van Nostran Reinhold.

ETSC, European Transport Safety Council (1995), *Reducing Traffic Injuries Resulting from Alcohol Impairment,* Brussels: Rue du Cornet.

ETSC, European Transport Safety Council (1997), *Safety Monitor – ETSC's Newsletter on Transport Safety Policy Developments in the EU,* Brussels: Rue du Cornet.

FILDES, B. (1995), "Driver Behaviour and Road Safety", in BREWER, N., WILSON, C. (Eds.), *Psychology and Policing,* Hove, UK: Lawrence Erlbaum Associates, Publishers.

FISHBEIN, M., AJZEN, I. (1975), *Belief, Attitude, Intention and Behaviour: An Introduction to Theory and Research,* Reading, MA: Addison-Wesley.

GREGERSEN, N. P., BERG, H. Y. (1994), "Lifestyle and Accidents among Young Drivers", *Accident Analysis and Prevention,* 26:297-303.

GROEGER, J. A. (1995), "Psychological Influence on Traffic Law Compliance", *Analise Psicologica,* 3 (XIII), 257-269.

GROEGER, J. A., CHAPMAN, P. R. (1997), "Normative Influence on Decision Making", *Applied Psychology: An International Review,* 46 (3), 265-285.

HARPER, J. G. (1991), "Traffic Violation Detection and Deterrence: Implications for Automatic Policing", *Applied Ergonomics,* 22 (3), 189-197.

HAUER, E., AHLIN, F. J., BOWSER, J. S. (1982), "Speed Enforcement and Speed Choice", *Accident Analysis and Prevention,* 14, 267-278.

HOMEL, R. (1988), *Policing and Punishing the Drinking Driver: A Study of General and Specific Deterrence,* New York: Springer Verlag.

HOMEL, R. (1994), "Drink-Driving Law Enforcement and the Legal Blood Alcohol Limit in New South Wales", *Accident Analysis and Prevention*, Vol. 26, No2, pp. 147-155.

KRETSCHMER-BÄUMEL, E. (1989), "Kraftfahrer im Konflikt ,Trinken und Fahren' Verhaltensmuster", *Blutalkohol,* 2:106-113.

KRÜGER, H.-P. (1995), *Das Unfallrisiko unter Alkohol*, Stuttgart: Gustav Fischer Verlag.

MANN, R. E. *et al.* (1986), "Sentence Severity and the Drinking Driver: Relationship with Traffic Safety Outcome", *Accident Analysis and Prevention*, Vol. 23, No. 6, pp. 483-491.

MATHIJSSEN, R., NOORDZIJ, P. C. (1993), "The Decline of DWI and Alcohol-Impaired Accidents in the Netherlands 1983-1991", *12th Conference of Alcohol, Drugs and Traffic Safety* (ICADTS-T92), Verlag TüV-Rheinland.

NÄÄTÄNEN, R., SUMMALA, H. (1976), *Road User Behaviour and Traffic Accidents*, Amsterdam: Elsevier.

NAU, P. A., VAN HOUTEN, R., ROLIDER, A., JONAH, B. A. (1993), "The Failure of Feedback on Alcohol Impairment to Reduce Impaired Driving", *Journal of Applied Behaviour Analysis*, 26, p. 361-367.

NOORDZIJ, P. (1976), *Influencing Road Users' Behaviour,* SWOV-Report 76-4e, Voorburg.

OPP, K.-D. (1971), "Einige Bedingungen für die Befolgung von Gesetzen", *Kriminologisches Journal*, 3. Jg. S. 1-25.

OECD (1990), *Behavioural Adaptation to Changes in the Road Transport System*, Paris: OECD.

PARKER, D., MANSTEAD, A. S. R., STRADLING, S. G., REASON, J. T. (1992), "Intention to Commit Driving Violations: An Application of the Theory of Planned Behaviour", *Journal of Applied Psychology*, 77 (1), 94-101.

REASON, J. T., MANSTEAD, A., STRADLING, S., BAXTER, J. S. (1990), "Errors and Violations on the Roads: A Real Distinction?", *Ergonomics,* 33:1315-1332.

REASON, J. (1994), *Menschliches Versagen*, Heidelberg: Spektrum Verlag.

RIEDEL, W., ROTHENGATTER, T., DE BRUIN, R. (1988), "Selective Enforcement of Speeding Behaviour", in ROTHENGATTER, J. A., DE BRUIN, R. (Eds.), *Road User Behaviour: Theory and Research,* Assen: Van Gorcum.

ROOIJERS, A. J. (1988), "Effects of Different Public Information Techniques in Reducing Driving Speed", in ROTHENGATTER, J. A., DE BRUIN, R. (Eds.), *Road User Behaviour: Theory and Research*, Assen: Van Gorcum.

ROOIJERS, A. J., DE BRUIN, R. A. (1990), "Selective Enforcement of Speeding Behaviour in Built-up Areas", *Proceedings of Enforcement and Rewarding Strategies and Effects*, OECD International Symposium, 19-21 September 1990, Copenhagen.

ROSS, H. L., MC.CLEARLY, R., KLETTE, H. (1983), "Liberalization and Rationalization of Drunk-Driving Laws in Scandinavia", *Accident Analysis and Prevention,* 16, pp. 471-487.

ROTHENGATTER, T. (1982), "The Effects of Police Surveillance and Law Enforcement on Driver Behaviour", *Current Psychological Reviews*, 2, 349-358.

ROTHENGATTER, T. (1988), "Risk and the Absence of Pleasure: A Motivational Approach to Modelling Road User Behaviour", *Ergonomics*, 31, 4, 599-607.

ROTHENGATTER, J. A. (1990), "Normative Behaviour is Unattractive if it Is Abnormal: Relationships between Norms, Attitudes and Traffic Law", *Proceedings of Enforcement and Rewarding Strategies and Effects*, OECD International Symposium, 19-21 September 1990, Copenhagen.

ROTHENGATTER, J. A. (1991), "Automatic Policing and Information System for Increasing Traffic Law Compliance", *Journal of Applied Behaviour Analysis*, 24, 85-87.

ROTHENGATTER, J. A. (1993), "Attitudes Toward Traffic Law Violations and Enforcement", *37th Annual Proceedings Association for the Advancement of Automotive Medicine*, November 4-6, 1993, San Antonio, Texas.

ROTHENGATTER, J. A. (1997a), "Psychological Aspects of Road User Behaviour", *Applied Psychology: An International Review*, 46 (3), 223-234.

ROTHENGATTER, J. A. (1997b), "The Effects of Media Messages, Police Enforcement and Road Design Measures on Speed Choice", *IATSS Research*, Vol. 21 No.1 pp. 80-87.

SALUSJÄRVI, M. (1982), "Speed Limits and Traffic Accident", *Proceedings of the OECD Symposium: The Effects of Speed Limits on Traffic Accident and Transport Energy Use* (pp.71-80), Paris: OECD.

SARTRE - *Social Attitudes to Road Traffic Risk in Europe* (1994).

SHINAR, D., MCKNIGHT, A. J. (1985), "The Effects of Enforcement and Public Information on Compliance", in EVANS, L., SCHWING, R. (Eds.), *Human Behavior and Traffic Safety*, New York: Plenum Press.

SIEGRIST, S. (1992), "Das Bedingungsgefüge von wiederholtem Fahren in angetrunkenem Zustand", *bfu-Report*, Nr. 19, Bern.

SÖDER, J. C. M. (1991), *Evaluatie onderzoek VVN-campagne alcohol in het verkeer 1986-1991*, VK 91-10, Haren: Traffic Research Centre, University of Groningen.

SPAN, D., STANISLAW, H. (1995), "Evaluation of the Impact of a Deterrence-Based Random Breath Testing Program in New South Wales", in KLOEDEN, C. N., MCLEAN, A. J. (Eds), *Alcohol, Drugs and Traffic Safety*, University of Adelaide, Adelaide, pp. 840-844.

TORNRÖS, J. (1994), *Atgädar mot rattfylleri Litteraturöversikt*, Report 384, Linköping: Väg- och Traffik-Institute.

VAN HOUTEN, R., NAU, P. A. (1981), "A Comparison of the Effects of Posted Feedback and Increased Police Surveillance on Highway Speeding", *Journal of Applied Behaviour Analysis*, 14, 261-271.

VAN HOUTEN, R., NAU, P. A. (1983), "Feedback Intervention and Driving Speed: Parametric and Comparative Analysis", *Journal of Applied Behaviour Analysis*, 16, 253-281.

WAARD, D., ROOIJERS, T. (1994), "An Experimental Study to Evaluate the Effectiveness of Different Methods and Intensities of Law Enforcement on Driving Speed on Motorways", *Accident Analysis and Prevention*, 25, 751-756.

WILDE, J. S. G. (1988), "Risk Homeostasis Theory Applied to a Fictitious Instance of an Individual Driver's Decision Making", in ROTHENGATTER, T., DE BRUIN, R. (Eds), *Road User Behaviour: Theory and Research*, Van Gorcum, 67-81.

WINKLER, W., JACOBSHAGEN, W., NICKEL, W. R. (1990), "Zur Langzeitwirkung von Kursen fuer wiederholt alkoholauffaellige Kraftfahrer, Untersuchungen nach 60 Monaten

Bewaehrungszeit", "Long-Term Effects of Drinking Driver Programs for Repeat Offenders, Analysis after 60 Months of Probation", *Blutalkohol,* 27(3):154-174.

Chapter 11

EFFECTIVENESS OF ROAD SAFETY CAMPAIGNS

Nicole De Vrieze
Psychologist, (IBSR) Belgian Road Safety Institute

1. FOREWORD

"People drive as they live." Hence, they have their own patterns of behavior, wants, habits, prejudices, frames of reference and ways of looking at things. The act of driving is just one of these patterns. Yet, in the context of road safety, the individual is too often regarded as merely a road user. The tendency is to forget that the individual is first and foremost a human being, with all that that implies.

We, the "experts," know (or think we know) the kind of behavior the road user ought to adopt and we look for a way or ways of convincing him/her. And yet, do we really pay enough attention to find out what the road users themselves need?

The following section focuses solely on mass media campaigns, whereas the other kind of initiatives, such as training, education and re-inclusion of road users, are dealt with elsewhere in this work.

2. BACKGROUND

Road safety experts from numerous developed countries decided during the 1960s to add safety campaigns to the array of measures already applied to make the roads safer.

As an example, in 1967, the Swedish authorities launched a large-scale safety campaign in a bid for ensuring that road users were aware that they were now supposed to drive on the right side of the road. Consequently, the stakes at play were very high indeed! (OECD, 1971).

Surveys carried out before the changeover helped determine the content, timetable and specific communication media used to target the different groups or categories of users. The organizers made use of advertisements, comic strips, cartoons, radio and television programs.

The road safety campaigns proper were seconded by radical legislation and technical changes. So it is difficult to determine how great their real impact was in changing the behavior of drivers or their attitudes towards the information being provided.

In 1970, the OECD decided to set up an international group of experts whose target was to consider the key factors involved in the design of road safety campaigns and the methods of evaluating those campaigns (OECD, 1971). This report was the state of the art. It focused in particular on the various types of campaigns, laboratory and field evaluation, the pros and cons of each method, the patterns of behavior most amenable to change. Also considered was the problem relating to the risk homeostasis.

The report was followed by a "Safety Campaign Handbook" in 1975, designed to help the relevant authorities organize such activities. Many surveys were then made on the campaigns and their effectiveness.

Over the years, road safety campaigns have increasingly drawn their inspiration from commercial advertising, health-education campaigns, theories of social psychology and, more recently, social marketing based on social-psychological principles.

3. DESIGN OF CAMPAIGNS

Before a mass media campaign is planned, the problem first has to be assessed so as to use the results as a basis on which the determination of the aims and the target population (or populations) can be based. This helps to decide which is the most appropriate media and the type of message to be used. The material considered has then to be tested before the campaign proper. The impact has to be assessed once it is over.

4. ASSESSING THE PROBLEM

An assessment of the situation involves defining the problems the campaign is supposed to solve. Accident statistics may be used to detect high-risk groups, black spots on the road network and the primary causes of accidents. Other indicators may be taken into account, such as near accidents, subjective insecurity (e.g., parents afraid of allowing their children to go to school by themselves, although no accident has ever occurred on the way to it).

Where an assessment reveals that the infrastructure is at fault, a campaign can only alleviate the problem, not cure it. The campaign can make the road users aware of the danger of a situation but cannot solve the origin of the problem. Nonetheless, campaigns have been conducted in such cases, due to the lack of funds required for infrastructural adaptations or technical measures, such as an automatic traffic inspection system.

Very often, however, the assessment shows that it is the user's behavior that is at the origin of the problem. The problems most frequently dealt with are the need to observe speed limits (mainly in built-up areas), the wearing of seat belts, the provision of adequate protection for children in cars, drunk driving and weekend accidents.

When the behavior of the driver is shown to be at fault, the question arises as to what determines such behavior. At this stage, experts tend to use behavioral models (see Hugenin on this topic).

Let's take as an example the Fishbein model (Ajzen and Fishbein, 1980), which is based on three key factors: The assumed advantages and disadvantages of a certain behavior, subjective standards (how the individual perceives the expectations of his social environment), and the possibility of adopting the behavior (behavioral intention). The three components may be used to try to exert some influence on behavior.

Where the individual regards his behavior as acceptable, he can see no grounds for changing it. It also has to be remembered that as far as the user is concerned "undesirable..." behavior often offers significant advantages and frequently has had a positive feedback. Let's take the example of a motorist who doesn't wear his seat belt. He has never been involved in an accident nor has he ever been sanctioned. Why would he change his behavior? Even more serious, the same motorist may be involved in a fairly severe accident without suffering any injury or traumatic effects. In this case, the "bad behavior" has been "rewarded."

Subjective standards also represent a major factor in determining behavior. Vogel (1985) has shown that colleagues and employers constitute a significant benchmark group for drivers of heavy good vehicles who normally drive fast. Hence a campaign designed to encourage drivers to drive more slowly will seek to influence fellow workers and the employer or to help drivers to resist peer-group pressure.

The possibility of adopting the recommended behavior must be present. For example, a driver may wish to wear a seat belt but find himself in a vehicle without one. A driver may want to observe the speed limits, then find himself at a spot where the maximum authorized speed is not credible and where observing the speed limit could place him or others in danger, given the speed at which the other drivers are traveling.

These are only a few examples. All too often, however, safety campaigns encounter such hurdles. Take as an example a campaign to discourage drunk driving. If messages seek to promote non-alcoholic beverages, such drinks have to be available at a reasonable price, at places where people go out to enjoy themselves. They need to have a pleasant taste and an eye-catching presentation (be attractive to look upon and be colorful).

The recommended behavior has to be presented not only as safer but also easy to adopt. The messages should also take advantage of contemporary social values, such as environmental protection (there are, however, grounds for wondering whether this is always a valid argument, now that countries can purchase the "right to pollute").

Appealing solely to the safety aspect obviously has little chance of success because people may regard road safety as important, but do not feel any need for "safety on the public highway." Generally speaking, everyone thinks he is a good driver and so does not always feel personally concerned by a campaign.

Furthermore, accidents always happen to other people. Rooijens (1986) wrote that "if one compares road safety campaigns with commercial advertising, it could be said that the campaigns are selling a bad product that the public does not need."

Any assessment of the problem has to reflect the social environment, the general climate and anything that may have a direct or indirect impact on the campaign (Järvinen, 1997).

5. DETERMINING THE TARGET POPULATION

To define the target population implies determining the group or subsidiary group of users most likely to "behave badly." The campaign may focus specifically on a particular group of road users or, on the contrary, on all users, and sometimes even on the entire population. For example, when the legal alcohol limit is lowered, all drivers have to be informed. However, the measure also affects owners of cafes, bars and restaurants as they can in principle be held liable if a driver leaves their establishment with more than the legal blood-alcohol limit.

A campaign targeted at individuals driving while under the influence of cannabis will focus more specifically on young people, whereas if it deals with driving under the influence of medication (benzodiazepines for example), it will target a more elderly population (BTTS, 1997).

The process of deciding which section of the population to target should not be based solely on time-honored social-demographic variables, but should also take account of others, such as lifestyle, prejudices, habits, etc.

An in-depth analysis of the target population provides a mean of determining the most suitable media and strategies. For example, if an investigation of the habits and lifestyles of the target population shows that the people involved are house music enthusiasts, there will be little point in running TV spots during commercial breaks in the middle of TV romances. Far better to use, say, a pop group in order to transmit the message.

6. CHOOSING THE MEDIA

There is a wide range of potential media: posters, radio advertisements or programs, TV advertisements or programs, leaflets, brochures, press releases or articles, books, conferences, advertisements in movie theaters, messages conveyed by pop groups. Road safety messages may even be put across during church services, or they may be featured on telephone cards, sugar wrappings, and lottery tickets (IBSR, 1990). In short, many objects in daily life may be used as a media for putting such messages across.

Rooijens (1988) considered the impact that the various media have on speed limits at bends on roads where visibility is very poor or even non-existent. He was able to highlight the positive, but short-live, effect of a personalized mail-shot addressed to drivers failing to observe speed limits. The medium that turned out to be the most effective was a poster, updated every week, showing the number of drivers sticking to the limit. The effect proved to last longer than was the case with the other media.

Whatever method is chosen to convey the message, the important thing is to reach the target population. Against this background, it has to be remembered that road users do not generally look for road safety information off their own bat, so it has to reach them in their normal environment, where they do not have to make any special effort to come into contact with or assimilate the information.

7. CHOOSING THE APPROACH

The approach has to be understood as the way the message is conveyed. Users have to feel concerned by the message so it has to take a form that attracts their interest.

As early as 1975, Mackie and Valentine were investigating the following approaches:

- The factual approach – an appeal to reason, a neutral approach informing about the facts without any emotional component;
- The "shock" approach;
- The approach seeking to appeal to family responsibility: An approach

highlighting the dependence of children and spouses (the current feminist movement is unlikely to give this approach its blessing!);
- Using well-known personalities (such as TV or sports stars);
- Prestige or authority: Using recognized personalities who may be described as opinion leaders (MPs, doctors, police officers, for example);
- Satire – people who fail to follow the recommended behavior are turned into objects of ridicule;
- Humor – the aim being to enhance the appeal of the communication;
- Sex appeal (this approach can also be expected to raise hackles in this day and age!).

Each one of these approaches has been assessed to determine its avowed effectiveness, and how long it is remembered. None, however, can be considered as ideal.

Approaches 4 and 5, it should be pointed out, are not without their dangers: They only need the "transmitting agent" to lose his credibility (ban on a sports personality, an MP's involvement in a political or financial scandal) for the campaign to collapse.

Over the years, several European countries have reported a change in approach: the focus used to be on the consequence of a given type of behavior (facial injuries for a seat belt campaign, child on crutches to combat drunk driving, for example), but the trend now is towards making drivers more aware of their responsibilities (IBSR, 1997). This could well be a consequence of ideas harking back to May 1968.

Let's face it: The jury is still out on the question of how effective the different approaches can be. Champions of shock campaigns continue to clash with the proponents of soft, positive campaigns.

A single approach is clearly unlikely ever to be discovered, because attitudes obviously differ considerably from nation to nation. The standards and values of citizens in northern Europe cannot be compared with those of the south, anymore than they can be compared with those of the United States or the Middle and Far East. Yet, all the industrial nations and developing countries are faced with road safety problems. Each country has its ways and means (in terms of financial and human resources), its values, its mentality, and culture, but can obviously benefit from the experience gained by other countries.

8. TESTING THE MATERIAL BEFOREHAND

It is rare for any tests to be done beforehand, even though they may prove extremely useful and avoid investing huge sums of money in campaigns doomed to fail. For example, officials in Belgium wanted to organize a campaign to encourage drivers to turn their headlights on in due time. They sought to discover why drivers were slow about turning on their lights. After much discussion, it was concluded drivers were perhaps anxious not to put too much of a strain on their batteries (compared with electricity consumption in the home). A poster was then designed on this theme and tested in advance. It was then found that apart from the "experts" none of the motorists questioned thought about the question of energy consumption when the headlights were switched on. Another argument was therefore being used!

9. EVALUATION

Assessing the impact is part and parcel of the different phases of a campaign. But the costs in terms of time and money no doubt explain why this stage is not undertaken on a routine basis (Rooijens, 1986). The author believes that it is perhaps better to conduct two well-assessed campaigns than three in which the assessment is inadequate or non-existent. It is certainly true that an effective assessment can provide valuable lessons for future campaigns.

The ultimate aim of a campaign is to minimize the number and seriousness of road accidents, so the effectiveness of a campaign can always be gauged by the accident statistics. However, this sometimes creates a problem, particularly when the time period covered by the campaign is so short that even if the campaign has been effective, this hardly shows up or is not reflected at all in the statistics. Moreover, there should be no suspicion that anything else (infrastructural measures, legislative amendments) has changed if one is to ascribe a decline in the number of accidents to the safety campaign that has been conducted.

The campaigns themselves prove to be convoluted processes, involving several factors and mechanisms. These generally concern long-term activities, for behavior patterns and attitudes change (if at all) only slowly and in stages (Järvinen, 1997).

Let's reconsider the question of drunk driving. This was not much of a problem in the early 1960s, as it was a type of behavior that did not excite much social opprobrium, unless it resulted in an accident. Nowadays, such behavior is deeply frowned upon by society and all countries are reporting sharp declines in the number of individuals driving once they have drunk more than the legal limit.

This is probably a result of the many campaigns organized on this theme and which have succeeded in producing a radical change in public opinion, even though, individually, they have failed to achieve spectacular results.

According to Allsworth (1997), further progress on this front is highly unlikely, because we should have now reached the irreducible hard core of chronic drinkers. For example a major survey conducted in Belgium to investigate the use of alcohol, legal and illegal drugs by drivers involved in road accidents revealed that 71 % had BAC under 0.5 ‰, 2 % between 0.5 and 0.8 ‰, but 11 % had a level over 2 ‰ (BTTS, 1997) (1.7 % between 0.8 ‰, and 0.9 ‰, 6.5 % between 1 ‰, and 1.4 ‰, 7.4 % between 1.5 ‰ and 1.9 ‰).

A change in the behavior of the target population is a relevant criterion for gauging the effectiveness of a road safety campaign. In the case of behavior such as the wearing of seat belts, child safety in vehicles, the heeding traffic lights by pedestrians, and the observing of speed limits, the best criterion is pre and post-observation in the field.

However, in the case of other campaign themes, objective criteria are much harder to determine.

For example, if the campaign is focused on drunk driving, a comparison may be made of the sales figures for alcoholic beverages before and during the campaign, the number of cars parked near to drinking establishments, the number of motorists booked for driving under the influence... However some evaluation criteria raise problems. In the above-mentioned example, the number of motorists booked is not necessarily a reliable indicator, as there is a good chance that the police carry out more, or better targeted spot-checks during campaigns of this type.

The assessment may also be made (and is often made for that matter), by way of surveys conducted before and after among a sample of the target population.

These methods encounter three types of difficulty:

First of all, it is rare for a control group to be created. Second, the answers given may be affected by what is regarded as socially desirable. For example, an individual is unlikely, particularly during a face-to-face investigation, to come out in favor of driving under the influence of alcohol, even if he regularly indulges in this kind of behavior. Third, surveys tend to explore an attitude or type of behavior to which lip-service is paid but is not necessarily followed in practice.

It is not the goal of this chapter to discuss the attitude-behavior relationship, as this is considered elsewhere in the work (see Åberg).

Other criteria often used include the higher level of awareness or the penetration of the campaign (have the users seen the campaign? What did they remember – spontaneous or assisted memorization? What did they think of it?). While often serving a very useful purpose such indicators raise the question of exactly what relation they have to the behavior exhibited in practice and therefore, indirectly, to road safety? Studies have proved that there is not necessarily any link between better knowledge and the adoption of the corresponding type of behavior.

The types of campaigns that have proved to be effective are those that signaled an increase in the objective risk of being checked and penalized (but this obviously has to match the subjective risk of being checked). If the latter did not increase, the impact of the campaign was blunted and its effectiveness substantially lessened. Unfortunately, the effect is short-lived.

It is also a fact that after a certain period, even if a campaign has made an impact, that impact tends to wear off over time. For example, two seat belt campaigns were conducted in Belgium a few months apart. The count gave the following results (they apply only to drivers):

April	June	October	December
Before the test	After	Before	After
42%	51%	48%	49%

They illustrate the point well: the campaign produces results (there is an increase in the number of people wearing seat belts), then the effect wears off a bit (but the number is still higher) and with the new campaign, the effect increases anew (Prigogine, 1995). It is nonetheless clear that not all the campaigns make a significant impact.

Example of a sensitization campaign: preparation, communication channels, evaluation.

From January 1st, 1991, wearing seat belts in the rear of a car is obligatory in Belgium.

The Belgian Road Safety Institute commissioned an opinion poll to prepare the campaign on this new regulation. The poll's aim was to decide in which tone the campaign was to be held: informative or challenging.

The poll concerned the actual habit of wearing rear seat belts, the public opinion about the new regulation and the arguments for or against wearing rear seat belts. 41 % of the polled persons who had been rear seat passengers wore seat belts, either always or sometimes. 80% of the polled persons had a positive opinion on the new measure.

Hindrance and lack of comfort were most often invoked as arguments against the measure. It is noteworthy that the same arguments were used against the wearing of front seat belts. On the whole, the public is not well informed on the purpose of wearing rear seat belts: many people do not have an opinion at all when they are confronted with arguments for or against rear seat belts.

With these results in mind, it was decided to run a purely informative campaign.

The following communication channels were used for the campaign: A government statement by the Transport Minister; a radio commercial and a TV commercial (both focused on the arguments that were most often cited against the measure); posters on billboards along roads and motorways, representing three people wearing safety belts on the rear seat of a car; an information folder.

A new poll was held at the end of the campaign. It revealed that the new measure was known by 99 % of the polled persons. 60 % of them knew about it from the television news, 30 % from the newspapers, 23 % from the posters along roads and motorways, 21 % from the road safety programs on TV, 19 % from the radio news, 19 % from the radio commercial and 14 % from the TV commercial. The folder had a very low score: Only 4 % of the polled persons knew of the campaign via this channel.

10. CONCLUSION

It may not be easy to speak with any confidence about the effectiveness of road safety campaigns, but no party claims to know what would happen on the road if the campaigns were not conducted. Rooijens (1986) wrote that: "the cost of road accidents is so high that even the limited effectiveness of campaigns justifies the cost." We would add that it is not simply a matter of minimizing the financial cost of road accidents, but also of minimizing the physical, moral and psychological suffering of road accident victims and of their families and friends.

11. REFERENCES

ALLSWORTH (1997), "La campagne de lutte contre l'alcool au volant au Royaume-Uni : Rappel historique, résultats obtenus et enjeux d'un succès durable", *Séminaire sur la communication en sécurité routière*, Varsovie.

BTTS (1996), *Belgian Toxicology and Trauma Study*, Brussels.

DEUTSCHE VERKEHRSWACHT (1973), Action Junge Fahrer.

FLENSTED-JENSEN (1997), "L'information sur la sécurité routière : Partie intégrante d'un tout", *Séminaire sur la communication en sécurité routière*, Varsovie.

JÄRVINNEN (1997), "Contraintes et limites des stratégies de communication. Methodes d'évaluation de l'efficacité des campagnes", *Séminaire sur la communication en sécurité routière*, Varsovie.

MACKIE, VALENTINE (1975), *Effectiveness of different "Appeals" in road safety propaganda,* TRRL Laboratory Report.

OCDE (1971), *Campagnes de sécurité routière – calcul et évaluation*, IBSR, Brussels.

OCDE (1975), *Manuel sur les campagnes de sécurité routière*, IBSR, Brussels.

PRIGOGINE (1995), *Dossier "Ceinture de sécurité"*, IBSR, Brussels.

ROOIJERS (1986), *Inventarisatie en analyse van voorlichting-campagnes op verkeersveiligheidsgebied,* Verkeerskundige studiecentrum-traffic research centre-Rijksuniversiteit Groningen.

ROOIJERS (1988), *De invloed van verschillende voorlichtingstechnieken op het snelheidsgedrag van automobilisten,* Verkeerskundig studiecentrum-Rijksuniversiteit Groningen, Haren.

VANSNICK (1997), *Campagne "Bob"*, IBSR, Brussels.

VOGEL (1989), *Voorlichting Studiedag Universitaire,* Campus Diepenbeek.

Chapter 12

SOCIAL INTERVENTION IN TRAFFIC SAFETY

Enrique J. Carbonell
University of Valencia, Traffic Research Institute, Spain

1. INTRODUCTION

Traffic accidents during the transport of merchandise or passengers on roads and highways have, from the beginning, taken by surprise those who have experienced them themselves or involving persons close to them, as well as those who have witnessed accidents or, in one way or another, helped to alleviate their effects. They say that Albert Einstein commented after the death of a friend: "The worst thing is not only the fact that traffic accidents occur, but also how little we know about why they happen and how little we do to prevent them..."

These two statements, with some nuances, contain the two key elements which define the problem of social intervention in Traffic Safety. The nuance refers to progress, given that, from the words of the Nobel prize winner, until now efforts have evolved which have aided in understanding why accidents occur. Some of these accomplishments have been gathered in this chapter.

Nevertheless, it is in the area of prevention where we can consider that our knowledge may have certain value (what we do to avoid them), and this chapter concentrates completely on this preventative dimension of social intervention. It would be difficult to think of driving or traffic as a solitary phenomenon. The influence of the social environment in which traffic is produced is decisive in the formation of safe driving styles among users of vehicles and roads. Numerous authors like Barjonet, Näätänen, Summala,

Rothengatter, Wilde or Huguenin, to name just a few, have dedicated an important part of their research to the influence of social factors.

On the other hand, it seems obvious that there exists a great degree of growing awareness about traffic accidents in what we could call the "immediate environment of the accident," to which we referred in the first lines of this chapter. But, from early times to Albert Einstein and continuing on to the present day, the other traffic users have remained distant from the concept of the accident, considering it something unrelated to their lives, definitely tragic, but highly improbable. This attitude has always been more marked among the groups defined by the objective indicators as high risk, and even today can be demonstrated. in any study.

So we can define the two main dimensions of Social Intervention to which we dedicate this chapter: Prevention based on the social environment of the traffic and the accident; and lack of knowledge about risk factors and a feeling of distance from the accident situation on the part of the users.

Using this problem as a springboard, in recent years and especially in Europe, great effort has been made to develop efficient preventative activities from the social environment. A clear example of these efforts is shown in a recent report of the OCDE (1993), designed to evaluate this type of intervention performed in European countries. The most generic conclusion of this work of numerous experts has been the strong tie that exists between the adequate analysis of the social setting of the group to which the intervention is directed and the effectiveness of the intervention. In other words, and as we have repeated constantly, magic recipes do not exist; on the contrary, the experiences which have achieved satisfactory results have counted on the participation of various sciences, professionals and social sectors that each have contributed their perspectives, experiences and knowledge to a group effort, and above all, have participated in an exhaustive analysis of the specific reality of each social setting.

2. THE NEED FOR CAMPAIGNS (SOCIAL INTERVENTION)

When justifying the need for social intervention, we must pause briefly to frame our arguments in the proper context. Tradition in the area of traffic safety says that prevention is achieved based on the three Es: Ergonomics – in some versions engineering –, enforcements and education (ergonomics – engineering –, supervision and control, and education). Based on this proposal, everything that is not engineering or ergonomics-vehicles (acquisitions like the turbo, the ABS, the Airbag, increased power, better soundproofing, power steering, etc.), or whose driving takes place in changing environments (atmospheric conditions, highways-motorways vs. roads, etc.).

The second process is much more consistent than the previous one and has an accumulative effect, which makes it much more dangerous. We are referring to the constant, from a probabilistic point of view, *absence of adverse consequences to imprudent behavior and the permanent presence of satisfactory elements* in these very same behaviors. In terms of learning processes, this constant implies the consolidation of imprudent behaviors.

Throughout the life of a driver (a similar process occurs with pedestrians) and once the initial period of lack of expertise in managing the vehicle has passed, circumstances will progressively arise (for example, pressure to arrive some place on time, distraction) that will lead the driver to transgress some norm. If on these occasions no unsatisfactory consequence is produced, (accident, a scare-almost accident, a fine...), and if on, the contrary, the events lead to meeting a goal or other reinforcing consequences (arriving on time, testing one's own skill, the thrill of risk...), we unleash a process of cognitive dissonance tending to generate attitudes contrary to the norm and favorable to accepting risk. Expressions like, "I always run this red light because no one is ever coming," "I made it to M in only X hours," "the speed limits on the highways are for those who drive hand-cranked cars," or other similar expressions, denote the confirmation that a deviant behavior proves more adaptive when them is an absence of adverse consequences. In a very graphic way, this bas been called Unrealistic Optimism in driving.

Both processes join to underline the transgression of the norm or its lack of acceptance as a certain type of "folkloric crime" for which there is no lack of socially accepted extenuating circumstances or even clearly exonerating reasons. This leads us again to the beginning of the circle, with a population of traffic users who experience the accident as something foreign to them and with an extremely low level of consciousness about the seriousness of the health problem represented by traffic accidents.

Numerous studies show that regarding the causes of traffic accidents, others or chance-destiny, both elements being unrelated to oneself or uncontrollable, are considered responsible. This attitude generates a tendency to consider one's own behavior as basically irrelevant in controlling the occurrence of accidents, which, far from producing defenselessness, given that the perceived probability of accidents occurring is extremely low, allows a freeing of the inhibitions of the behavior emphasized by the cognitive consistence of maintaining attitudes contrary to safety, or at least to a specific norm.

The previous reflections on attitudes and behaviors lead to the need for interventions which counter-balance the effects of these processes from the framework in which they are produced, which is social.

Finally, there is an even more definitive argument for considering Social Intervention imperative, and that is the epidemiological situation. The numbers of accidents, incidents and conflicts which occur on the road require urgent measures to try to mitigate their effects. Thus, social interventions are one of the possible ways, and one of the most effective, of achieving this. In fact, the interventions that focus on areas like formal education (Road Education and Road Training) are too slow to be considered efficient in the short term, even though predictably they will be more effective in the long run.

In conclusion, the transmission of new information, the modification of attitudes and behaviors, the improvement of the average level of consciousness about the difficulties and risks of traffic and the importance of the statistics, constitute sufficiently strong arguments to justify the need for social intervention.

3. LIMITATIONS OF SOCIAL INTERVENTION IN TRAFFIC SAFETY

In recent years, we have frequently demonstrated publicly the envy we feel towards the marketing and communications specialists when they are presented with the "terrible" problem of selling more trousers, soft drinks or detergents of certain brand names. These specialists usually begin with a relatively acceptable position for their brand within the market, with a type of product the public already buys habitually, with a pretty well-defined sector of the population as the target of the communication and with certain ease in evaluating the effectiveness of their interventions. The most outstanding problem seems to center on the novelty of the strategy of communication or, at most, optimizing the other parameters. If any specialists are reading these lines, you will have to excuse me for the simplification, above all because of the complexity that follows.

In the area of Traffic Safety, social intervention may not seem viable if we focus for a moment on its limitations, or rather on the difficulties arising from its origin. In rater sections of this chapter we will demonstrate its possibilities. For the time being, we will analyze its problems.

3.1. The Safety Product

The first and primary consideration regarding a social intervention in Traffic Safety centers on the very product we want the traffic users to acquire: Safety. Safety is not necessarily a desired product. All too frequently, professionals in this area accept, perhaps influenced by a learned tradition, the idea that people hold safety as one of their priorities. This is not true for everyone as has been demonstrated in repeated studies on

Temerarious Attitudes, on Risk Seeking when Driving, etc. On the other hand, it would be acceptable to assume that the majority of users desire safety if it were not for the large variety of connotations this term has for distinct groups and its diverse levels of counter-position to (the opposite?) Risk.

Furthermore, other communicative environments, with products to a greater or lesser degree related to traffic tend to somewhat consistently emphasize the positive aspects of Risk like Freedom, Power, Excellence, Expertise, Satisfaction, Sensations and Emotions, and a long etcetera. And, on the contrary, the positive elements of Safety are not mentioned, but instead this concept is ignored, often as out-of-date or reactionary. The satisfactory note in recent times in our country is the increasing approach by the automobile industry to the use of Safety as a value to be kept in mind. This change is not far removed from the advances introduced in this industry and the normative pressure exerted from different petitioners, like the DGT in our country and in a wider spectrum, the CEMT resolutions in this sense (see for example: CEMT Resolution 56/1989 On Publicity Harmful to Traffic Safety). At other levels, however, the elements of risk mentioned before continue to be associated with a certain "Life Style," especially among young people.

All of this allows us to refine the sentence we highlighted before indicating that Safety is not necessarily a desired product in all groups, nor does it have the same significance among those who consider it a relevant value.

3.2. The Behavior we Want to Achieve

If the problem were stationed on the previous level, the solution could be reduced to an adequate selection of the target audience before an intervention. But the difficulties go even farther when we mention the word Safety, what type of behavior: (a) individual or group; (b) generic or situational; (c) only one or multiple.

(a) The distinction between safe individual behavior vs. collective is controversial at the present. A good definition of safe individual driving would be that which observes the limits imposed by the characteristics of the road and the atmospheric conditions, the vehicle and the capabilities of the driver. But it would only be valid in the absence of other participants in traffic, in solitary driving, like, for example, on a closed test circuit. Quite the contrary, the option of safe driving implies a normative arrived at by agreement which maximizes the benefits of driving, while minimizing its risks for the population as a whole. This carries with it some clear and problematic consequences related to accepting the formal normative, on many occasions contradictory to what the individual holds as his behavioral guide, that is, his subjective or informal norm. This is a difficult springboard for the intervention, to establish what the level of the informal norm is and its discrepancy with the format norm, as well as the reasons that lead to this disparity.

(b) Should we understand that passing where it is permitted by the normative frees us from risk? Obviously not. It will depend on whether the other vehicles also comply with this norm, whether our own capability permits it, and whether our vehicle meets the proper conditions to do so, for example. The safe behavior is very near, on a large number of occasions, to the lever of the situation, or to the operational level, as some authors have called it. Nevertheless, the option of generic communication seems the most satisfactory. The other alternative becomes confusing in a process of social intervention, but it is necessary, on this generic plane, to transmit how greatly dependent safety is on individual decisions. The reasons vary from the relevance of the situation to constantly attributing the responsibility for accidents to destiny or to others, as we referred to a few pages back.

(c) Only one or multiple? The answer is obvious: multiple. The "safe behavior" is related to speed, distance, time, turns... or more broadly, emotions, thoughts, perceptive organization, information processing, or, in prevention language: high-risk and low-risk behaviors. Every social intervention process must necessarily count on the multi-behavioral character of the product called Safety.

Therefore, the social intervention must be based on the reality that the behavior of Safety has a collective and therefore normative definition, must limit itself to a generic consideration while emphasizing the importance of individual decision-making, and is multiple. Without a doubt, this sounds somewhat different from "buy these trousers".

3.3. To whom do we Direct the Intervention?

The segmentation of the market into different target groups is a key task for the success of an intervention. This is well-known by the specialists who meticulously analyze the groups they want to convince of the excellence of their products. Frequently, one of the greatest worries is finding new segments of the market by approaching probable new audiences. In our case, the problem is the opposite of this latter example and closer to the former.

The entire population of a country, except in rare exceptions, participates in one way or another in traffic. At a lesser level, the majority of the population is susceptible to carrying out behaviors relevant to safety and/or risk in the road setting. And finally, an important part of the population can be influenced through social intervention.

From this broad block, we must extract the target groups for our interventions, by carrying out the appropriate segmentation. In this task, we will encounter at least the following difficulties:

(a) *Confusion of roles in traffic.* For some sectors of the population it will be easier than for others, but the reality is that given the characteristics of the flexible traffic system, each individual or group will assume different activities, functions and responsibilities in different moments. Let's think, for example, of a young person of the age of 20 who goes to his place of work or study on a motorcycle several days a week (driver of a two-wheeled vehicle), other days uses public transportation (bus passenger, for example), and walks a few hundred meters (pedestrian), some weekends his parents lend him the car (car driver), and others he rides with a friend (car passenger), and finally, on occasions, he drives the family car when the family travels at vacation time or on weekends (driver of the same car as before, but in different contexts), etc. If the previous example seemed complicated to the reader, stop to evaluate the complexity of the activities, functions and responsibilities that a professional driver (bus, lorry, or driving school instructor) can generate.

This reflection leads us directly to considering the social interventions that must be taken into account when segmenting the population, the roles they play in the system of traffic. All too frequently, this elemental consideration is forgotten in some groups (e.g., young people) while it is emphasized in others (e.g., professional drivers).

(b) *Behavior as segmentor.* The analysis prior to the intervention reappears as having great relevance based on the study of the behavior. High-risk practices seem, pretty well-defined, as in the case of alcohol consumption and driving. However, the results of very different research studies conclude that the type and evaluation of a high-risk practice, like the motivations for doing it and the practice of non-risk alternatives, vary throughout the population according to well-defined sectors.

Let's consider the example of the consumption of alcohol and driving. Young drivers that consume alcohol and then drive, do so in a context of pleasure, preferably at night and concentrated on the weekends. Their motivations are tied to their style of recreation and in general they consider exceeding the legal limit slightly as of little seriousness for their safety as exceeding it by a lot. By contrast, middle-aged drivers consume alcohol during or after meals, drive after consuming it in a context where the euphoric effect is decreasing and consider exceeding the limit slightly as not very serious but exceeding it by a lot as very serious. The latter group perceives the possible consequences of this behavior more clearly (lack of control over one's own behavior, fines, accidents, etc...), while the former group values the consequence of lack of control as positive and views the possibility of fines or accidents as distant or very improbable. For the first group, alternative transportation like buses or trains could be promoted (less cost and inter-urban use), while the second group would prefer the taxi (greater privacy and urban use).

We could go on indefinitely, but it seems sufficient to illustrate what we want to emphasize: the clear differentiation that the behavior imposes in determining target audiences. Keep in mind that an example was deliberately chosen in which the age characteristic has designated two groups from the start. But this is not necessarily clear for "young people" in general nor for "middle-aged adults" in general.

In conclusion, the analysis for determining target audiences should not only include the more traditional criteria of socio-demographic order, but also an approximation according to activities, functions and responsibilities (roles), and a differentiation based on the behaviors and their determinants, which would, without a doubt, improve the effectiveness of the social intervention. Again this sounds somewhat different from "the potential purchaser of these trousers is a young man of such and such age with this socio-economic level", and at best "interested in..."

3.4. How we Evaluate the Effectiveness of an Intervention

The last problem amongst those we dare to present in this chapter is perhaps one of the most controversial. Are interventions in Traffic Safety effective? How can we know whether they are or not?

The problem ceases to exist if we present it in its just terms. In fact, an intervention is effective if it achieves, in some measure, the actual objective for which it was designed. In reality the difficulties arise when we analyze the supposed objectives for which we plan an intervention, and we contrast them with the results, something which certainly is not always done.

The interventions in Traffic Safety may contain one or more of the following objectives for a certain sector of the population:

a. Improve the position, the image, of an institution or business;
b. Transmit information;
c. Change attitudes toward Safety or any of its concrete aspects;
d. Change behaviors;
e. Reduce accidents.

All of the previous objectives are completely fair and desirable in our area of work. We can analyze each of the objectives and consider the relevance of its evaluation.

The image of a business or institution is not only a fair objective, but it is also very often desirable that they be interested in improving their positions by transmitting safety. It would be of great benefit if Safety as a value and the behaviors of traffic safety were used as arguments by a large number of companies: The accumulative effect of these interventions would generate a positive net result for safety in traffic. Companies or industries like those of automobiles, insurance, service station chains, some associations of discotheques and pubs, etc., have already begun to develop campaigns of this type in Spain. On its side, the public institutions responsible, totally or in part, for traffic and its safety like Traffic, Health, Public Works, Transport, Education (whether in the central administration or in autonomies), and the Municipalities, must make sure to participate in interventions which not only improve their images, but also address high priority objectives.

The transmission of information is enough, as we stated in section 1 of this chapter, to justify a social intervention. Often, the aspects of an intervention which are evaluated are the improvement of image and the quality and clarity of the transmission of the information, and at best, the acceptability of the message transmitted.

The changing of attitudes is a somewhat more complex process that cannot be considered from a simple "I think this campaign has influenced my opinions and will change my behavior." The evaluation of attitudes has been shown to be one of the most complex tasks from a psychological point of view and requires a more refined analysis, and in fact, good working models exist in this respect (see Montoro, Carbonell, and Teiero, 1993). At the same time, achieving results in attitude changes implies entering into the analysis of the antecedents and consequences of the behavior, in conclusion, carry out a motivational analysis in order to optimize the intervention and its results. A broad scientific literature on persuasion strategies exists, and specialists in social intervention have definite experience with these strategies. From a generic point of view, being what fits in this short chapter, the most effective strategy is that which uses emotional arguments which tend to produce cognitive resonance in the receiver, either because of the content of the arguments, or because of the subject emitting them (a social prescriptor of reference, prescription by peers, etc.).

Behavior is a word close to attitudes only on some occasions. In fact, some publicity geniuses have achieved, or rather have accelerated, changes in behavior in certain social groups. But in general, publicity, in the broad sense of the word, proves ineffective in being able to create new behaviors not found previously in the repertoire of the individual, and relatively effective in changing the direction or objective of a behavior already available to a subject.

If the objective of the social intervention is the change of behavior, we see ourselves as obligated to carry out actions which complete at least one of these requisites: Act on the attitudes in a persistent manner, expecting an effect on the behavior in the medium or long run, or act directly on the behavior and its consequences. For example, changes in the system of fines and, above all, increases in vigilance and control have a direct effect on the behavior, although they sometimes produce undesired collateral effects on the attitudes which end up propitiating unfavorable alternative behaviors.

The evaluation of effectiveness based on the behavior is the level of analysis which is carried out less often. It is somewhat different from calculating the reductions in the numbers of accidents. In this sense, how many people drive now under the influence of alcohol (a very generic but valid question to use as an example) is very different from how many people have had accidents while under the influence of alcohol. If we confuse both planes of analysis, we can fall into causal interpretations of simple streaks of luck. The problem usually resides in that evaluating the behavior is often difficult and almost always implies a high economic cost. However, from our point of view, it is in reality less expansive than the evaluation of attitudes.

The relationship between behavior and accidents is somewhat more sophisticated than that existing between attitudes and behavior. Suppose we carry out an exceptional intervention on a stretch of road used for going out at night and in one year we eliminate 50 % of the accidents and 70 % of the deaths on this extension. We can conclude that our actions were excellent based on these results as long as we are not producing a deviation of the access to the night spots and an increase in the accidents on other roads. Nevertheless, and in the best of possibilities, it is possible that these results have no effect on the general indicators of accident rates in the country as a whole, because a day of intense fog on a fast road or the over-turning of a bus offsets the positive results. We would not use this data to conclude that the intervention was ineffective. In conclusion, the analysis must be limited to the area of the intervention and its repercussions.

On the other hand, consider a global action in a much wider area, for example, an autonomous community, with an apparent result of a reduction of 5 % in the accident rate. Again, to draw conclusions on the relationship between the intervention and the objective indicators, we would have to make sure that parallel reductions in the exposure to risk had not been produced, for example.

In conclusion, the reduction of the rates is dependent on the reduction of the probability of conflicts in the sense analyzed in a previous chapter in this book, and to the extent that the intervention has acted on the probability of conflicts, it will have been effective in reducing the accident rate.

4. APPROACHING THE PROBLEM: THE MARKETING OF TRAFFIC SAFETY

Social intervention in Traffic Safety is as old as the existence of motor vehicle traffic itself. From the earliest times, and using the means available to them, traffic specialists and public institutions have tried to transmit to their fellow citizens the importance of traffic and its dangers, explaining the pertinence of certain norms. We can find all types of examples during the long and fruitful tradition of creating posters in our country. During the beginning stages, marketing specialists and psychologists were also involved, as is demonstrated in various historical documents (see examples in Tortosa, Montoro and Carbonell, 1989).

The general strategy commonly developed during this long tradition, with notable changes in orientation to which we will later refer, was always what we could call "Social Publicity." In contrast, and more recently, interventions based on the principles of "Social Marketing" have begun to increase. Both lines of action have had their successes and failures, their supporters and opponents. From our point of view, as we expressed in the third section pf this chapter, their methods, limitations and objectives, make them complementary. For now, we will analyze their characteristics and some of their most outstanding results.

4.1. Operational Marketing Interventions

The first line of intervention, that which is referred to as Social Publicity, basically contains what we commonly call informative campaigns of a general character which, in fact, have enjoyed a long tradition both within and outside of our country, the orientations used to transmit the messages having evolved and fluctuated with the passage of time.

Its basic strategy is operational marketing directed towards a social end, or acquiring a social value complementarily. There are two kinds of final objectives for these types of interventions: Selling a product which advocates safety as a value, or as an alternative, selling a product called safety. The prototypical examples of the former are the Volvo campaigns, white examples of the latter are General Traffic Direction campaigns. Amongst the characteristics which unify both types of interventions, the following stand out:

(a) Apriority in the segmentation. Customarily, these kinds of intervention do not provide an excessive definition of the sectors of the population to which they are directed. Instead, it is the public itself which identifies its greater or lesser tie to the communication strategy selected and its contents. On some occasions, as has occurred with some examples of DGT interventions, a segmentation based on the criteria of accident rate is established, or, at most, high-risk behaviors as in the case of "use of the seat belt" and "use of the helmet."

(b) Use of the mass media. The star is the television, although with some frequency other means are used like the newspaper, and, to a lesser degree, the radio. This tactic coincides with the intention to reach the greatest volume of public possible, within the declined sectors.

(c) Closed messages. This is one of the most outstanding characteristics. We move in the area of very well-executed advertising projects, designed by prestigious professionals who elaborate materials which are perfectly structured and adapted to the medium in which they are presented, and which occupy a time slot (or a space in the case of the written press) deliberately selected to achieve maximum effectiveness in terms of the reception of the message. Nevertheless, and due to all of the above, they are closed projects.

(d) Short term effectiveness measured. The fourth and last of the most relevant characteristics is the evaluation of the effectiveness of these interventions in terms of level and quality of message reception and improvement of the product's image. The variations are round between the intention, for example, of an automobile company (determine if the type of intervention bas improved the image of the product – a model – and if so, has increased sales), and the objective of an institution like the DGT (determine whether the intervention has improved the image of the institution and the product – safety – and if so, if it has been able to reduce numbers of accidents). The experience in our country shows that the relationship between an intervention designed to reduce accidents and the actual reduction is not usually rigorously analyzed.

One aspect of the complete spectrum, which could very well be included as a fifth characteristic, is that each intervention is usually financed by only one company or institution.

We have had many examples of these types of interventions in our country in the last few years. Campaigns whose objective is to sell a product which advocates safety have fortunately proliferated recently. They include almost all of the activities related to driving like: Automobile manufacturing and sales, a sector in which together with the traditional proposals of Volvo and Mercedes, the majority of companies have carried out campaigns offering safety, although the recent efforts of Ford Spain and BMW Iberian have been especially outstanding for very different reasons; the gasoline supply networks related to petroleum companies like BP and to a degree Repsol; tire manufacturer like Michelin, more traditional in these types of proposals, or Firestone; insurance or mutual benefits companies for which

Mapfre and some Mutual Benefit Associations of Taxis provide an outstanding example in our country; etc.

This generalization of the safety value in driving as an element of the market has certainly helped to improve the general level of consciousness of a society distanced for too long from this problem. It is because of this, that this type of interventions should not be rejected in any case, but rather should be encouraged more openly. If we reflect for a moment on the strong counter-productive effect automobile advertising had in the 1980's and the beginning of the 1990's by stressing high-risk factors like speed, we can understand even better why the current emphasis on safety is more desirable.

In the same block, we have included the campaigns of the DGT because of their basic strategy, and we could also include examples developed in autonomous communities (e.g., "Mixing alcohol and gasoline kills," Cataluñia, 1989 – symbolic and forceful; "You've got the key," Valencian Community, 1994 – mild and innocuous). At another time, we carried out sufficient analyses of these examples (see Montoro and Carbonell, 1994); nevertheless, it may be appropriate here to make some comments on this campaign model.

The first reflection cannot be other than to affirm the high quality publicity of the traditional (before 1992) line of the "soft" campaigns of the DGT, their low impact on the memories and behaviors of drivers, as well as their presumably low effect on the improvement of accident rates, together with their high economic cost. This same problem had been discussed previously in countries like the United States, England, Canada, Japan, France, Italy and, especially, Australia which had opted for very forceful and realistic campaigns.

The second reflection concerns the emotional character of the DGT campaigns from 1992 to the present. Studies carried out in our Institute (see Tejero and Chóliz, 1995) show that these campaigns produce an emotional response at both the cognitive as well as physiological levels, superior even to advertisements having a marked erotic character (colognes and condoms). Therefore, the activation mechanism intended by this advertising orientation works, which together with the elevated memory index, points toward considering on a basic level that the chosen line is effective.

In the third place, together with the short term effect, based on the emotional response, a medium and long term effect was sought: Reflect on the seriousness of accidents and their consequences and, from that point, deduce the relevance of safe and prudent driving. This element of reflection arose from the social debate produced by the controversy over the campaigns. This was an authentic effect among those sought by the other strategy to which we will later refer: Social marketing. Arguments for reflection were not provided by the campaign itself, but rather by the debate about it. The controversy, without a doubt, made it possible for many agencies involved in traffic safety to appear in the media discussing accidents and their consequences, and, at the same time, provoked the same kind of debate in small areas of influence around each driver. The consequence was a notable improvement in the social climate against reckless driving behaviors.

As a fourth reflection, we must add the reasonable qualifier that the effectiveness of the campaigns begun in 1992 has not yet been proven. However, its communicative impact, its social impact and the new base situation it created for the debates on accidents have been proven. Furthermore, it is necessary to stress that the acceptance of the campaigns was not equal in all of the collectives, the group of young drivers having felt less impact from this line of intervention.

In conclusion, this proposal has had a large number of positive components as shown by the previous reflections, but it seems that we must complement these efforts. We will analyze this question in the next section.

4.2. Strategic Marketing Interventions

The second line of intervention, usually referred to as Social Marketing, includes what we commonly tend to think of as social dynamiting. Its basic premise is of a strategic nature, that is, analyze and comprehend the markets, the potential safety consumers, in order to bring our product nearer to the user. The final objectives of this type of intervention are: Socially activate the desire for safety in groups or well-defined sectors of the population, and transmit high-risk and safe practices in the traffic system.

It is more difficult to find prototypical examples of this type of intervention in Spain. Perhaps the most consistent approximations to this model have been produced at the municipal level, with the Alcoy Project developed by the DGT in collaboration with a fair number of institutions, amongst them, in particular, the city hall of the town of Alcoy (Manso and Vilaplana, 1994), and at the autonomic level, with the Campaign for the Prevention of Accidents in Young People, developed in 1993 by the Valencian Generalitat, the DGT and Aumar (Montoro, Carbonell and Tejero, 1993).

The most relevant characteristics of this type of intervention can be synthesized into the following epigraphs:

(a) Profound analysis, before the intervention, of the high-risk practices and behaviors. In general, this type of intervention requires an obligatory exhaustive analysis of one or various high-risk practices in a well-defined collective. This previous evaluation implies a profound knowledge of the group's habits (not only those related to driving) and the motivations that sustain these habits. The point of support for the change in high-risk behaviors is the intervention on these motivational factors (and cognitive on many occasions) which maintain the high-risk practices, and motivation of the safe practices.

(b) Use of alternative means of communication. The television tends to occupy a place of exile among the means chosen for the communication, and, at most, is useful for sustaining a closed communication which informs as to the existence of the campaign or establishes a certain "umbrella" for the challenge of communication. The media plans center on other more agile means like the radio. They develop alternative communication proposals with greater emphasis (posters, stickers, pins, etc.) which include some quite original examples (e.g., disc jockeys in the discotheques, direct peer communication in places of leisure or study – in the case of young people –, diskettes for personal computers sent by mail, etc.).

(c) Open messages and communication from authorities or peers. This is one of the characteristics that best defines this type of intervention. Along with some elements which articulate the communication, these interventions provide a wide margin of flexibility in the messages. This seems logical if we keep in mind that the objective is that persons of very different professional and cultural backgrounds participate in sending messages. For example, it is of great interest when a social authority from the sector of the population chosen uses his own communicative structure to transmit the basic communication core of the campaign, either in a structured interview or in an informal setting. The level of acceptance of the message, if the authority is properly chosen, is greatly elevated.

(d) Medium and long term effectiveness. The evaluation of the effectiveness of these interventions in terms of level and quality of the reception of the message is, without a doubt, more problematic, given the diversity and complexity of the messages. The influence on cognitive and motivational aspects also complicates the evaluation of the effectiveness of these interventions, which usually require short term and long term analysis levels. Evaluations of this type of campaign leave much to be desired in our country due to the lack of evaluation of behavior changes, the effect on accidents, and the systematic forgetfulness regarding the necessary continuity in the intervention and its evaluation.

The financial aspect completes the list of characteristics, in this case with two distinct notes: Low budgets and co-financing. This model of intervention usually springs from a cooperative effort between diverse institutions, associations and businesses, very near to the objectives drawn up in the National Pact.

We can find interventions of this type at the municipal level, in Torrejon de Ardoz or Talavera de la Reina, and in a much more consistent way in Alcoy, in the Valencian Community. The Alcoy Project was organized as a collaboration between the DGT and the Town Hall of Alcoy, whose Neighbors' Associations proposed an initiative in 1989 to develop programs which would act on persons and infrastructures with the objective of improving the road setting and reducing the accident rate. Participating in this project were distinct departments of the Town Hall, representatives of the educational sector in the town, the local media and, especially, the citizen movement represented by diverse associations. The educational aspect took precedence over the social at many times in the process, but this project can still be considered a more than acceptable example of social intervention at the municipal level, and is, without a doubt, the most complete.

There are other examples at the level of the autonomous communities, including several of the interventions whose initiatives lie in the Catalan Institute of Traffic Safety in Cataluñia, or like the modest effort developed in Castilla-La Mancha in the year 1994, initiated by the Junta of the Communities of Castilla-La Mancha, or in the Community of Madrid in 1994 (don't lose your head). The most thoroughly planned effort is, without a doubt, that which was carried out in the Valencian Community during 1993, although it was a short-lived effort and should even be considered a mere pilot experience.

The intervention was directed in this case toward young drivers of two and fourwheeled vehicles, a group which, as the objective indicators show, can be considered high-risk. The campaign was sponsored by the General Traffic Direction Agency, the Valencian Institute on Youth and various General Directors of the Valencian Generalitat together with private companies like Aumar, a highway concessionaire, and the Vitalicio Group, an insurance company. This group, clear and most dynamic antecedent of the National Pact against Traffic Accidents (signed in April of 1994), was able to generate an authentic social dynamic of communication and action, which led to uniting forces as apparently disparate as those of the Federation of Driving Schools of the Valencian Community, the owners of specially constructed amusement centers, youth associations with all kinds of purposes and tendencies, Valencian artists, professional drivers, athletes, etc., from one common broadcaster of messages: "La Taula against traffic accidents." The central axes of the communication were three intertwined messages: "It's in hour hands," "Accidents, I say no," and "Say no too." The channels or media were very diversified, because of which, together with the necessary, but insufficient, televised communication, the effort focused on radio, newspapers and magazines, posters, pins, distributed at times and places involving direct contact with the target group. At the same time, a parallel project was carried out with great success where the co-operating entities developed measures which consisted of providing alternative forms of transportation, objective risk indicators to be used by young people before driving, creating instruments to facilitate safe habits (places to keep helmets in places where young people spend their leisure time, for example), etc. All of this helps to reduce the cost of safe practices.

The experience was, as pre- and post-test evaluations prove, highly fruitful, with a 73 % level of acceptance of the credibility of the communication among young people, which greatly exceeds the results the best campaigns in our country directed to this collective. At the same time, 89 % of the young people interviewed in the post-test considered the content and communication technique to be positive.

In general terms, it is fairly obvious that a social objective like the reduction of traffic accidents among young people, can and must be undertaken from proposals like those presented by the application of social marketing. In this sense, and following the indications of the OCDE in its recent report, we must emphasize that the social interventions in this area must share, together with the necessary segmentation of the target audience and the election of a closer, non-contaminated speaker, at least a series of requirements to be able to anticipate its effectiveness (OCDE, 1993):

(a) Center the intervention on creating a Social Conscience favoring Safety, at various levels: a) internally, planning measures to activate the specialists, persons responsible for making social and political decisions and multipliers (e. g., journalists), and b) externally, motivating the target group to modify its behavior and attitudes.

(b) Favor the reasons for safety that may be present to a greater or lesser degree in all of the collectives, minimizing their psychological costs (time, effort, economics, loss of freedom and independence, etc.).

(c) Increase the social pressure against high-risk behaviors, through interventions by persons who may be close to the subject in the same situation in which the problematic behavior could be produced, not only using the partners or friends of the subject, but also the personnel of the places of leisure, as promoted in the campaign in the Valencian Community.

(d) Maintain a permanent analysis of the target group and adapt the actions to the fluctuations of this group. It would be highly recommendable to obtain more specific information restricted to the groups of high-risk, especially to quantitatively determine what the perceived power of the factors is that these subjects regard as obstacles at the moment of avoiding high-risk practices.

(e) The permanent evaluation of the target group undergoes interventions whose effects are appropriately analyzed. In this way, and again following the recommendations of the OCDE, a pre-test and post-test evaluation becomes imperative in the more massive interventions esteemed necessary, in order to obtain the short, medium and long term feedback required for adjusting the actions to the desired objectives.

In any case, the initiation of this type of program, given its high economic cost, the responsibility it implies and its social importance, should be preceded by a rigorous study of its objectives with a well-developed psycho-social planification, and end with a rigorous evaluation and follow-up of its effectiveness and a study of its level of social penetration.

5. INTERVENTION PROPOSAL: A MODEL FOR INTEGRATING THE EFFECTS

Now that the need for social intervention, and its limitations, have been analyzed as well as the important lines developed until now, it does not seem appropriate to end this chapter without daring to take a stand.

The proposal, based on the considerations made up to this point tries to unite the successes of the diverse types of intervention, integrating and coordinating their advantages to increase effectiveness. Using these references, the proposal can be synthesized into five points:

1. Use of the preventative health models;
2. Maintenance of Social Publicity;
3. Development of Social Marketing;
4. Combination of the previous interventions with supervision and control measures;
5. Anticipation of behavioral and technological changes in traffic system users.

The justification, which should be obvious in some cases, is synthesized in the following lines.

5.1. Preventative Health Model

From the perspective of the Psychology of Health, it has been reiterated that a person needs to know the health risks, as well as the possible sources of danger, concretely the so-called high-risk behaviors or practices, so as to avoid accidents or illness. This argument is based on the well-known fact that the perceptions of susceptibility to health or risk problems predict preventative behavioral action, so that the more conscious a person is of the possible or actual risk factors – and driving is one of the tasks which involves the most danger – the more preventative types of actions he/she will carry out, or in other words, the fewer decisions he/she will make involoving an objective risk. From this position of realism, traffic safety can be intensified, increasing people's desire to avoid accidents.

Therefore, in the content of these safety campaigns, complete information on the causes of accidents – in which human error plays a starring role – must be included, as well as the most relevant preventative measures like periodic vehicle revision or the necessity of using seat belts, insisting in the case of the high-risk behaviors like drug or alcohol consumption, distractions, fatigue, sleepiness, etc.

All of this must be based on an adequate identification of the beliefs-attitudes individuals maintain regarding the behavior, the subjective norms and the social references sustaining them, and the mechanisms of control and self-control over the behavior which the individual has and recognizes.

5.2. Maintenance of Social Publicity

The line of intervention we have called Social Publicity has had definite benefits in our country which have been highlighted above. With some modifications and evolution, it seems logical to recommend its continued use. Both types of intervention, those of an informative character and those involving social dynamiting, can and should be complementary.

Without a doubt, the actions developed in Spain in the past few years have coincided with a slight reduction in traffic accidents with victims. Nevertheless, it would not be very precise to think that this reduction is due only to the action of these campaigns. During this time period, the new Traffic Regulation was also put into effect, the network of roads improved somewhat, the Strategic Traffic Safety Plan was put into effect and there was an improvement in police supervision. Some even point to the fact that due to the economic crisis, the number of kilometers-year per vehicle has descended in our country, and, therefore, the exposure to risk. However, the appreciable improvement in the climate of social consciousness related to the seriousness of traffic accidents and the consolidation of an increasing tendency to consider them avoidable, allows us to think that the chosen line should be maintained, with the opportune corrections.

5.3. Development of Social Marketing

The question could be asked if the use of a successful means of communication in a series of publicity campaigns is enough reason to assure, as a minimum, the adequate communication of the relevant messages. We find this statement goes too far. The alternative which best fills this gap is the use of Social Marketing. Nevertheless, its actual application in the field of Traffic Safety is not that obvious.

It has been shown that Social Marketing has not been used frequently in the field of Traffic Safety. The reasons offered have been partial ignorance, prejudices, fears, etc. That is, social marketing applied to Traffic Safety has not been used, not because of lack of faith in the benefits it offers, but rather due to fear of and even resistance to the change implied in undertaking a little known discipline. Solid experience in the use of marketing allows us to state that the extrapolation of its principles and techniques to the field of Traffic Safety could yield positive results.

We must consider the marketing of Traffic Safety from a strategic point of view. To this end, all the variables intervening in a Marketing Plan must be considered: Carry out analyses of the market and the segments which comprise it, mark objectives to be reached, design an action strategy, and control and evaluate the results.

In any case, the objective of the social mobilization against traffic accidents is an unwavering objective which can be sustained more effectively in the long run by social marketing than by mere social publicity.

5.4. Combination of the Previous Interventions with Control and Supervision Measures

Numerous investigations and social interventions (Le Frise, Holland, 1984; Salzburg, Austria, 1992; Valencian Community, Spain, 1993) have demonstrated that programs of prevention of socially undesirable behaviors are more effective when they penalize the undesirable behavior, at the same time that they modify the attitudes toward it in the direction of avoiding the behavior. In this sense, we must repeat the need for multi-factorial interventions which unite the efforts to teach and convince with demonstrations of the effectiveness of the normative pressure.

It is very clear that, for example, a large number of the young people who use helmets do so because of the possibility of a fine. This is a necessary path, but dangerous when dealing with something more essential than putting an object on your head to avoid a fine. Thus, consumption of alcohol is more tied to – and more defended as – a pastime, which must be counteracted effectively by actions which co-ordinate objective pressure with subjective pressure.

5.5. Anticipation of the Behavioural and Technological Changes in Traffic System Users

From the first pages of this chapter, we have insisted on the importance of the evaluation of the problem, the process and the results of the intervention. Furthermore, we have made a constant bid in these pages in favor of the Purposeful interventions and against the Reactive.

This last point in the proposal must serve to recall the limitations of the interventions in traffic safety when the changes in the behaviors of the subjects in traffic that may be caused by the intervention itself are not considered. In a more generic sense, the changes – deliberate (publicity, intervention), or not (technologic changes in roads and vehicles) – of values and behaviors must be anticipated with a two-fold end: Prevent them in the case of increase of risk, or use them, providing incentives if they favor safety.

We can find clear examples of both cases. We referred to the first case when we considered the temporary increase in risk acceptance caused by a certain improvement in safety mechanisms or even the sound-proofing of the vehicle. At the second lever, we find the growing movement among young people in favor of healthy habits and against the consumption of some addictive substances like tobacco and, in part, alcohol, confluent with the increase in ecological concern.

6. IN CONCLUSION

The social intervention in Traffic Safety must come out of the traditional media plan (radio, written press, television, etc.) to add to and provoke an authentic social debate, supplying arguments, instruments and moral support to the sector which defends our interests, Traffic Safety. We must favor the resurgence of a culture which diminishes social tolerance of infractions and reckless driving behavior. It is a cultural transformation which has already begun in other European countries.

At this time we need to favor a culture which begins to react to the idea of traffic related death as a "simple almost aesthetic individual act," as expressed by our good friend Pierre Barjonet, as the prosaic analyses show the growing level of risk that exists on roads and cities forming part of the traffic system.

Safety can be considered a cardinal value on which ultimately rests the legitimacy itself of the Human Society. The research on safety is growing and examining all aspects, in the same proportion as progress, something that would undoubtedly satisfy the saddened Albert Einstein with whom we began this chapter. Nevertheless, it still remains for this zeal for Safety to impregnate the society as a whole without producing a feeling of loss of Freedom. To achieve this, we must all cooperate in bringing the term Freedom closer to Safety, distancing it from Risk, as we keep in mind that only from a position of Safety is it possible to experience Freedom in a social framework.

Chapter 13

THE EFFECTS OF ROAD DESIGN ON DRIVING

Jan Theeuwes
Vrije Universiteit, Amsterdam, The Netherland

1. INTRODUCTION

The design of many roads and their typical appearance reflects the way these roads developed through history. Because the road design is not considered as a system property optimally adapted to the human capabilities, there is by no means a structurally safe traffic system. The crucial question is how potential errors occurring in traffic can be reduced by designing a road environment that is optimally adapted to the goals and needs of the road users. It is commonly estimated that over 90 % of the traffic accidents are related to human error. Because better education, information and enforcement only have marginal effects on accident reduction, it is absolutely crucial that the road environment is designed in such a way that human errors are reduced to a minimum (see e.g., Theeuwes and Godthelp, 1993, 1995a, 1995b). The crucial question is how design principles can reduce the probability and consequences of an error during driving.

In order for road design measures to have the desired effect on the behavior of the driver, the road environment should be perceived by the driver in the way it was purposely designed. Note however that perception is an *active construction* process: it is the result of an interaction between sensory information and expectations of the driver. Therefore, it is quite feasible that drivers may perceive the road environment different from what the road initially was designed for. For example an urban four lane road may be designed for a driving speed of 50 km/h; yet, if drivers *perceive* the road as "a type of highway" even though it is in the built-up area it is likely that they will behave and treat the road as a highway.

In general it has been estimated that over 90 % of the information that a driver has to process is visual (Hills, 1980). Since the visual system is limited and the driving environment is relatively complex, in many circumstances perception of the road environment will rely on top-down expectations. In other words, drivers will perceive those events that are in line with their expectations and will overlook events that are not in line with their expectations. Typically, accidents occur because drivers did not expect these events to happen and did not anticipate adequately. Studies investigating errors occurring in actual traffic show that expectations do play a crucial role in the occurrence of accidents. Accident data show that a large portion of drivers involved in automobile crashes do not act too late but do not act at all to avoid the collision (Sussman, Bishop, Madnick and Walter, 1985). In addition, Malaterre (1986) indicates that 59 % of all accidents are the result of inappropriate expectations or interpretations of the environment.

This chapter discusses some theoretical issues regarding road design. The first section discusses how the road environment will be perceived by the driver depending on the level at which the driving task is performed. The second section discusses the role of saliency and conspicuity of road elements in designing a road environment. The third section addresses the role of expectations and how this affects our perception and interpretation of the road environment. The final chapter gives some hands-on advice and rules on how to design an optimal road environment.

2. PERCEIVING THE ROAD ENVIRONMENT

Visual perception should be considered as part of the driving task. In general it is recognized that human behavior is intrinsically goal directed (Duncan, 1990). In order to understand how people perceive the road environment, it is important to consider both the current state and the state the person tries to accomplish (i.e., "goal state"). In order to accomplish a particular main goal (e.g., traveling from location A to B), the main goal is divided into smaller subgoals (e.g., Laird *et al.*, 1987). Each smaller goal can be accomplished by a particular set of actions. For example, to accomplish a main goal such as "being at a particular time at a particular place," at lower levels, drivers may decide whether or not to pass a car, drive a particular speed or whether or not to stop for a traffic light.

To understand the effect of road design on driver's behavior we developed a theoretical framework based on various existing psychological models (e.g., Michon, 1985; Rasmussen, 1985) that allows one to appreciate the possible effects particular road measures might have. For example, at the control level of the driving task (i.e., keeping the car on the road) actions are performed within a very brief time constraint and at a so-called skill-based level (i.e., Rasmussen, 1985) within which actions take place without conscious control. This implies that road measures that want to induce certain changes at this level of the driving task (e.g., forcing drivers to change lanes at in work zones) should do so by providing information at the appropriate level. In other words, just putting up a sign which indicates the driver should change lanes without actually providing lane markings on the road to guide the delineation manoeuvre would be a an example of providing information at the wrong level. Signs to warn the driver that a delineation is coming up are important; yet, to let the driver negotiate the lane change manoeuvre at the appropriate level of the task hierarchy it is absolutely crucial to also provide information at the control level (i.e., by providing lane markings).

The driving task can be divided into three dimensions: *task hierarchy, task performance,* and *information processing.* Figure 1. gives the structure of the road user task in three dimensions. The X- and Y-axis give the stages of information processing and the task hierarchy, respectively. The Z-axis gives the level of task performance which is related to the experience with the driving task.

2.1. Task-Hierarchy

With respect to the task hierarchy, the highest *strategic* level entails the general planning of a trip, including the determination of trip goals, route, and modal choice, plus an evaluation of the costs and risks involved. For example, someone may want to travel from location A to B (for reasons of work or leisure), following a particular route (for reasons of time or pleasure), with a particular time schedule (utility). The strategic level determines the constraints within which the task at the *manoeuvering* level will be performed.

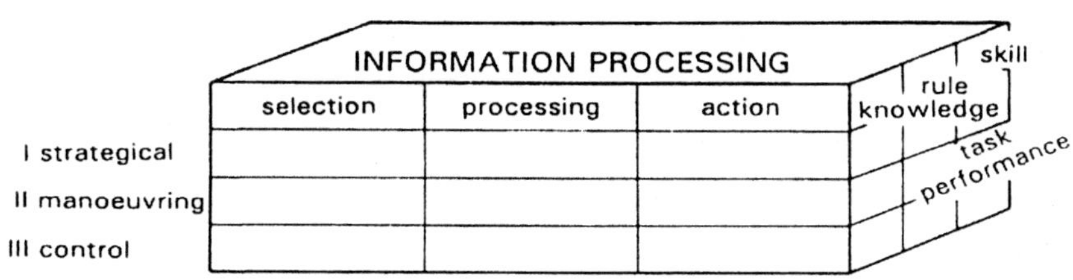

Figure 1. Structure of the driver's task in three dimensions.

Given the constraints of the strategic level, performance at the manoeuvering level is largely constrained by the actual environmental input. This level includes manoeuvres such as overtaking, stopping, parking, crossing, give way, etc. The *control* level is the lowest level in the task hierarchy and entails tasks dealing with vehicle handling such as the control of the vehicle on the road, steering, shifting gears, etc. Control actions are performed within a relatively short time constraint and automatically, that is, as soon as a particular stimulus is perceived a particular response is given.

2.2. Task Performance

The efficiency with which a driver can perform the driving task depends upon the task level and upon the proficiency of the driver with a particular task. Rasmussen (1985) recognizes three levels of task performance: tasks performance can be knowledge-based, rule-based or skill-based. *Skill-based behavior* is performed automatically and represents sensory-motor performance during activities which take place without conscious control as smooth and highly integrated patterns of behavior. Shifting gears and steering control are examples of this type of behavior. *Rule-based behavior* consists of a sequence of subroutines controlled by a stored rule or procedure which is developed over time during repeated practice with a particular situation. When a particular situation occurs, the rule or procedure is retrieved from memory and is executed similar to the execution of a cookbook recipe. The rule or procedure is retrieved from memory simply on the basis of previous successful experiences. *Knowledge-based* behavior occurs in unfamiliar situations when there is no rule for control from previous encounters. The behavior is performed at a higher conceptual level involving deductive reasoning and understanding of the situation. For example, finding the appropriate road to a particular destination. Knowledge-based behavior is required when faced with unusual situations or/and when the person does not have much experience with a situation.

2.3. Information processing

The different stages of the information processing sequence play a key role in the driving task. Information processing occurs at each hierarchical level and its arrangements will depend on the level of task performance. With increasing experience with a particular task, the level of task performance will move from knowledge-based to rule-based, thereby changing the type of information processing occurring. Due to the kind of task, however, most tasks will never become fully automatized up to the level of skills.

2.4. The Effect of the Road Design on Different Driving Task Levels

Strategic level: The design of the road environment plays a minor role at the strategic levels of the driving task. As noted, at the strategic level aspects such as the goal of a trip (e.g., work or leisure) and the route choice are important. For example, drivers may choose a certain route not because it is fast but because it is a nice road with a lot to see along the road. In this sense the design of the road environment does affect the driving task at a strategic level. Note since the levels are hierarchically organized, choosing a road for certain reasons at the strategic level will determine the constraints at the manoeuvering level. For example, when choosing a road for reasons of leisure, people will look more at the road surroundings instead of the road, will drive less fast, will pass cars less frequently, etc.

Manoeuvering level: At the manoeuvering level the road environment has a large influence on driving behavior. At this level, the execution of a driving task is controlled by previous experiences with the road environment in which such a manoeuvre was successful. Given the occurrence of a particular environmental condition, a manoeuvre will be performed at a rule-based level by executing a more or less fixed set of actions similar to the execution of a cookbook recipe. This rule-based manoeuvering behavior is based on memory representations which develop through experience with the driving task in a particular road environment setting. For example, when overtaking a car on a (Dutch) freeway, several implicit assumptions stored in memory are activated, such as, the assumption that there are no oncoming cars, that there is no slow traffic, that there are no driveway exits, no traffic lights etc. On the other hand, there are presumptions that other cars might be overtaking your car, or that other cars might drive much faster. When overtaking the car in front all these assumption control the execution of the manoeuvre (e.g., checking the outside mirror, increasing speed, etc.).

Control level: At the control level, road design also plays a crucial role. Since the total performance is smoothly integrated and rolls along without conscious attention or control, it is absolutely crucial that the road design fits well with what drivers need to perform this fully automated task. For example, course control is based on a combination of correlates of lateral speed, lateral position and heading rate (Riemersma, 1987). This type of information is rather "primitive" and enters the visual system at an early perceptual level. If for some reason information regarding any of these parameters is not adequate (e.g., the line markings go off the road or are not clearly visible), it is almost sure that an accident will occur. Since the behavior runs off in a more or less automatic fashion the wrong input into the system will immediately result in the wrong output, in particularly because there is no active feedback. Again because the driving task is hierarchically organized, problems at performing the task at a lower level will also influence the performance at a higher level. For example, leaner-drivers which have problems with course control, might decide not to overtake another car or might decide to look less frequent in the inside mirror. On the other hand, there is also an influence from top to bottom levels. For example, driving under high time constraints (a planning aspect) will result in frequent overtaking, and will require a stricter course control.

3. THE ROLE OF CONSPICUITY OF ROAD ELEMENTS

In a road environment there is an enormous influx of visual information and the appropriate sampling and integration of information is critical for the driving task. In optimizing the information acquisition process, the extent to which driving-relevant objects and events are capable of attracting the attention of the driver is considered to be crucial since failure to notify these objects and events may result inappropriate actions and behavior on part of the driver. The efficiency to which an object is capable of attracting attention is commonly referred to as conspicuity or saliency (Hughes and Cole, 1984). In this view, what is noticed in the environment is solely determined by the physical properties of the objects and its background.

Typically when designing a road, the driving relevant objects are made as conspicuous as possible. For example, relevant signs such as stop signs and yield signs are large and red while less relevant signs such as street names are small and blue. The underlying notion is that drivers will immediately notice the conspicuous signs. When examining the definitions of conspicuity in the literature two properties are mentioned in nearly all definitions. First, conspicuity is determined by the object-background characteristics; Second, a conspicuous object exerts control over the visual system in such a way that attention automatically is attracted by the object. The first property is an environmental aspect whereas the latter is the behavioral consequence of the presence of such a salient object. It is important to note that conspicuity should always be considered in relation to the background. A red sign is conspicuous in a road environment in which there are not many red signs. Yet, a red sign among very many other (red) signs (as for example in a busy business district with many billboards and flashing lights) will become inconspicuous. Also, adding more signs to warn a driver is usually not a good solution because the addition of more signs will render basically all signs somewhat less conspicuous.

3.1. Definitions of Conspicuity

Formally conspicuity has been defined operationally as that sensory attribute of a visible object in its surroundings by which it is able to control sensory selection via the visual system (Engel, 1977). Cole and Jenkins (1980) define an conspicuous object as one that would – for a certain background – be seen with certainty within a short observation time, regardless of the location of the object in relation to the line of fixation. The concept of the conspicuity lobe is related to these formal definitions and is defined as the visual field in which the relevant object can be discovered in its background during a brief presentation of the stimulus pattern (Engel, 1977). This measure is based on the presumption that an inconspicuous object should be close to the fixation point to be detected whereas a conspicuous object will be noticed even when it is presented in the far periphery. Important in Engel's definition is the control the object exerts over the visual selection system, i.e., the object exerts involuntary control over the next saccade; it demands to be looked at (Jenkins, 1979).

From a road design perspective considerable research effort has been directed to ensure that drivers notice information relevant for the driving task. It has been argued that the main function of driving relevant objects is to attract attention independent of the state of the driver. For example, with respect to traffic signs, Forbes (1939) defined the effectiveness of a traffic sign as the likelihood of whether a person's eye fixation would be attracted to a particular sign. Odescalchi (1960) argued that a sign should be made so conspicuous as to attract attention before the driver comes within reading distance of it. Conners (1975) stated that conspicuity was concerned with attracting attention to hazards when not actively searching for them. Gerathewohl (1954) considered conspicuity as “attention getting” effectiveness. Also, in order to ensure that other road users are noticed, measurements are taken to increase their conspicuity, i.e., flashing lights on slow moving vehicles, road crossings and on emergency vehicles.

In order to understand what is meant by conspicuity and the conspicuity area, you can perform the following test: First, fixate an object; then move your eyes a little bit away from the object, and try to fixate a location near the object (e.g., move your eyes in small steps to the right of the object). After a few steps, the object has disappeared in its background, that is, you moved your eyes so far into the periphery that you cannot discriminate the object from its background anymore. When you have to move very far into the periphery before the object disappears, then the object is very conspicuous. If after a few steps the object already disappears then it is an inconspicuous object. The eccentricity (the visual angle) at which an object disappears gives a measure for the conspicuity of the object. The obtained eccentricity is the radius of the conspicuity area or lobe. When, in situations of free search, an eye fixation falls within the conspicuity area of an object, it is assumed that the object attracts attention followed by an eye shift towards the object. Objects which are very conspicuous (e.g., a red traffic sign against a green background) do have very large conspicuity areas. For example, when you are driving along a road and fixate around the vanishing point, you are still capable of detecting a red sign against a green background because your point of fixation falls within the conspicuity area of the sign. It is clear from this example that the conspicuity of an object is directly related to the speed at which an object can be detected.

3.2. The Effect of Conspicuous Road Elements

From a road design perspective it is important to address the role of conspicuity of road elements. As indicate above, if a sign is conspicuous and you are searching for a sign then you will be able to find the sign very fast. Yet, as note above in many definitions of conspicuity it is assumed that people will look at conspicuous objects and events even when they have no intention to look for them. When a particular road design is inadequate and several accident have occurred at that site (i.e., black spot) often road designers will try to improve the situation by increasing the conspicuity of the sign. For example, they add flashing lights to the signs or choose larger signs to warn the driver. The reasoning is that drivers cannot miss these signs because they are so salient.

In a set of laboratory studies the question was addressed whether people always look at the most conspicuous object (Theeuwes, 1990, 1991, 1992). In these studies subjects had to search for a particular target (a small line segment placed in the middle of a small circle). There were sixteen circles on the computer display and the line segment subjects were looking for was placed randomly in any of the sixteen circles. One of the circles was very conspicuous because it was red while the others were green. The question addressed was whether people always would start searching at the most conspicuous object (i.e., the red circle). The time it took to find the target was measured. The results showed that subjects started searching randomly at any of the sixteen circles, basically ignoring the conspicuous red circle. Yet, when the line segment subjects were searching for was always positioned in the only red circle, subjects were very fast in detecting it. In other words, people do not automatically look at the most conspicuous object. When it is not relevant for the task they simply ignore it. Yet , when it is conspicuous and relevant for the task, then conspicuity helps a lot because people can find the conspicuous object very fast.

Hughes and Cole (1984) introduced the distinction between attention and search conspicuity. In their field study the conspicuity of target disks erected along an experimental route was determine under two different instructions to the subjects. One group of subjects had to report verbally "all objects or things that attracted their attention." These subjects did not know about the presence of the experimental disks. The other group of subjects received the instruction to locate and report the target disks along with all other traffic signs. The frequency of reporting the target disks was a measure of their conspicuity. For instructions given to the first group this measure was called attention conspicuity since the target disks had to attract attention in order to be noticed. For instructions given to the second group the measure was called search conspicuity, since drivers were actively searching for the target disks. As expected the hit rate for attention conspicuity was much lower than the hit rate for search conspicuity. In line with the earlier mentioned laboratory studies (Theeuwes, 1990, 1991, 1994) this field study shows that people do not necessarily look at (or in Hughes and Cole's study "report") conspicuous objects when they are considered irrelevant for the task at hand. Cole and Hughes (1988) further confirmed these finding showing that eye movement patterns of driver's watching a movie displaying a road scene depended very much on the instructions given to the drivers.

In conclusion, these findings indicate that drivers do not necessarily look at the most conspicuous road elements. For many years, it has been erroneously assumed that drivers will always notice road elements as long as it is ensured that they are conspicuous. The findings above suggest that our eyes do not necessarily go to these conspicuous road elements. In order for conspicuous objects to be noticed it is important that the road environment is designed in such a way that these conspicuous elements are expected and considered to be relevant for the driving task.

4. ROAD DESIGN AND THE ROLE OF EXPECTATIONS

As note above drivers do not necessarily look at objects in the road environment that are conspicuous. Only those events and objects that are considered to be relevant for the driving task at hand and are in line with the on-line expectations will be perceived and acted upon. Since perceiving the road environment is the result of this interaction between sensory information, task demands and expectations of the road user, it is crucial to design a road environment that takes into account the expectations and task demands of the driver. Driving is often a visually demanding task indicating that it is impossible to exhaustively scan the road environment for possible clues of what the appropriate behavior is. Drivers will therefore rely on their experience with the driving task and perceiving the road environment will rely greatly on top-down expectations.

Based on the general notion that people try to structure their world, it is assumed that road users will categorize the traffic environment. It is not the individual objects, nor the individual environments that will be stored in memory but an abstract representation of the world which contains a basic set of typical properties. These prototypical representations develop through experience. In order to ensure unity in the way people structure their world, it is required that there is a large consistency in the physical appearance of an object or environment and a large consistency with respect to the behavior displayed in relation to that object or environment. When these conditions are fulfilled, it can be expected that the prototypical representation of certain road environments will be more or less the same for everyone (Theeuwes, 1995b, 1998a, b; Theeuwes and Diks, 1995a, b). For example, Theeuwes and Diks (1995a) showed that there is a large consistency among a representative group of Dutch drivers regarding the prototype of a freeway. All people more or less agree on what a motorway in the Netherlands looks like and how to behave on such a type of road. However, there was hardly any consistency among drivers with respect to other types of roads outside the build up area (see also Theeuwes, in print; Kaptein and Theeuwes, 1996; Theeuwes and Godthelp, 1993, 1995a).

The prototypical representation of the road environments which is the basis for the categorization process contains "information" regarding the typical spatial relationships between the road elements and road users, so called *schemata* (Bartlett, 1932; McClelland and Rumelhart, 1981); and 'information' regarding the typical sequences of events in time, so called *scripts* or *frames* (Minsky, 1975). Classification of a road environment activates particular scripts and schemata which, in their turn, induce where – in place and in time – particular road users and elements can be expected. If the environment induces inappropriate expectations, errors are likely to occur.

The nature of contextual effects on the processing of road environments is thought to be the result of an interaction between incoming perceptual information and the higher level memory representations (i.e., schemata and scripts). For example, it has been demonstrated that objects that are obligatory in the schema are encoded more or less automatically (with a minimum use of the processing resources), whereas objects which do not fit in, require more resource-expensive encoding processing involving active hypothesis testing (Friedman, 1979). Loftus *et al.* (1983) argue that scenes are processed in two stages. Holistic information is extracted first, followed by search for specific features. The holistic information can be assessed within a single fixation of the scene (Potter, 1975). This information is thought to activate the *scene schema* which is held in a presumed pictorial memory system (Paivio, 1971). A search is then initiated for specific objects as held in temporal storage.

Riemersma (1988) investigated subjective road categorization as represented in the "heads" of the Dutch road users. For the built-up area, the results indicate that the official objective criteria of road categorization as used in the design of road environment is only marginally present in the subjective categorization. In addition, the study shows that the estimated safe speed depends only on the effort it would require to keep the car on the road. The probability of the occurrence of an encounter with another road user (e.g., pedestrian) did not have an effect on the estimated safe speed. Outside the built-up area, Riemersma (1988) demonstrated that the emergency lane which, in the Netherlands discriminates freeways from other types of fast speed roads, is not used in the subjective categorization. In addition, roads which were not freeways but which allowed a high speed were often erroneously classified as freeways (15-20 %).

Mazet and Dubois (1988) claim that the categorization of road environments occurs only on the basis of the behavior displayed in these environments. This implies that different categories of roads that generally require the same type of behavior will subjectively be represented by the same prototype. A residential area where everybody drives 80 km/h (although the speed limit is 50 km/h) and a city highway where the same speed is utilized will be categorized in the same way, although they look quite different. Inadequate categorization is dangerous because the inadequate categorization will induce inadequate expectations.

4.1. The Role of Expectations in Detecting Road Elements

Theeuwes and Hagenzieker (1993; Theeuwes 1995a) demonstrated the effect of contextual information on visual search in road environments. More specifically, the study explored the effect of the object-context relation "position" as defined by Biederman *et al.* (1982). This property refers to the fact that objects which are likely to appear in a given scene often occupy specific positions in that scene. This effect on "position" information is particularly important because this relation might be violated in every-day life traffic situations. Theeuwes showed that errors evolved when road users had wrong expectations regarding the location appearance of particular target objects. Figure 2 gives an example of the stimulus material used in this study.

With respect to this example, subjects were instructed to search for a traffic sign and respond "yes" when they found it and "no" in case they thought that no traffic sign was present. In the upper picture, the traffic sign is positioned at an expected location which gave a search time of 1.1 sec. and 6% errors. In the lower picture, the traffic sign is positioned at the left side of the road which is an unexpected location given the overall lay-out of the scene. In this condition, search time was 1.7 sec. and in 33% of the cases subjects thought that no traffic sign was present. Since both pictures are physically identical, that is, the conspicuity of the traffic signs is exactly the same for both pictures, the difference in performance can only be attributed to the top-down driven search strategy induced by the lay-out of the scene.

Figure 2. Example of stimulus material (In the top picture the traffic sign is positioned at an expected location; in the bottom picture at an unexpected location).

As is evident in Figure 3, visual search behavior is dependent upon the expectations induced by the environment. The lay-out of the environment activates a particular schema which induces expectations where to expect the target object. As is clear from this study, search behavior is biased towards those portions of the visual field were the target is expected. It should be realized that the effects of contextual driven search might be much stronger in real driving especially in conditions in which there is a relatively high visual load, i.e., driving in busy traffic in urban environments, or under reduced sight conditions, for instance when driving in the dark or in twilight. Especially in these situations, rapid resource-inexpensive and conceptually-driven feature detection is advantageous. The study shows that objects at unexpected locations are not seen too late but, in most cases, not seen at all, i.e., when searching for objects at unexpected locations subjects tend to think that the target is not present.

Given these considerations, it is clear that extremely dangerous situations may occur when the design of the traffic environment induces certain expectations regarding the spatial arrangement of objects in that scene, which are not correct. The importance of inducing the correct expectations is supported by studies discussed earlier showing that visual selection does not so much depend upon the conspicuity of the target objects but more on the demands of the search task (Theeuwes, 1989, 1990, 1993). It is quite dangerous when a traffic environment induces incorrect expectations because conspicuous signs and/or other infra structural measures can hardly correct these expectations because these signs may not be perceived.

Although there is no empirical evidence regarding the time frame of these expectations, it is likely that once expectations are set they cannot be changed easily. When initially the lay-out of a road indicates that it is a freeway, the driver will keep on interpreting the road as a freeway. Gradual changes to another type of road will lead to confusion and incorrect expectations. Consequently, the same type of road should connect a section which psychologically is interpreted as one unit: For example, a road connecting to cities, a road from a shopping to a residential area. Because people interpret a connecting road as a single psychological "unit," it should be designed as such, that is, one type of road.

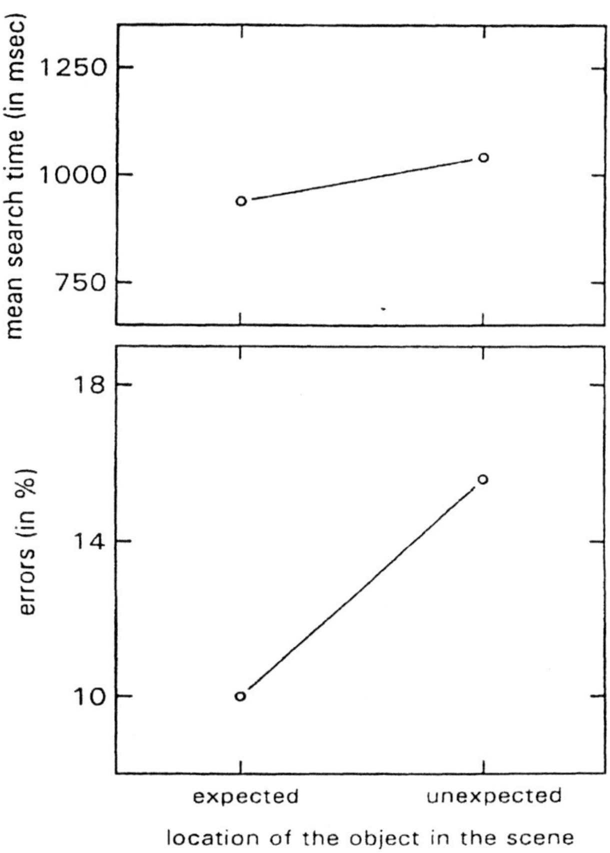

Figure 3. Mean reaction time and error percentage for searching for a target object placed at an expected or unexpected location within a particular road scene.

4.3. Subjective Road Categorization

The idea that people categorize the road environment is based on the general notion that people try to structure their world. From studies investigating representations in memory of objects, it is known that people classify objects as belonging to a particular category (Rosch, 1978; Rosch, Mervis Gray, Johnston and Boyes-Bream, 1976). Through experience, internal representations develop, which contain the typical characteristics of a category. The category to which an object belongs tells something about the characteristics of the object as well as the behaviors associated with it.

Categorization is the grouping of our experiences into categories (groups, classes, rubrics). This is done by treating different units (stimuli, objects) the same: To call them the same, to put them in the same pile, or to react in the same way to them. To categorize a stimulus means to consider it not only equivalent to other stimuli in the same category but also different from stimuli outside that category. Our categorical knowledge is organized in a hierarchical structure, containing a basic level which is the first level learned by children, the most informative and the most widely used in our language. One object is a better exemplar than the other, a phenomenon called "goodness of example" or "graded structure." One can think of a hierarchical structure containing the concepts *animal-bird-robin*, with *bird* at the basic level. If we know that something is a robin, we know also that it is a bird and that it is an animal. Also, we know that it can fly and has wings (information stored at level bird) and a red breast (stored at level robin). Furthermore, a robin is a good example of a bird, while a chicken is not such a good example.

There are different categorization models, each making different assumptions regarding the representation of categories in memory and about the way we produce categorizations (see Barsalou, 1992 for an overview). Besides models which are based on rules (classical models) and similarity (exemplar and prototype models), there are also mixed models in which categorization is based on rules as well as similarity. It seems that humans use a lot of ways to categorize: Next to rules, exemplars and prototypes we can also categorize on the basis of goals and theories.

There are some differences between categories of environments and those of objects. Environments have less limitations in the way the attributes are arranged. For example, the attributes of a school (tables, books, chairs) can be arranged in various ways while the legs of a table can only be placed in certain locations (Tversky and Hemenway, 1983). Also, unlike with objects, we are part of the environments. On the other hand, it has been claimed that environments are perceived and categorized analogous to objects (Ward, 1977; Ward and Russell, 1981a).

In many respects categorization of environments shows the same phenomena as that of simple objects. The categorization of environments also shows a hierarchical structure with a basic-level (Tversky and Hemenway, 1983; Russell and Ward, 1982). In addition, people not only distinguish environments on the basis of physical characteristics, but also on the basis of goals to go there or the behaviors that take place in it (Genereux, Ward and Russell, 1983). Finally, there are indications that a "graded structure" is present within environmental categories; i.e., one environment is a better example than an other for a particular category.

Rosch (1978) described two general principles for the formation of categories, cognitive economy and perceived world structure. Cognitive economy refers to the function of category systems and asserts that the task of category systems is to provide maximum information with the least cognitive effort. Perceived world structure refers to the structure of the information and asserts that the perceived world comes as structured information rather than as arbitrary or unpredictable attributes. Thus maximum information with least cognitive effort is achieved if categories map the perceived world structure as closely as possible. Applying the first principle of cognitive economy to the road environment this entails that road users try to reduce the large number of roads that exist in the "real" world to a few behaviorally and cognitively relevant road categories. It is to the road user's advantage to only differentiate among road categories when the difference is relevant. To categorize a road as belonging to a certain category implies that it is similar to all other roads within that category and different from roads outside that category. For the road environment, the perceived world structure suggests that road users see the environment consisting of a set of attributes that are highly correlated. The world does not consist of a set of attributes that are randomly picked. Thus, through experience with the road environment, road users develop a perceived world that contains attributes that are likely to occur in combination (Mazet and Dubois, 1988). Thus, if one sees a four lane freeway, one expects some road markings, an emergency lane and fast traffic that moves in the same direction.

A category represents a set of objects or environments which are considered to be equivalent (Rosch, 1978). In order to determine what road users consider as a road category, it is necessary to determine the perceived similarity among the pictures of the different road environments. Similarity data are typically analyzed by multidimensional scaling (MDS), which is a set of procedures in which the judged or otherwise assessed similarities between members of a set of stimuli are used to produce a geometrical representation of the stimuli (Ward and Russell, 1981b). More similar stimuli are represented as closer to each other in geometric space; while dissimilar stimuli are represented farther apart. Road environments which are considered to be close in geometric space can be considered as being part of the same road category while road environments which are far apart are thought to belong to different road categories.

Theeuwes and Diks (1995a) used this MDS technique to determine the similarity among road environments outside the built-up area. Participants sorted pictures of the various existing road environments in piles with the instruction to put pictures of environments which they thought were similar in the same pile and pictures they thought were different in different piles (see e.g., Gärling, 1976). Participants sorted pictures of existing roads outside the built-up area with respect to the behavior they would show and expect on these roads. These roads belonged to one of the four "official" main road categories outside the built-up area in the Netherlands. If the official categorization is clear and self-evident, it was expected that participants would have been able to sort the pictures in coherence with the four official categories. If, however, there is not enough clarity and homogeneity in the official categories and not enough diversity between the official categories, one would expect that the subjective categories are different from the official ones.

Figure 4 shows the results of this study. The four official categories are given at the bottom of the figure. The results are based on a two dimensional solution of the MDS analysis based on the similarity matrix of the roads. The boundaries around the data points are based on a subsequent cluster analysis. As is clear from this Figure, the freeways ("A" roads indicated by open circles) all seem to cluster together quite well suggesting that people thought that these type of road were quite alike. In fact, it suggests that there is a coherence between the official road categorization and the way people categorize them at least for freeways. However, as is clear from the figure other official clusters of roads (B, C and D roads) do not cluster together. For example, road types B and C were classified in four different clusters. Road type D is clustered basically in two groups: The photos D1 to D4 (all clustering together) are small rural roads without a center lane marking, while roads D5 to D8 are road with a center lane marking. Note that officially these road belong to the same category.

This study demonstrates that the subjective categorization (the way people categorize the roads in their "heads") does not match the official road categorization. This study supports the idea that road behavior is related to the appearance of the road and that road characteristics and traffic behavior are cognitively integrated by drivers into subjective categories. In addition, the results indicate that the four "official" categories used as stimuli are only partially reproduced in the subjective categorization. The fact that the "official" categories which do not correspond well to the subjective categorization are those road categories with the highest rate of accidents suggests the importance of an adequate road design. If too many accidents occur on a road, putting up of a traffic sign with a lower speed limit will not have any effect as long as the road itself is not changed. If, on a particular road, everyone is driving too fast, this is probably caused by the road design, i.e., the road looks like a road where you can drive fast. If this speed is unwanted, because cyclists have to cross or because of the presence of slow traffic, putting up warning signs will probably not show the desired effect, because those signs will have a low credibility as long as the road appearance remains the same. Only by changing the road design so that the design corresponds to the subjective categorization will help to elicit the correct expectations and behaviors.

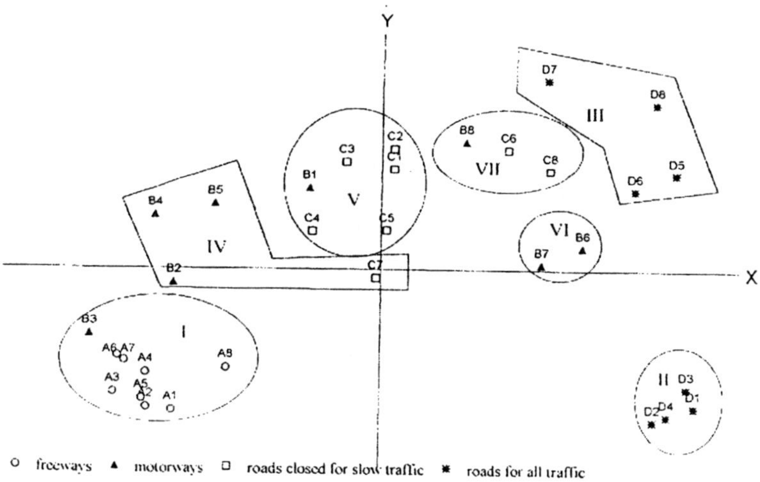

Figure 4. Two-dimensional solution of MDS analysis regarding subjective road categorization. The results of the cluster analysis are represented by group boundaries.

5. DESIGN RULES FOR AN OPTIMAL ROAD DESIGN

Studies investigating errors occurring in actual traffic show that expectations do play a crucial role in the occurrence of accidents. As note before there is evidence that a large proportion of the accidents are the result of inappropriate expectations or interpretations of the environment (e.g., Malaterre, 1986). Note that accidents occur seldom indicating that errors during driving are not fed back to the driver. On the other hand, correct expectancies, i.e., finding an object where you expected it, are consistently reinforced because the traffic environment is reasonably predictable.

Because expectations play such an important role it is crucial that the design of roads is adjusted to these expectations. Purely by its design roads should elicit safe behavior (Godthelp, 1990). By taking into account the constraints and the limitations of the driver (all drivers including the elderly), road design can reduce the number of errors occurring in traffic (Theeuwes and Godthelp, 1992;

Theeuwes, Godthelp and Riemersma, 1992).

The type of roads which elicits by its design safe behavior are recently classified as Self-Explaining Road (SER) (Theeuwes and Godthelp, 1992). In the Netherlands, the design of freeways and woonerfs are to some extent self-explaining and inherently safe. On the other hand, a very large extent of the Dutch roads – for example the 80 km/h rural roads – are not designed according to the safety principles mentioned above. These type of roads are not easily classified because they do not have any prototypical recognizable properties, nor do they compel the traffic behavior required on these roads. For example:

- The probability of the occurrence of slow traffic cannot be inferred from the road design.
- The probability of the occurrence of oncoming traffic is often unclear.
- The location and the presence of crossings and exits of driveways is not well marked.
- The estimation of the location and the required speed of curves is often difficult and inaccurate.

Figure 5 and 6 give some examples of the issues raised above.

Figure 5. The probability of the occurrence of slow traffic cannot be inferred from the road design: *Should you expect bicyclists on this road?* (the answer is "yes").

The use of consistent and easily understandable codes can to some extent reduce these problems. The design of roads should reflect the probability of encountering particular road users. Along the same line is the concept of "Positive-Guidance" as developed by Alexander and Lunefeld (1986). They also suggest that the traffic situation should be in line with the expectations of the road users.

Figure 6. The probability of the occurrence of oncoming traffic is often unclear: *Should you expect oncoming traffic?* (the answer is "yes").

In addition to the development of a road which is self-explaining, is the development of a modern traffic control system which can add some "intelligence" to the road environment. For example, navigation systems in the car with variable directional signs along the road can guide traffic, can reduce uncertainty in finding the optimal routes, and can remove instabilities (e.g., traffic jams) in the traffic flow. Variable speed advice dependent on the local circumstances (intensity, rains, fog) can optimize the traffic flow.

On theoretical grounds we identified some criteria which will increase the self-explaining character of roads. When developing the "road of the future" one should start with a few easy recognizable and distinguishable road categories. These types of roads should be designed in such a way that high speed differences and directional differences are not possible. Four categories can be distinguished: Freeways, highways connecting larger regions, rural roads connecting residential and shopping areas, and woonerfs, i.e., roads going from door to door. For these four categories, self-explaining roads should fulfill the following tentative criteria:

- Unique road elements (homogeneous within one category and different from all other categories).

- Unique behavior for a specific category (homogeneous within one category and different from all other categories).
- Unique behavior should be linked to unique road elements (e.g., woonerfs: obstacles-slow driving; freeway: smooth concrete-fast driving)
- The lay-out of crossings, road sections and curves should be linked uniquely with the particular road category (e.g., a crossing on a highway should physically and behaviorally be completely different from crossing on a rural road).
- One should choose road categories with are behaviorally relevant.
- The same road category should connect a section which psychologically is interpreted as a single unit (e.g., a road connecting two cities).
- There should be no fast transitions going from one road category to the next.
- When there is a transition in road category, the change should be marked clearly (e.g., rumble strips)
- When teaching the different road categories, one should not only teach the name but also the behavior required for that type of road.
- Category-defining properties should be also visible at night.
- The road design should expel speed differences and differences in direction of movement.
- Road elements, marking and signing should fulfil the standard visibility criteria.
- Traffic control systems should be uniquely linked to specific categories (e.g., on freeways, systems that regulate traffic flow and on rural road, systems that restrict driving speed).

6. REFERENCES

ALEXANDER, G.J. & LUNENFELD, H. (1986). Driver expectancy in highway design and traffic operations. Report No. FHWA-TO-86-1, U.S. Department of Transportation Federal Highway Administration Office of Traffic Operations, Washington, D.C. 20590

BARTLETT, F.C. (1932). Remembering. Cambridge: Cambridge University Press.

BARSALOU, L.W. (1992). Cognitive Psychology, an Overview for Cognitive Scientists. New Jersey: Lawrence Erlbaum.

BIEDERMAN, I., MEZZANOTTE, R.J. & RABINOWITZ, J.C. (1982). Scene perception: Detecting and judging objects undergoing relational violations. *Cognitive Psychology 14, 143-177.*

COLE, B.L. & JENKINS, S.E. (1980). The nature and measurement of conspicuity. *Australian Road Research Board, 10, 99-107*

COLE, B.L. & HUGHES, P.K. (1988). Drivers don't search they just notice. D. Brogan (Ed.): *Visual Search*. Amsterdam: North Holland.

CONNERS, M.M. (1975). Conspicuity of target lights: The influence of color. NASA Technical Note TN-D7960

DUNCAN, J. (1990). Goal weighting and the choice of behaviour in a complex world. *Ergonomics 33, 1265-1279.*

ENGEL, F.L. (1977). Visual conspicuity, visual search and fixation tendencies of the eye. *Vision Research 17, 95-108.*

FORBES, T.W. (1939) The normal automobile driver as a traffic problem. *Journal of General Psychology, 20, 471-474.*

FRIEDMAN, A. (1979). Framing pictures: The role of knowledge in automatized encoding and memory for gist. *Journal of Experimental Psychology: General, 108, 316-355.*

GÄRLING, T. (1976). The structural analysis of environmental perception and cognition, a multidimensional scaling approach. *Environment and Behavior, 8, 385-415.*

GENEREUX, R.L, WARD, L.M. & RUSSELL, J.A. (1983). The behavioral component in the meaning of places. *Journal of Environmental Psychology, Vol. 3, 43-55.*

GODTHELP, J. (1990). Naar een beheerst wegverkeer. *Verkeerskunde 41, (3), 112-116.*

GERATHEWOHL, S.J. (1954). Conspicuity of flashing light signals of different frequency and duration. *Journal of Experimental Psychology, 48, 247-254.*

HILLS, B.L. (1980). Vision, visibility, and perception in driving. *Perception, 9, 183-216.*

HUGHES, P.K. and COLE, B.L. (1984). Search and attention conspicuity of road traffic control devices. *Australian Road Research, 14, 1-9.*

JENKINS, S.E. (1979). An investigation into the nature and physical determinants of visual conspicuity. Melbourne: University Press (PhD thesis).

KAPTEIN, N.A. & THEEUWES, J. (1996). Effecten van vormgeving op categorie-indeling en verwachting ten aanzien van 80 km/h wegen buiten de bebouwde kom. Rapport TNO-TM 1996 C-10. TNO Human Factors Research Institute, Soesterberg.

LAIRD, J.E., NEWELL, A & ROSENBLOOM, P.S. (1987). Soar: an architecture for general intelligence. *Artificial Intelligence 33, 1-64.*

LOFTUS, G.R., NELSON, W.W. & KALLMAN J.J. (1983). Differential acquisition rates for different types of information from pictures. *Quarterly Journal of Experimental Psychology, 35A, 187-198.*

MALATERRE, G. (1986). Errors analysis and in-depth accident studies. *Ergonomics 33, 1403-1421.*

MAZET C., and DUBOIS, D. (1988). Mental organizations of road situations: Theory of cognitive categorization and methodological consequences. Paper presented at the SWOV conference: Traffic safety theory and research methods, Amsterdam.

MCCLELLAND, J. & RUMELHART, D. (1981). An interactive activation model of context effects in letter perception; Part I. An account of basic findings. *Psychological Review 88, 375-407.*

MICHON, J. (1985) A critical view of driver behaviour models: what do we know, what should we do. In L. EVANS & R.C. SCHWING (Eds.): Human Behavior and traffic safety (pp.485-520). New York: Plenum.

MINSKY, M. (1975). A framework for representing knowledge. In P.H. WINSTON (ed.): The psychology of computer vision. New York: McGraw-Hill.

PAIVIO, A. (1971). Imagery and verbal processes. New York: Holt, Rinehart & Winson.

POTTER, M.C. (1975). Meaning in visual search. *Science, 187, 965-966.*

RASMUSSEN, J. (1985). Trends in human ereliability analysis. *Ergonomics 28(8), 1185-1195*

RIEMERSMA, J.B.J. (1987). Visual cues in straight road driving. Ph.D. thesis. TNO Institute for Perception, The Netherlands.

RIEMERSMA, J.B.J. (1988). An empirical study of subjective road categorization. *Ergonomics 31, 621-630*

ROSCH, E. (1978). Principles of categorization. In: E. ROSCH & B.B. LLOYD (Eds.). Cognition and Categorization. Hillsdale, NJ: Erlbaum.

ROSCH, E., MERVIS, C.B., GRAY, W.D., JOHNSON, D.M., & BOYES BREAM, P. (1976). Basic objects in natural categories. *Cognitive Psychology,. 8., 382-439.*

RUSSELL J.A. & WARD, L.M. (1982). Environmental psychology. *Annual Review of Psychology, 33, 651-688.*

SUSSMAN, E.D., BISHOP, H., MADNICK, B. & WALTER, R., (1985). Driver inattention and highway safety. *Transportation Research Record 1047, 40-48.*

THEEUWES, J. (1989). Visual selection: exogenous and endogenous control: a review of the literature. Report IZF 1989 C-3. TNO Institute for Perception, Soesterberg.

THEEUWES, J. (1990). Conspicuity is task dependent: evidence from selective search. *Acta Psychologica 74, 81-99.*

THEEUWES, J. (1991). Visual selection: exogenous and endogenous control. In A. Gale et al.: *Vision in Vehicle III* (pg. 53-62). Amsterdam: North Holland.

THEEUWES, J. (1993). Visual attention and driving behavior. In SANTOS, J.A. (ed.): *Human Factors in Road Traffic*. Lisboa: Esher.

THEEUWES, J. (1994). Stimulus-driven capture and attentional set: Selective search for color and visual abrupt onsets. *Journal of Experimental Psychology: Human Perception & Performance, 20, 799-806.*

THEEUWES, J. (1995a). Visuelles Orientierungsverhalten an Kreuzungen. *Zeitschrift fuer Verkehrsicherheit, 41, 151-156.*

THEEUWES, J. (1995b). Towards a structurally safe traffic system: The development of Self-Explaining Roads. *Analise Psicologica, 3, 287-290.*

THEEUWES, J. (1996). Visual search at intersections: An eye-movement analysis. In A.GALE et al.: *Vision in Vehicle V. Amsterdam: North Holland.*

THEEUWES, J. (1998a). Self-Explaining Roads: subjective categorization of road environments. In A. GALE: Vision in Vehicle VI. (pp. 279-288). Amsterdam: North Holland

THEEUWES, J. (1998b) Introduction. In P. ALBUQUERQUE, J.A. SANTOS, C. RODRIGUES & A.H. PIRES DA COSTA (eds.). Human Factors in Road Traffic II: Traffic Psychology and Engineering. Braga: Universidade Do Minho.

THEEUWES, J. (in press). Commentary on RÄSÄNEN and SUMMALA, "Car Drivers' Adjustment to cyclists at Roundabout". Transportation Human Factors.

THEEUWES, J. & DIKS, G. (1995a). Subjective road categorization and speed choice. Rapport TNO-TM 1995 B-16. TNO Human Factors Research Institute, Soesterberg.

THEEUWES, J. & DIKS, G. (1995b). Categorisering van omgevingen: een overzicht van de literatuur. Rapport TNO-TM 1995 B-2. TNO Human Factors Research Institute, Soesterberg.

THEEUWES, J & GODTHELP, H. (1993). Self Explaining Roads. In DE KROES & STOOP (eds.) Safety of Transportation (pg. 56-66). Delft: University Press

THEEUWES, J. & GODTHELP, H. (1995a). Self-Explaining Roads. *Safety Science, 19, 217-225.*

THEEUWES, J. & GODTHELP, J. (1995b). Self-Explaining Roads: How people categorize roads outside the built-up area. In Proceedings of "*International Conference Road Safety in Europe and Strategic Highway Research Program*" 26-28 September, Lille, France

THEEUWES, J. & M.P. HAGENZIEKER (1993). Visual search of traffic scenes: On the effect of location expectations. In A. GALE et al.: *Vision in Vehicle IV (pg. 149-158). Amsterdam: North Holland.*

TVERSKY, B & HEMENWAY, K. (1983). Categories of environmental scenes. *Cognitive Psychology, 15, 121-149.*

WARD, L.W. (1977). Multidimensional scaling of the molar physical environment. *The Journal of Multivariate Behavioral Research, Vol. 12, 23-42.*

WARD, L.M. & RUSSELL, J.A. (1981a). The psychological representation of molar physical environments. *Journal of Experimental Psychology: General*, Vol. 110, 121-152.

WARD, L.M. & RUSSELL, J.A. (1981b). Cognitive set and the perception of place. *Environment and Behavior*, Vol. 13, 610-632.

Chapter 14

THE EFFECT OF ROAD TRANSPORT TELEMATICS

Lena Nilsson, Lisbeth Harms and Björn Peters
Swedish National Road and Transport Research Institute (VTI), Sweden

1. INTRODUCTION

The introduction of road transport telematics (ITS, the most commonly used abbreviation for the topic, standing for "Intelligent Transport Systems") may turn out to be the greatest and most comprehensive transformation of road transportation since the invention of the combustion engine. General objectives of ITS are to increase transport efficiency, to improve traffic safety and to reduce traffic related pollution (Franzén, 1993; Perrett and Stevens, 1996). The way ITS will influence future traffic, and how it will affect driving behavior is only to a certain extent predictable from current knowledge, therefore much research has been devoted to this topic during last decade. Obviously, the addition of advanced technology to the traditional driver-vehicle-environment interaction will increase the complexity of the road traffic system (Rumar, 1990a). Presumably also the task of driving will gradually change as the implementation of ITS functions proceeds. One possible development is that electronic communication between infrastructure components and vehicle components will replace several current driver functions, for example, detection of relevant traffic and roadside information and speed choice recommendations. Also certain parts of the vehicle-to-vehicle interaction, for example, distance keeping, may change from being performed manually to being performed by more or less automatic ITS.

Considering the negative impact of human errors on traffic safety, for example, the traditional view of human related factors as frequent crash causes, it may seem attractive to replace as many human functions as possible with automatic ones. Thereby the driver would play a less critical role in future traffic. However, experience from related areas like aviation and process control suggests that human operators are required in the control-loop even with a high degree of automation (Wickens, 1992; Endsley and Kiris, 1995). The introduction of advanced technology has a potential to change both the task structure and task allocation between the human and technical agents. Thereby the role of the human operator has been transformed mainly from manual to supervisory control. Usually the demand on human cognition has thereby increased, while the demand on human action has decreased (Wickens, 1992). Provided that a similar development will take place in road traffic, the impact of driver behavior in future traffic may become more rather than less critical. The frequency of driver errors leading to an accident may decrease while accidents due to technical shortcomings in combination with driver errors may occur seldom but their consequences may be severe.

In the prospect of such a development of road traffic, the influence of technical support on driver behavior has become a core problem in the development of ITS. Extensive effort has been devoted to investigating how different ITS functions influence driving performance. Most of this ITS research has been carried out within the framework of large research programs such as DRIVE and PROMETHEUS in Europe, and IVHS in the United States.

Several ITS functions are intended to directly influence the way the driving task is performed. Examples are cruise controls, navigation and route guidance systems, crash-avoiding systems, vision enhancement systems and systems for automatic vehicle operation. This type of support is directed towards a specific part of the driving task, which does not mean that other parts are left uninfluenced. On the contrary, it is reasonable to assume that use of an ITS system will also influence the performance of sub-tasks other than the one supported. The effects on individual driving behavior of other ITS functions, such as trip planning, travel information, road-pricing and fleet-management, are expected to be more indirect and therefore the impact of those support systems are also less clear. Similarly unclear are the net-effects on the task and the behavior of drivers when a number of ITS functions are used together. These issues are difficult to approach with empirical research methods, and they are therefore often subjected to theoretical assessments (Brand, 1994; Carsten, 1995; Perrett and Stevens, 1996).

The present chapter is primarily concerned with user related effects of the most prevalent in-vehicle ITS functions, aiming at direct driver support mainly at the operational and tactical levels. In the discussion of foreseen positive and negative effects of ITS implementations, crucial issues are: The drivers' need for different types of support, the design of the human-machine interaction (HMI), the influence of a driver support system on the way the specific driving task is performed, and last, but not least, the general effects of ITS on driving behavior including behavioral adaptation (Grayson, 1996). Driver acceptance of different types of support has also to be considered. As the breakthrough of ITS is still to come, most of the reported effects result from experimental studies including few drivers using a single support system during a short period of time.

2. THE INFLUENCE OF ITS ON TASK STRUCTURE

The assumption that driver support systems will affect driving performance more generally draws attention to the relationship between the different sub-tasks of driving. The inherent structure of the driving task is hierarchical with strategic, tactical and operational components demanding different levels of driver control (Allen *et al*., 1971). Human control structures on the other hand are highly flexible and highly dependent on practice and experience. Human control behavior is well described in the skill-based, rule-based, and knowledge-based framework (Rasmussen, 1974, 1986). The main issue in this framework is the hierarchical nature of human control, replacing the serial model used in early models of human information processing. Current modeling of driving performance takes into account both the task structure and the human control structure by combing the two frameworks mentioned above (Hale et al., 1990; Ranney, 1995).

The combination of the two above mentioned frameworks (Figure 1.) reveals that human control is not inherently in correspondence with task requirements: Although a sub-task may be best performed at a certain control level this may not necessarily be the level a human driver uses for that task.

ABS = Anti Blocking System
CC = Cruise Control (Constant)
ACC = Adaptive Cruise Control
CAS = Crash-Avoidance System

Trip Planning = Information System for Trip Planning
Navigation = Navigation Support and Route Guidance

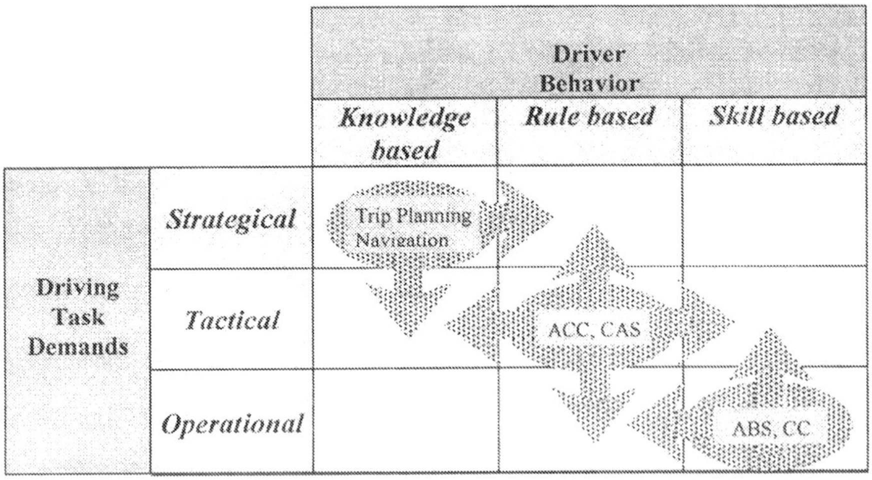

Figure 1. Matrix combining the three levels of task demand and the three levels of behavioral control.

For instance, unsupported navigation is a strategic task, which may be performed by the driver either as knowledge-based, rule-based or skill-based behavior. Likewise a novice driver may use rule-based behavior for vehicle handling while this develops into a skill as driver experience increases. Thus the level of driver control is dependent on factors like driver expectations, driving experience and prior experience of the actual road network. Figure 1. relates the flexibility of human control behavior to the demands of driving with some current ITS support systems available.

The use of some ITS devices may increase the resource demand from the driving task by adding new task components. Moreover, current feedback loops may be disregarded and the driver may be left to supervisory control, being a task humans are least suited for (Wickens, 1992). On the other hand the fact that human control structures may inherently differ from the task requirements suggests that the current design of road traffic is far from optimal. Most likely, ITS devices can help drivers to meet the actual requirements of the driving task more appropriately. Support systems could for instance improve driver anticipation and decisions by providing necessary information which is not available (e.g., visible) in the traffic environment. Moreover, support systems could prevent slips and lapses (Reason, 1990) due to feed-forward control, for example, preventing skill-based behavior where this is not appropriate. Another aspect of great interest is the possibility that ITS support at a certain task level could promote behavioral changes at other task levels. It is still possible that positive feedback from a system supporting speed-keeping (operational task) may result in a better overall speed perception and speed choice, and thereby improve driving performance also at the tactical level.

3. THE NEED OF DRIVER SUPPORT FOR INCREASED TRAFFIC SAFETY

The need for driving support is an important but also a delicate question. Driving is normally considered a self-paced task, and drivers are used to driving without assistance. Thus, individual drivers seldom experience an explicit and urgent need of support. Instead, the need of driver support systems may be judged from the fact that driver *errors* and inappropriate driver *behavior* are major problems in the field of traffic safety (Parkes, 1995). According to Reason (1990) erroneous behavior in traffic can be of different types. Apart from intended violations and mistakes, due to the application of strong but wrong rules, errors may also result from unintended actions (slips and lapses). As a matter of fact drivers seldom experience their own errors, probably because they act in accordance with their intentions. At the low level of attention associated with skill based behavior errors may pass unnoticed unless they result in a conflict or in an accident.

A detailed analysis of accident files has revealed that selected ITS (more precisely, Prometheus functions) could meet certain identified but not necessarily experienced driver-needs, and thereby contribute to reducing many types of traffic accidents (Malaterre and Fontaine, 1993). The needs identified in rural areas were judged to be relatively well covered by the conceptual ITS solutions (Prometheus functions), while the proportion of needs appearing in urban areas that can be satisfied by ITS were judged to be lower. In both environments the need for detection and estimation support was reported to be high and also possible to solve to a large extent by ITS. Driver support through vehicle control was judged to be a much smaller need. Yet, many of the support systems that are closest to market introduction aim at supporting vehicle control. It is worth mentioning that Malaterre and Fontaine (1993) considered that 20 to 25 % of the identified driver needs would not be possible to fulfill utilizing ITS.

Another point of departure for analyzing drivers' need of support is the inherent human limitations that may impede appropriate driving performance. Such human limitations can be used to guide the development of ITS functions. Data limitations refer to the limited ranges of the human senses, and resource limitations refer to limited cognitive capacity.

Data limitations may be overcome by ITS supporting the detection of vital information, objects and road alignment. For instance vision enhancement systems (VES) may facilitate the driving task under conditions of low visibility and reduced sight. Nilsson and Alm (1996) investigated the effects of vision enhancement on driving performance in low visibility. During driving in fog, with a visibility distance of 50 meters, a "clear" picture of the road scene in front of the driver was presented on a monitor mounted on the bonnet (simulated head-up). The access to the high-visibility picture made the subjects in the study drive faster compared to unsupported driving in fog. The observed increase in driving speed was accompanied by an increased variation of lateral position, suggesting that vehicle control was somewhat affected by visibility conditions. Presumably both the small size of the picture and its lack of vehicle reference contributed to the more unstable road tracking. It was also found that the drivers' ability to respond quickly to unexpected visual stimuli was improved, to the level in clear sight, when the VES was used. An interesting finding in this study was that drivers adopted different strategies for coping with the two parallel information sources (the road and the monitor picture). Apparently some drivers looked at the monitor most of the time, while others were less consequent in their division of attention. Also, some drivers used the VES for the detection of hindrances along the road, but looked at the road in situations requiring maneuvers like overtaking and meeting oncoming cars. Thus, the introduction of the vision enhancement system made driving in fog somewhat easier, but still the information acquisition task was more complicated compared to driving with normal visibility. Apparently the VES put the drivers in a choice situation, making divided attention a potential risk factor. Knowledge about possible consequences of VES use, like those described above, should be added to the fact that VES devices still suffer from severe technical shortcomings. Available techniques, used for instance in aviation, are less suitable to road traffic. According to Bossi *et al.* (1995) no VES systems can currently be recommended for use in road traffic.

Drivers' *limited cognitive resources* were a recognized problem in road traffic long before introducing ITS. Drivers that have passed the novice stage normally show a relatively automatic driving performance, carrying out the driving task at a low level of attention, and being capable of monitoring several sources of information in parallel. In spite of that, their cognitive load is sensitive to variations in the driving environment (Harms, 1991; Verwey, 1993). High cognitive load presumably cause late detection and in turn late responses to critical events (Rumar, 1990b). Therefore it is a concern that many in-vehicle systems aimed at driver support may instead increase the demand on the driver by making the task more complicated or adding a new component to the task (Summala, 1997). One obvious change related to ITS introduction is the increasing amount of information available to the driver. In a situation of an extended number of sources of information it is reasonable to expect increased safety risks due to divided attention, distraction, incomplete processing or deteriorated situation awareness.

The additional ITS information is usually visual, as the user-system interface of most in-vehicle ITS consists of a display or other types of visual feedback. The displays are normally dashboard mounted, and require drivers to look away from the road scene to detect the information (Lansdown, 1997). Thus, a serious drawback of vision based support systems is that they increase the time drivers spend with their eyes off the road. Display glances between 1 and 2 seconds are not unusual (Wikman *et al.*, 1997; Wierwille, 1993) and looking away from the traffic scene for more than 2 seconds is generally not recommended (Gale, 1997). To make better use of human resources, it is sometimes proposed that in-vehicle driver support systems should direct information to other sense modalities than vision, the main alternative being acoustic information. Of importance for the driver receiving the information is that the characteristics of visual and audible information differ concerning time and space aspects. Audible information is omnidirectional, but it is not persistent in time. Besides, some acoustic messages can be experienced as annoying, not leaving the drivers any possibility to disregard them.

Visual information is persistent in time, but most often requires direct viewing. Visual information has the advantage that it can be presented for a long time interval, thereby being less interruptive to ongoing driver activities than for instance verbal messages. It can be argued that this aspect makes visual information better suited to support a self-paced task, which driving is usually considered to be. On the other hand it can also be argued that visual information should not be presented in situations where it is especially important that the driver keeps his eyes on the road. Effects of presenting tactile information via the gas pedal and the steering wheel have been investigated (Reichart *et al.*, 1996), but generally the tactile channel has been considered to a lesser extent in the development driver support systems. However, contrary to both a warning light and a warning buzzer, a counterforce in the gas pedal was found to improve distance control. The improvement was obtained without generating undesired side-effects like increased driving speed, harder accelerations/decelerations, and increased left lane driving (Janssen and Nilsson, 1991). From the findings above it is obvious that addressing different human senses can result in unintended (unwanted) side effects to a varying extent. What sense to address with an ITS system in order to distribute human resources most beneficially, and generate an optimal driver-ITS interface, depends on several factors, for example the type of message to be conveyed, the traffic environment, driver preference and capability. However, regardless of mode of presentation, most information demand processing resources and may also increase driver workload considerably (Verwey, 1993). It has been reported that distance keeping is markedly impaired when drivers look at the road ahead and perform a mental calculation task (Summala, 1997; Brookhuis *et al.*, 1991). Also, decision making and memorizing activities communicated over a hands-free mobile telephone have been found to negatively influence choice reaction time, distance keeping, and mental workload in a car following situation (Alm and Nilsson, 1995). Despite being aware that the mental activities were demanding resources, the drivers did not increase their safety margins in driving. So, it is reasonably evident that an increase in cognitive load can impair also information-based judgments. Visual information is an obvious but not the only source of distraction and increased mental load during driving. This fact should be seriously considered when designing in-car driver support systems.

4. DRIVER ACCEPTANCE A PREREQUISITE FOR ITS INTRODUCTION

Driver acceptance of ITS functions has been elucidated at an early stage of ITS development by describing ITS functions to potential users and letting them rate their stated acceptance of the different solutions (Brand, 1990). The in-car support systems can be designed to prevent accidents by informing or warning the driver, or even by taking over the control and act automatically. They can also be used for avoiding unsafe driving behavior, such as a too high speed, or facilitating parts of the driving task, like route finding. It could be suspected that some of these functions may not be welcomed by drivers since they could reduce the feeling of freedom and the joy of driving by interfering with the drivers' driving habits and wish of being in control of the situation. An early study by Alm and Rumar (1991) also showed that drivers were positive to ITS functions described to them as service functions (e.g., general help function, route guidance, trip planning), while they were skeptical to ITS functions interfering with their actual control of the car (e.g., car following, speed and lane keeping). As an example, on a seven-point scale the difference between the mean ratings of a speed guiding function and a speed keeping function differed two steps in favor of the guiding alternative. However, more recent research indicates that drivers may accept intervening ITS (like ACC) and find them attractive and useful in particular after having had an opportunity to try them in real traffic (Fancher *et al.*, 1998).

It is important but not sufficient to get an idea of driver opinions about prospective ITS at an early stage. However, the acceptance concept includes not only subjective estimates of system descriptions, but also adoption of the system as manifested in driver behavior must be considered. Therefore, both subjectively expressed opinions and performance while using ITS have to be studied together. In-vehicle devices of limited relevance to the driving task (radios, tape recorders, CD-players and telephones) have become popular and their user acceptance manifested by frequent use, even though these devices can disturb the driving performance (Wikman *et al.*, 1997; Alm and Nilsson, 1994; Wierwille and Tijerina, 1996).

The first driver support systems in modern cars are systems supporting navigation and route guidance. A number of systems are already on the market and the drivers' preferences are important also from commercial point of view. Several functions and interfaces have been evaluated throughout the past decade. The relevance and reliability of the presented information have been found important to drivers' acceptance (Bonsall and Parry, after Kantowitz *et al.*, 1997). Also the relationship between system reliability and drivers self-confidence is important for route finding information to be used. Thus, Kantowitz *et al.* (1997) found that drivers being familiar with the surrounding road network used route guidance information less efficiently compared to unfamiliar drivers, even when the information was highly reliable. The interface design, the experienced relevance and usefulness of the information compared with the driver's experience of own capability influenced the outcome.

Considering the user-system interface, many navigation systems are based on electronic maps. However, findings from research suggest that the optimal configuration of navigation aids is short standardized voice instructions with a redundant screen displaying visual information, preferably in the form of symbols (see e.g., Parkes and Burnett, 1993; Verwey, 1993; Burnett and Joyner, 1996). Voice instructions reduce the demand on visual inspection, and the presence of a visual display helps drivers maintain their spatial orientation. However, individual drivers may still have their own preferences and have been reported to perform better with an interface they prefer (Wochinger and Boehm-Davis, 1997). In this investigation elderly drivers preferred electronic maps to alternative navigation systems. This preference may suggest that elderly drivers preferred the mode of support, which was most similar to a presumably life-long habit of reading paper maps for navigation, and/or to physiological issues, for example, reduced hearing. Drivers with specific needs concerning the design of ITS support (like elderly or disabled drivers) are currently an important issue in European Transport Telematics Research and have been addressed in a number of EC funded projects (Hakamies-Blomqvist and Peters, 2000). One of these projects, TELSCAN (TELematic Standards and Co-ordination of ATT systems in relatioN to elderly and disabled travellers – within the 4th Framework) has produced a design guidelines handbook with the objective to make ITS more useful to travelers with specific needs (Nicolle and Burnett, 1999).

Inferring from current driver attitudes, automatic limitation of the driving speed is not what most drivers would ask for. However, constant cruise control (CC) is a frequently used device, highly appreciated by many drivers. One reason may be that the drivers set the cruising speed themselves, and experience a gain in driving comfort. The CC design is best suited for use on rural roads and motorways under low traffic conditions. Likewise positive results have been reported from a full-scale, long-term study of driving with an automatic "speed limiter". This device is designed mainly for use in built-up areas. Speed control is performed manually by the driver up to the signed speed limit. The "speed limiter" only prevents the driver from driving faster than the signed speed. Twenty-five volunteers used the "speed limiter" for a three months period, during which their acceptance was found to increase (Almqvist and Nygård, 1997).

Another function, which seems relatively close to market introduction, is adaptive cruise control (ACC), which is an extension of constant cruise control (CC). ACC has the possibility of automatically adjusting the speed to the prevailing conditions, for instance signed speed limits and slower traffic ahead. In the latter case the speed is controlled so a "safe" following distance is kept. Surprisingly enough, drivers expressed positive opinions about ACC even when situations not possible to manage by the ACC were included in the evaluation (Nilsson, 1995). At the time of this experiment the speed control was designed so as to be sensitive to moving objects but not to stationary ones (including stationary vehicles). In some cases, when a stationary convoy was presented in front of the driver, the required driver intervention was so late that crashes actually occurred. Thus, the drivers' acceptance of ACC seems to be based on their experience of system functionality in frequently occurring "normal" situations (moving vehicles) rather than by experienced system limitations in critical situations appearing more seldom (stationary convoy). On the other hand it has been reported that driver opinions about ACC were negatively influenced when another system (mobile telephone) was used simultaneously (Nilsson and Nåbo, 1996).

Acceptance of ACC in terms of comfort and convenience qualities has been demonstrated also from more "long-termed" use in real traffic (Fancher, *et al.*, 1998). Volunteer drivers (108 people) used vehicles equipped with ACC functionality as their own cars for two or five weeks. The volunteers were found to appreciate the ACC and were strongly disposed to let the ACC "drive", as they tended to utilize it over a wide range of driving conditions. Furthermore, they adopted a way of driving that prolonged the duration of each ACC engagement. A high preference for ACC control across many different driving environments as well as a reluctance to manually intervene upon ACC control after a speed reduction supported the conclusion that the volunteers liked the function. In spite of the participants' initial concerns, their total amount of successful operation of the ACC covered a total of some 56 000 km of system engagement.

In the design of time-critical support, such as collision avoidance systems (CAS) the timing of the system activation is extremely important, both to gain driver acceptance and to effectively avoid crashes. The criteria for system activation have to be carefully specified, considering at least the main alternatives "early" and "late" system activation. Support initiated late reduces the prerequisites for effective collision avoidance. Collisions may occur in spite of a correct (but too late) CAS action, thus undermining the trustworthiness of the system. Support initiated early may instead produce what drivers may experience as false alarms or unnecessary actions that can be experienced as very annoying (Janssen and Nilsson, 1991). In the latter case drivers may get used to ignoring the support system, or to not use the system at all. Both reactions, rejection and behavioral adaptation, are threats to future traffic safety if ITS are not well adjusted to serve their very purpose: Driver support.

A CAS can support safe driving either by notifying the driver (collision warnings) or by automatic interventions (mainly braking). Different CAS configurations have been investigated. It seems as drivers prefer warnings (i.e., being in the loop) before automatic system interventions, even though evaluations of driving performance suggest that automatic system control produce superior driving performance in terms of driving safety (Nilsson *et al.*, 1992). In the case of collision warning, the timing of CAS initiation must allow enough time for the driver to process the message and to act. Thus, the shorter reaction time of the system compared to that of an average driver, and the consequences of this in terms of driver reactions, have to be considered in the system design. If the driver is about to take action, an information may lead to distraction and divided attention ("I already know!"), an advice about how to behave may lead to confusion ("I have already made up my mind!"), and a warning may lead to stress ("I have recognized the danger!"). An activation criterion based on the time-to-collision (TTC) concept seems to coincide well with human control strategies and preferences (Janssen and Nilsson, 1991), while a worst case criterion assuming leading cars to brake maximally at every moment would be optimal from a safety point of view. However, a criterion combining time headway (i.e., time gap) and TTC could be a realistic solution, as headway is a more sensitive indicator in distance keeping, whereas in approach situations the speed difference, indicated by the TTC measure, is more important to safe driving.

5. CO-OPERATIVE DRIVING

Future driving will probably be a co-operative activity based on the interaction between human and technical components. Co-operation is mediated by trust (Muir, 1994). Therefore, the relationship between the drivers' confidence in a support system and in their own capabilities is crucial to appropriate use of ITS. A system that cannot be well suited to the actual complexity of road traffic is a threat to traffic safety. It is a recognized and also a serious problem that an ITS system could fail to perform its function. However, in co-operative driving the occurrence of false alarms may influence drivers' confidence in the system profoundly, and may lead to sub-optimal system performance. The driver may then cease to rely on the system and may choose to ignore it or to not use it at all. Another negative side effect of false alarms could be that the system would disturb normal driving by distracting the driver.

An important precondition for a successful implementation of future ITS is formal standards for system design and system functions. In order to obtain a general agreement on standards, they need to be generated in an international and governmental framework. Recently a task force has presented guidelines for HMI (Human Machine Interface) for in-vehicle information and communication systems (European Commission, 1998). Also numerous working groups are currently concerned with development of HMI standards for different design features relating to driver support systems. Visual presentation of information, dialogue management, audible symbols are among the issues for which standards are currently considered (see also Noy, 1999a).

The absence of formal standards for system functions could lead to a large number of different ITS systems being available on the market. Presuming that some systems may be very similar though not identical in their design and functions, human reliability will probably be far from optimal since ITS support could appear very confusing. Most drivers do not have detailed knowledge about how ITS systems are designed, or what exactly they are designed for. Adaptive cruise control (ACC) for instance is an extensions of automatic speed keeping (CC) only with the possibility of adjusting speed to the current traffic situation, for instance when catching up another car. Negative effects have been found for an ACC system being sensitive to moving but not to stationary road objects (Nilsson, 1995). ACC supported drivers collided more often than unsupported drivers with vehicles in a simulated stationary queue. The outcome could not be explained by a change in workload or alertness during ACC use. Instead it illustrates that it can be extremely difficult for drivers to identify situations that require human intervention when using an ITS system, operating within a limited range of driving sub-tasks. In the case of the ACC system sensitive only to moving objects, the lack of inappropriate driver actions resulted from a "transparency problem", that is, from incorrect driver expectations and interpretations of the system functionality. Several current support systems relate to the driving speed. All the same, they can be designed as comfort systems or as safety systems. "Speed limiters" prevent speeds higher than the speed limit on specific road segments, automatic ACC devices control the speed taking traffic and road conditions into account, and CAS systems inform or warn the driver or act automatically in due time to avoid a collision. It is an obvious threat to safe driving that drivers, using differently equipped cars, can be confused by these different functions.

In a comparison of different systems for longitudinal control Rothengatter and Heino (1994) found that advisory visual information neither affected headway, subjective workload, nor on driver acceptance. Tactile warnings via the gas pedal lead to prolonged headway (time gap), to an increase in subjective workload, and a decrease in perceived comfort. Thus, the potential safety benefit of the tactile warning was obtained by the observed increase in headway, while the visual warning did not show any gain in safety. Driving with a distance controlling cruise control (ACC) lead to decreases in headway, as well as in subjective workload and perceived support. Thus, the possible potential safety benefit was related to the reduced workload. The obtained effects are supported by the results in another simulator study (Janssen and Nilsson, 1991). A warning pulse (counterforce) in the gas pedal combined with a TTC activation criterion (4 sec) was the only HMI design yielding a positive effect on the car following performance itself, that is, on proportion of headways (time gaps) < 1 sec without suffering from side-effects which could influence safety negatively. Both a warning light and a warning buzzer were associated with increased driving speed and increased speed variation. Besides, the studied visual warning, in the form of a red lamp on the dashboard, resulted in an increased proportion of short headways.

The "problem of transparency" relates not only to in-car devices, but also to the application of infrastructure components. Some roadside ITS may be present – and also best suited – only for certain road standards. This implies that the driver could meet different levels of ITS support during a single trip. Even though drivers may know about the different system functions and/or about the different levels of support on different roads, they may not be currently aware of them (Reason, 1990). Being used to a certain system may prevent appropriate behavior, and even encourage erroneous behavior, when a similar system with slightly different functionality is used.

Behavioral adaptation (Grayson, 1996; van Winsum, 1996) is also a major concern in ITS development. There is much evidence that potential safety benefits of conventional safety devices (studded tires, safety belts, etc.) are not obtained due to drivers giving precedence to comfort, mobility and time saving. Behavioral adaptation in road traffic is usually – but not always – associated with increased driving speed. ITS support systems might not leave car drivers with that possibility. However, facilitating the driving task could make drivers direct their attention to other issues and to other car equipments than those being relevant to safe driving. Talking to passengers or telephoning are very popular side activities among drivers, and there are reasons to believe that the tendency to entertain oneself during driving could increase with ITS. Unless wisely used, this practice could jeopardize traffic safety.

6. ASSESSMENT OF ITS FUNCTIONS

A basic question, when assessing driver behavior, acceptance, workload etc., is what variables should be selected and used as safety related indicators. A set of variables generally agreed upon and measured in a consistent way would enable the realization of harmonized procedures, and comparisons of results obtained in different studies. Such standards for assessing driving performance, both in general and related to ITS introduction, are highly desired but unfortunately not yet available. A recent attempt to define common standards for human machine interaction (HMI) on the basis of a general "Driver Model" did not lead to encouraging results, implying that it is most likely that "soft" standards based on consensus will develop gradually (Parkes, 1997).

As driving inherently is a self-paced task, assessments of driving performance must consider a variety of performance aspects. Current driver support systems, being designed for supporting specific sub-tasks, are generally found to influence also other aspects of driving. Speed is usually considered a most sensitive indicator of driving task variance, and both reduced visibility and increased cognitive load is known to reduce speed. Thus, speed changes are almost always assumed relevant for indicating effects due to ITS use. Usually also "side effects" such as mental workload, choice reaction time and variation in lateral position are taken into account and measured. But as assessment standards are not available, reported assessments of driving performance in relation to ITS use differ considerably in terms of the indicators used. For instance, in a comparative study of route guidance systems Zaidel and Noy (1997) used a score combined of eight different measures for assessing the quality of driving.

Also, the driving task can be described as a complex and dynamic control process. Therefore it is reasonable that the assessment of factors intended to influence driving performance (e.g., ITS) is carried out in a dynamic environment. However, a certain level of control with respect to test conditions is necessary in order both to interpret the obtained empirical data and to ensure reliability assessment. Advanced driving simulators offer a sufficient approximation to the driving task for being valuable test-beds for ITS devices. A dynamic traffic environment, in combination with experimental control, provides a possibility both to simulate and to assess an ITS system even in the conceptual phase of its development. Thus, driving simulators, test tracks and field studies in real traffic compose a battery of complementary methods for ITS assessment, including the complete driving task. The assessment methodology issue has been brought up as a European research effort in Transport Telematics as well as a world-wide issue (Noy, 1999b). Examples of European efforts are the GEM (Generic Evaluation Methodologies for Integrated Driver Support) project within the DRIVE II program, and the ongoing projects RESPONSE (Vehicle Automation – Driver Responsibility – Provider Liability – Legal and Institutional Consequence) within the 4th Framework and ADVISORS (Action for advanced Driver assistance and Vehicle control systems Implementation, Standardisation, Optimum use of the Road network and Safety) within the 5th Framework.

The methods considered in this chapter relate solely to ITS effects at the individual driver level. It is equally important to gain pre-implementation knowledge about possible safety effects of ITS at an aggregate level. Traffic simulation is a suitable technique, capable of demonstrating the most likely effects of different degrees of ITS penetration and distributions of driver performance on the traffic process (Lind, 1998). A prerequisite for a reliable utilization of traffic simulation is the availability of valid traffic and network models, and of relevant input data, for example, behavioral data also at the single-car level. If these requirements are met, traffic simulation becomes a powerful tool for estimations of safety related effects of various ITS solutions, without the necessity to wait for long term effects and a corresponding collection of field data (usually accidents). A somewhat weaker method than traffic simulation is expert assessment, for example, letting independent traffic safety experts judge the safety impacts of different ITS applications (Lind, 1998).

7. CONCLUSIONS

This chapter has been focused on one important aspect of ITS development, the effect of driver support systems on driving performance and driver behavior. Up to now most studies of behavioral effects of ITS have been short term studies of behavioral responses to a single and relatively simple ITS device. However, the point of departure in developing intelligent driver support systems is not an urgent need for driver support in contemporary traffic. The need for technical innovation of road traffic stems from societal needs and technical feasibility. Three actors are important to the future development and implementation of ITS applications: Society, industry and individual drivers (Figure 2.). Their motives differ but their interests are overlapping. In the perspective of societal control of road traffic with a still increasing traffic volume to manage safety, environmental protection and efficiency are strong motives for improvement. To the individual driver mobility, comfort and safety are important determinant of driving behavior. Presumably the industrial actors want to develop products possible to manufacture and put on the market within a relatively short time frame. It can be assumed that these different perspectives of the involved actors, shown in Figure 2, result in different expectations, attitudes and uses of ITS. In the initial phases of the ITS development the industry has been relatively strong. Consequently, the system development has to a large extent been more driven by a technological push than by a pull from end-user needs.

As the breakthrough of ITS is still to come, the effects of ITS devices on traffic safety and driver behavior are still relatively uncertain in many respects. Also, most of the current knowledge is derived from experimental studies of driving performance with a single support system. So, the knowledge is still sparse about driving performance when using combinations of systems, or systems integrating several functions is sparse. Likewise, the knowledge about behavioral changes when larger vehicle fleets are utilizing ITS systems is limited. So far, ITS equipped vehicles studied in real traffic have constituted a very small proportion of research. However, the preparation of a comprehensive implementation of driver support systems requires not only assessments of the effects of single-task support systems on driving behavior. In addition, allocation of tasks between technical and human components must be considered. Moreover, it is important to also ensure the coherence of the different sub-tasks of driving by giving due consideration to their timing, scheduling and priority.

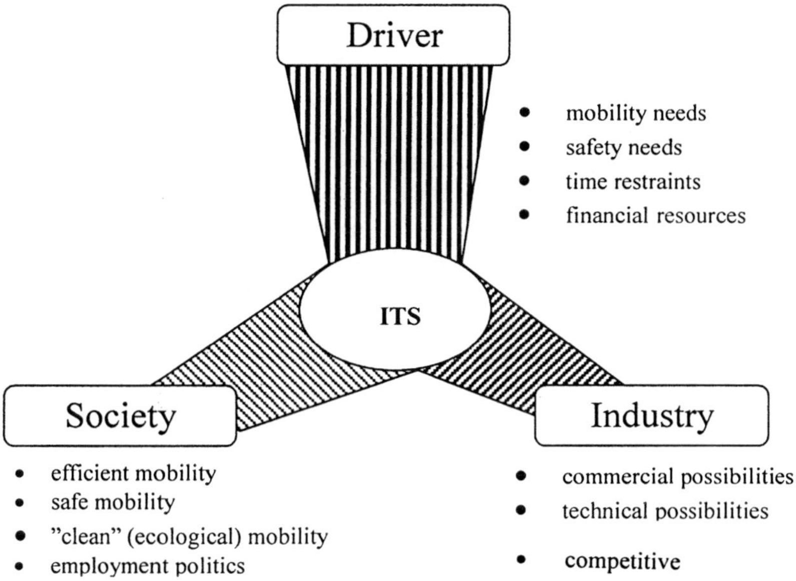

Figure 2. Different perspectives of actors involved in ITS implementation

Most current studies of ITS effects are concerned with driving performance, which indicates what drivers can do. Driving behavior on the other hand is about what drivers usually do. Assessments of the latter can hardly be carried out unless drivers are given the opportunity to get used to driving with ITS functions. Thus, long-term studies are needed in order to clarify persistent effects on driver behavior of different driver support systems.

Apart from driver support systems other types of ITS functions may be of more benefit to traffic management than to the individual driver. Future research may to a larger extent be directed towards these topics.

8. REFERENCES

ALLEN, T. M., LUNEFELD, H., ALEXANDER, G. J. (1971), *Driver Information Needs*, Highway Research Record 336.

ALM, H., NILSSON, L. (1995), "The Effects of a Mobile Telephone Task on Driver Behaviour in a Car Following Situation", *Accident Analysis and Prevention,* 27 (5) 707-715.

ALM, H., NILSSON, L. (1994), "Changes in Driver Behaviour as a Function of Handsfree Mobile Phones – A Simulator Study", *Accident Analysis and Prevention,* 26 (4) 441-451.

ALM, H., RUMAR, K. (1991), *Driver Attitudes Concerning Eleven Different Prometheus Functions*, Prometheus, PRO-GEN 74 000, Behavioural aspects, Swedish National Road and Transport Research Institute, Linköping, Sweden.

ALMQVIST, S., NYGÅRD, M. (1997), *Dynamisk hastighetsanpassning. Demonstrationsförsök med automatisk hastighetsreglering i tätort*, Bulletin 154, Department of Traffic Planning and Engineering, University of Lund, Lund, Sweden.

BOSSI, L. L., WARD, N. J., PARKES, A. M., HOWARTH, P. A. (1995), "The Effect of Vision Enhancement Systems on Driver Peripheral Visual Performance", in NOY, Y. I. (Ed.) *Ergonomics and Safety of Intelligent Driver Interfaces*, Mahwah: Lawrence Erlbaum.

BRAND, D. (1994), *Criteria and Methods for Evaluating Intelligent Transportation System Plans and Operational Tests*, Transport research record 1453 (Intelligent Transportation systems: Evaluation, Driver Behaviour, and Artificial Intelligence) TRB, 1-15.

BRAND, J. E. (1990), *Attitudes Toward Advanced Automotive Display Systems: Feedback from Driver Focus Group Discussions*, Report UMTRI-90-22, The University of Michigan Transportation Research Institute, Ann Arbor, USA.

BROOKHUIS, K. A., DEVRIEZE, D., DEWARD, D. (1991), "The Effects of Mobile Telephoning on Driving Performance", *Accident Analysis and Prevention,* 23 309-316.

BURNETT, G. E., JOYNER, S. M. (1996), "Route Guidance Systems: Getting it Right from the Drivers' Perspective", *Journal of Navigation,* 49 (2), 169-177.

CARSTEN, O. (1995), "A Framework for Safety Evaluation", in PAUWELUSSEN, J. P., PACEJKA, H. B. (Eds.), *Smart vehicles*, Lisse: Swets & Zeiglinger.

ENDSLEY, M. R., KIRIS, E. O. (1995), "The Out-of-the-Loop Performance Problem and Level of Control in Automation", *Human Factors,* 37 (2), 381-394.

EUROPEAN COMMISION (1998), *European Statement of Principles on Human-Machine Interface for In-Vehicle Information and Communication Systems*, DGXIII.

FANCHER, P., ERVIN, R., SAYER, J., HAGAN, M., BOGARD, S., BAREKET, Z., MEFFORD, M., HAUGEN, J. (1998), *Intelligent Cruise Control Field Operation Test (Final Report)*, Report DOT HS 808 849 (UMTRI-98-17), Ann Arbor, MI 48109-2150, USA.

FRANZÉN, S. (1993), "Introduction", in FRANZÉN, S., PARKES, A. M. (Eds.), *Driving Future Vehicles*, London: Taylor and Francis.

GALE, A. (1997), "Drivers' Visual Search of In-Vehicle Devices", in ROTHENGATTER, T., CARBONELL VAYA, E. (Eds.), *Traffic and Transport Psychology, Theory and Application*, Pergamon.

GRAYSON, G. B. (1996), *Behavioural Adaptation: A Review of the Literature*, TRL Report 254, Crowthorne, England.

HAKAMIES-BLOMQVIST, L., Peters, B. (2000), "Research on Older Drivers in Europe since 1985", *Accident Analysis and Prevention,* 32 (4) 133-139.

HALE, A. R., STOOP, J., HOMMELS, J. (1990), "Human Errors Models as Predictors of Accident Scenarios for Designers in Road Transport Systems", *Ergonomics*, 33 (10/11) 13377-1388.

HARMS, L. (1991), "Variation in Drivers' Cognitive Load, Effects of Driving through Village Areas and Rural Junctions", *Ergonomics*, 34 (2) 151-160.

JANSSEN, W. H., NILSSON, L. (1991), "An Experimental Evaluation of In-Vehicle Collision Avoidance Systems", in Proceedings from *24th ISATA International Symposium on Automotive Technology and Automation*, Florence, Italy.

KANTOWITZ, B. H., HANOWSKI, R. J., KANTOWITZ, S. C. (1997), "Driver Reliability Requirements for Traffic Advisory Information", in NOY, Y. I. (Ed.), *Ergonomics and Safety of Intelligent Driver Interfaces*, Mahwah: Lawrence Erlbaum.

LANSDOWN, T. C. (1997), "Visual Allocation and the Availability of Driver Information", in ROTHENGATTER, T., CARBONELL VAYA, E. (Eds.), *Traffic and Transport Psychology, Theory and Application*, Pergamon.

LIND, GUNNAR (1998), "Prediction of Safety Impacts of Non-Existing ITS Applications – Experienced Problems and a Way Forward", Paper presented at *Road Safety in Europe*, Bergish Gladbach, Germany.

MALATERRE, G., FONTAINE, H. (1993), "Drivers' Safety Needs and the Possibility of Satisfying them Using RTI Systems", in FRANZÉN, S., PARKES, A. M. (Eds.), *Driving Future Vehicles*, London: Taylor and Francis.

MUIR, B. N. (1994), "Trust in Automation, Part 1: Theoretical Issues in the Study of Trust and Human Intervention in Automated Systems, *Ergonomics*, 37 (11) 1905-1922.

NICOLLE, C., BURNETT, G. (1999), *TELSCAN Code of Good Practice and Handbook of Design Guidelines for Usability of Systems by Elderly and Disabled Travellers*, TELSCAN Deliverable 5.2, Thessaloniki, Greece.

NILSSON, L. (1995), "Safety Effects of Adaptive Cruise Controls in Critical Traffic Situations", in VERTIS (Ed.), Proceedings of *Steps Forward*, the Second World Congress on Intelligent Transport Systems, Vol. III, Yokohama, Japan.

NILSSON, L., ALM, H. (1996), "Effects of a Vision Enhancement System on Drivers' Ability to Drive Safely in Fog", in GALE, A. E. *et al.* (Eds.), *Vision in Vehicles V*, Amsterdam: Elsevier.

NILSSON, L., NÅBO, A. (1996), *Evaluation of Application 3: Intelligent Cruise Control Simulator Experiment, Effects of Different Levels of Automation on Driver Behaviour, Workload and Attitudes*, VTI Särtryck Nr 266, VTI, Linköping, Sweden.

NILSSON, L., ALM, H., JANSSEN, W. (1992), *Collision Avoidance Systems - Effects of Different Levels of Task Allocation on Driver Behaviour*, VTI Särtryck Nr 182, Linköping, Sweden.

NOY, Y. J. (1999a), "Status Report", *International Harmonized Research Activities (IHRA)*, Working group on intelligent transport system (ITS WG), unpublished manuscript.

NOY, Y. J. (1999b), "ITS Safety Test and Evaluation", *International Harmonized Research Activities (IHRA): Intelligent Transport*, Workshop April 14-15, 1999, Washington D.C., Report Transport Canada, Road Safety, Ergonomics Division.

PARKES, A. (1997), "The Role of Standards for In-Vehicle MMI", in NOY, Y. I. (Ed.), *Ergonomics and Safety of Intelligent Driver Interfaces*, Mahwah: Lawrence Erlbaum.

PARKES, A. (1995), "The Contribution of Human Factors Guidelines and Standards to Usable and Safe In-Vehicle Systems", in PAUWELUSSEN, J. P., PACEJKA, H. B. (Eds.), *Smart Vehicles,* Lisse: Swets & Zeiglinger.

PARKES, A. M., BURNETT, G. E. (1993), "An Evaluation of Medium Range 'Advance Information' in Route Guidance Displays for Use in Vehicles", in Proceedings from the *4th International IEEE Navigation and Information System Conference*, Ottawa, Canada.

PERRETT, K. E., STEVENS, A. (1996), *Review of the Potential Benefits of Road Transport Telematics*, TRL Report 220, Crowthorne, England.

RANNEY, T. A. (1995), "Models of Driving Behaviour: A Review of their Evolution", *Accident Analysis and Prevention*, 26 (6) 733-751.

RASMUSSEN, J. (1986), *Information Processing and Human-Machine Interaction, an Approach to Cognitive Engineering*, New York: North Holland.

RASMUSSEN, J. (1974), *The Human Data Processor as a System Component, Bits and Pieces of a Model*, Report Risø-M-1722, Risø National Laboratory, Denmark.

REASON, J. (1990), *Human Error*, Cambridge: University Press.

REICHART, G., HALLER, R., NAAB, K. (1996), "Driver Assistance: BMW Solutions for the Future of Individual Mobility", in ITS America (CD), Proceedings of *Intelligent Transportation: Realizing the Future*, Third Annual World Congress on Intelligent Transport Systems, Vol. III, Orlando, USA.

ROTHENGATTER, J. A., HEINO, A. (1994), "Safety Evaluation of Collision Avoidance Systems", in ERTICO (Ed.), Proceedings of *Towards an Intelligent Transport System*, the First World Congress on Applications of Transport Telematics and Intelligent Vehicle-Highway Systems, Vol. 4, Paris, France.

RUMAR, K. (1990a), "Driver Requirements and Road Traffic Informatics", *Transportation,* 17, 215-229.

RUMAR, K. (1990b), "The Basic Driver Error: Late Detection", *Ergonomics,* 33 (10/11) 1281-1290.

SUMMALA, H. (1997), "Ergonomics of Road Transport", *IATSS Research,* 21 (2) 1-9.

VERWEY, W. B. (1993), *First Test of Man-Machine Interface Adaptation to Driving Situation*, TNO-report IZF 1993 C-44, Soesterberg, The Netherlands.

WICKENS, C. D. (1992), *Engineering Psychology and Human Performance* (2nd edition), New York: Harper Collins publishers.

WIERWILLE, W. W. (1993), "An Initial Model of Visual Sampling of In-car Displays and Controls", in GALE, A. *et al.* (Eds.), *Vision in Vehicles IV*, Amsterdam: Elsevier North-Holland.

WIERWILLE, W. W., TIJERINA, L. (1996), "An Analysis of Driving Accident Narratives as a Means of Determining Problems Caused by In-Vehicle Visual Allocation and Visual Workload", in GALE, A. *et al.* (Eds.), *Vision in Vehicles V*, Amsterdam: Elsevier North-Holland.

WIKMAN, A. S., NIEMINEN, T., SUMMALA, H. (1998), "Driving Experience and Time-Sharing during In-Car Tasks on Roads of Different Width", *Ergonomics,* 41 (3) 358-372.

WINSUM, VAN, W. (1996), *From Adaptive Control to Adaptive Driver Behaviour*, Dissertation, Traffic Research Centre, University of Groningen, The Netherlands.

WOCHINGER, K., BOEHM-DAVIS, D. (1997), "Navigational Preference and Driver Acceptance of Advanced Traveler Information Systems", in NOY, Y. I. (Ed.), *Ergonomics and Safety of Intelligent Driver Interfaces*, Mahwah: Lawrence Erlbaum.

ZAIDEL, D. M., NOY, Y. I. (1997), "Automatic Versus Interactive Vehicle Navigation Aids », in NOY, Y. I. (Ed.), *Ergonomics and Safety of Intelligent Driver Interfaces*, Mahwah: Lawrence Erlbaum.

Chapter 15

USERS' ACCEPTANCE AND SOCIETAL ACCEPTABILITY OF NEW TRAFFIC TECHNOLOGIES

Safety Implications of In-Vehicle Devices

Stefan Petica
National Institute for Transport and Safety Research (INRETS), France

1. INTRODUCTION: NEW TRAFFIC TECHNOLOGIES - NEW PROBLEMS IN THE HUMAN FACTORS DOMAIN

In the Traffic Technologies field, a great variety of information systems or other types of applications are already available, while many new systems for driver assistance are being designed. Traffic information and management, user guidance, navigation systems, autonomous / intelligent cruise control, for example , or more complex systems, which are monitoring the driver or take over driver tasks in hazardous situations, has now entered a decisive development phase.

The competition between all of the industries of the developed countries in this field is generally very constructive and imaginative, but the consensus on different approaches concerning the real implications, is far from being established. For the manufacturers or for the engineers, system performances might be sufficient, but this is not necessarily the case for the ultimate users (drivers) or for different social key-actors (intermediate users).

For the information or advisory systems, the most important criteria determining the feasibility is not the technical performance, but the social context for implementation (Rothengatter *et al.*, 1991).

One of the latest characteristics of the development of information/assistance technologies is to provide the information/help to the drivers – not only through different types of messages located in the road environment, but by their integration inside the vehicle, through different types of devices.

Various psycho-physiological factors, the specificity of the driving task and the great variety of information, give rise to a new problem concerning the agreement of those devices by the users. This agreement or acceptance becomes, in fact, a real need for protecting safety and promoting a device's utility. Moreover, poorly designed or misused information systems inside vehicles can adversely affect driver behavior and safety by distracting attention from driving itself.

It is quite astonishing how little attention was affected until some years ago to users in the published literature on evaluation of the implications of the in-vehicle devices. But since Common European Programs began, the new prerequisites are unanimously required by different authorities, in order to anticipate the level of societal and economic benefits.

The needs, the acceptance degree, the general or particular opinions of the users are analyzed, almost every time new developments are envisaged. Those different measures focus especially on a general agreement of proposed systems on the basis of real or supposed needs. Is it that sufficient, while a society as a whole could be influenced in its own functioning? Which concept is most adapted to be used in this field, "acceptance" or "acceptability?" Is the first, sufficient and satisfactory or is the second more appropriate?

This paper intends to illustrate, even if not exhaustively, different aspects involved potentially by some existing systems or those which planned to be developed, their implications for the drivers or for the social point of view. It could be also considered as a critical analysis concerning different approaches in solving problems of new in-vehicle devices implications.

2. THEORETICAL AND CONCEPTUAL BASIS

2.1. Social, Societal and Systemic

Even if by convenience the concept of *"social"* is adopted in many situations and contexts, in this field of the development of new technologies it should be more appropriate to use the concept of *"societal"* implications rather than *"social."* Indeed, in order to distinguish the meanings of those concepts, the connotation of *social,* underlines and is pertaining to the *relations* between human beings or to the interactions in a given society. It is also pertaining to the welfare of the members of an organized society. However, the *societal* concept points out larger social groups and especially their *activities, customs, cultural differences, etc.*

Probably less precise than *"social,"* but more accurate in our field, the word *"societal"* designates members of the larger social organization, stresses better the welfare of the society as a whole, not only of its members or some categories of users. All things considered, the implications and the impact of any action or modification in the society are better distinguished.

In order to solve the overall challenge of the development of new technologies, jointly with real societal benefits, science needs to open in several directions. Principally, towards ethics and the sacred, and secondary, toward social action. The general method which seems to be more adapted in this field is the *systemic approach* which rejects the reductive materialistic, mechanical and utilitarian approaches and tempt to re-found the society. This approach expresses also, "the concept of generalized cognition, in the sense of a collective art of living, conscious or unconscious," during the development of the societies (Lorigny, 1996). That is the other reason to focus our approach toward societal acceptability rather than toward user acceptance.

In practice, the most common requirement for the introduction of in-vehicle technology is the *Acceptance* by the public, because "it is unproductive to invest effort in designing an intelligent co-driver if the system is never switched on, or even disabled" (Jinke *et al.*, 1997). But there are not yet standard ways of assessing users' acceptance even if many convergent or divergent methods are already proposed in the literature.

Even if this approach of users' acceptance assessment is necessary, it seems to be unsatisfactory, because it solves only part of the problem of adequate implementation of in-vehicle devices (in our case). That presumes essentially subjective or empirical opinions of users opinions who are not necessary aware other potential effects.

In this framework and through a systemic approach, the concept of societal *acceptability* which takes into account the social consequence, seems to be more adapted.

2.2. Acceptance and Acceptability: Conceptual Demarcation and Boundaries

The concept of *acceptance*, which can be considered the *positive aspect of attitude*, has previously been used in several domains and studies. This use comes within a theoretical framework which is relatively well defined. It is a question of evaluating the subject's degree of approval of such and such an action in relation to the possibilities offered by a given system or product.

"Social acceptance of a system can be seen as behavior directing judgment of that system or a behavioral intention" (Rothengatter *et al.*, 1991).

2.2.1. Main Mediate Method of Assessment: Attitudes and Opinions Evaluation

Adopting such a framework has several advantages, one of the most important of which is the strict methodology of measuring attitudes it provides. However, for the purpose of the preliminary and inevitably *a priori* assessments, we must widen the scope of action, for both practical and theoretical reasons.

The practical reasons which could be involved – the assessment of needs, of acceptance and of general awareness, HMI details – are to be carried out on the concepts which underline the bases of the design of a product which *does not yet exist.*

As for the theoretical reasons, these are the result of the disparity of definitions of "attitude," of which acceptance, as we have already said, is the positive aspect. This semantic diversity is all the more problematic when "one observes the increasing use of this concept in the field of traffic safety" (OCDE, 1994), without being able to determine which of the various definitions is the most relevant.

A brief history will help us to see where we stand. In 1931, Thurstone defined *attitude in terms of affect*: "Affect for or against a psychological object." In 1935, Allport placed attitude in a more general framework. It is "a state of mental and neural disposition, organized by experience, which exerts a directive or dynamic influence on the individual towards all the objects and situations with which he finds himself in contact." This definition necessarily implies a theory which deals with the *structure* and *function* of attitude, particularly in its relation to the real behavior of the individual.

Numerous contradictory theories subsequently emerged in connection with these structural and functional aspects of attitude. However, most authors agree on certain key notions concerning attitude:

- An individual's internal disposition to an object, relatively stable state from one situation to another;
- Is an acquired disposition and has some connection with what the object evokes in terms of affect and/or judgment, intention to act or actions. Psychological theories on the latter (the evocation of the object and the terms of this evocation) diverge.

On the one hand, for the defender of a "restricted" concept, attitude must be confined to the affective domain (unfavorable/favorable feelings) and distinguished from the cognitive (judgments, beliefs, knowledge) and "impulsive" (tendency to act). On the other hand, most advocates of a broader conception agree that assessment and memory are the principal aspects of attitude as well as the affective, cognitive and behavioral components (OCDE, 1994). Within this broader framework, theoretical problems are raised in terms of relationships between these different components.

Attitude is not a directly perceptible object and can be approached only in a mediate manner through various indicators: Opinion scales, dispositions, behaviors, affects, etc. (Lemaine, 1972; Scott, 1968). But these different indicators do not give necessarily identical results. In fact, behaviors are often different from what words might suggest or from what surveys could have brought out. These differences result from conflicting elements of the attitudinal structure, or from too much conformism.

Research into attitude would seem to favor the broader concept, since it can in that case examine both the affective aspects and beliefs and opinions, (different ways of replying to a questionnaire may correspond to the different components of attitude).

According to de Montmollin (1984), "there would seem to be no reason, or few valid reasons, either methodological or theoretical, for maintaining clear-cut distinctions between concepts of attitude, belief, representation, image or even value." In short, let us not forget that "attitude" is first of all a conceptual tool comprising several components that reveal a variety of predisposition's perceived with the help of various indicators and expressed through opinions.

In consequence, for not yet existing systems, acceptance is not the attitude in the strict sense of the term which is assessed, but rather general tendencies and intentions towards a "mental representation" at a given moment, of a product, tool or service expressed by the individual or by a sum of individuals (users).

2.2.2. Extending Reference System: From Individual to Overall Viewpoint

On the other level of the discourse of the social consequences of different advanced technologies implementations, it could be observed easily that the user acceptance (even if, ultimate users) are not enough to assess the real repercussions for this level. Other different judgments must be involved and other types of implications must be assessed. On the one hand, the benefits or the inconveniences of the elements of a given system/product are not necessarily the same for all the system/product. On the other hand and similarly, the benefits/inconveniences in a frame of reference are not the same in another frame of reference.

In this context, the concept which must be promoted is social *acceptability* which is more adapted and realistic because it allows one to take into account other aspects as: Technical feasibility, economical development, behavioral side effects, social safety benefits, etc. Evidently we could speak about social acceptance, but that means an utopian social perfection, impossible to be achieved at least theoretically, because there are too many contradictory claims and intersections in a given society. And *vice versa,* it is possible to speak about a user's acceptability; but is that sufficient and adequate, when the focus of interest is the safety and the social benefits?

Semantically, *acceptability* means "the quality or state of being acceptable" (College Dictionary) and acceptable involves being capable or worthy of being accepted. And especially that involves *only minimum requirements, barely adequate.* However, the users a*cceptance* is the *act of accepting, the approval, the quality or state of being accepted or acceptable.* It is the favorable reception according to the individuals point of view, their needs and requirements, while acceptability takes into consideration all of the stakeholders requirements and especially allows to focus or to privilege such and such domains, which is in our case *safety.*

Another concept, less "hard" than acceptance, is the term *acceptation-* which is only a *favorable regard,* or a belief, and frequently produces confusion in this field. Finally, the user acceptation and acceptance could be suitable for public agreement only, but not for social implications. To focus the approach at least towards the *minimum requirements* for the social "interest" and benefits becomes absolutely necessary in this field.

2.3. Combine Risk with Acceptability

It is generally agreed that any human intervention on a given state of a given system could produce risks; that is, some benefits to the detriment of other benefits already accumulated.

Consequently, taking into consideration the great number of factors involved in the implementation of advanced technologies means to associate the societal acceptability with the social, economic or safety risk. In this field it is evident that the risk of any action is narrowly associated to the acceptability of the minimum requirement for a given point of view.

The acceptability of a risk depends on many factors and there are several elements in the process of the determination of risk.

On the one hand, the measurement of risk may be viewed objectively and probabilistically. On the other hand, the measurement or estimation, of the amount of overall risk is usually followed by an evaluation of the acceptability of that risk. The nature of this phase implies subjective value judgments of different actors. The distinction between those two phases, acknowledges that individuals may agree with the results of one phase without necessarily agreeing to both.

2.3.1. Users' Acceptance but not Societal Acceptability

Generally, the people or the users don't accept or reject hazards in isolation. Rather, they make choices between potential direction of actions, whose consequences may include risks. If users accept a given action, such as deciding to drive fast, despite knowing about risks, then those risks might be acceptable in the context of the other positive results of this action. They need not be acceptable in any absolute sense of consequences (Fischhoff, 1997). At the social level, the mechanism is similar; the society makes even a risky decision, if there are other real benefits for other members. And sometimes not necessarily for all of their members. The problem is to evaluate and choose an impartial and reasonable balance. In several domains this problem begins, step by step, a real concern for different actors.

"Those same individuals might choose a riskier course of action (e.g., deciding to pass a slow car), if it brought a compensating benefit. Or, they might choose a less risky course of action, if that could be done at reasonable cost. A level of risk that is acceptable for one activity might seem horrendously high or wonderfully low in other contexts" (Fischhoff, 1997).

In this field, a technology should be accepted by a potential user if it is assessed as an acceptable balance of personal risks and benefits. Consequently, if a traffic technology is acceptable for each ultimate user of a given society, then it should be satisfactory to society as a whole. In this vision, it is considered generally and wrongly that a risk of "technology is societally acceptable (considering its benefits), just because the benefits are considered (by the individuals), societally acceptable (considering its risks)" (*ibid.*).

This philosophy is in our opinion restraint, because a confusion is kept between the direct ultimate users (who are considered as representatives of the society as a whole) and other stakeholders not directly involved in the usefulness of this technology but concerned by other type of implications (socially, politically, environmentally...), not necessary evident for the first category.

Even if the risk (and acceptability) could be considered as a result of the objective and subjective assessment, a good part of this assessment is a problem of perception of different actors involved. In this context, improving the communication of risk information among lay people, technical experts and policy makers (Slovic *et al.*, 1982) could be considered as an empirical but adapted principle for a societal acceptability. This approach, which we support, also should be circumscribed in the general frame of a Safety Culture and Societal Precaution Principle.

2.3.2. Societal Acceptability but not Users' Acceptance

Symmetrically, it is not rare to observe that some systems seems to be acceptable from the social point of view and even accepted " a priori " by the users because it solves some problems of basic needs. But after implementation, the same systems are rejected by the same users or other stakeholders. For example, in the domain of the navigation systems it is found that despite the initial and a priori overall positive assessment of systems, users feel that improvements are urgently required in other areas (Winkler and Nowicki, 1997). Thus, in a specific study the authors found that 72 % of persons interviewed suggested a total of 486 improvements. They requested an easier way of entering the destination; they considered the destination programming difficult and time-consuming, and they commented the poor recognizability of the characters because of their small size, etc.

The same study shows that self-sufficient navigation systems have positive effects on traffic safety only *in principle*. The main factors supposing enhancing safety are facilitation's for drivers and saving time but several deficiencies that have to be removed concern major obstacles to the success of the systems on the market. The unusual price and the lack of system flexibility via current traffic relationships and connections are other important factors *for an a posteriori inacceptance.*

In the other studies also (confidential and not yet published), the navigation systems are widely accepted by the vast majority of drivers but they greatly ignore the final efficiencies of the systems. In order to maintain users acceptance, the manufacturers are under duty to improve the systems constantly and systematically. Implicitly, new "customization", new needs and "conditionings" are generated.

Several requirements from the perspective of the users are "the inclusion of current traffic information in route calculation and a reasonable choice of routes, safe and well-developed main streets, additional touring information and a further digitalization of roads for smaller and medium-sized towns. Finally, it is important to reduce the prices of navigation systems at least three times" (*ibid.*) to make them lucrative for the mass market.

2.4. Acceptability as a Risk Balancing

In the field of transportation technologies in general and of in vehicle devices, in particular, a focus on *safety and comfort traffic actions,* as a main criterion, may produce the best balance of benefits/ potential inconvenient and so, societal acceptability. That could be considered as a proposal of the definition of Societal Acceptability. "A technology has a societally acceptable level of risk if its benefits outweigh its risks for every member of society" (Fischhoff, 1997).

2.4.1. Winners and Losers

From the ethical viewpoint this proposal may be seen in opposition with the pragmatism approaches (that look at the total benefits provided to a society by a technology), when we take into consideration the acceptability of risks. A rough method for doing so, is to perform a cost-benefit analysis, summarizing economic measures of a technology's total benefits and total costs (including the risks that it imposes).

"A central ethical presumption of many such analyses is that one should look at the overall balance of consequences for society, while ignoring the balance actually experienced by individuals. Under this hypothesis, one would not care if a technology made society as a whole better off, at the price of making some of its members miserable" (*ibid.*). Nor would one care if a few stakeholders perceived very big advantages, while several others had small losses, or *vice versa*.

For ethical reasons too, if the sharing between the stakeholders is made only in two categories, as ultimate users (safety winners) and intermediate users (economic winners) the balance in favor of the first category must be assisted by the honest contribution of the second category even if the benefits of this one are smaller and conventional. The potential losers in these "dealings" may not know or not be aware of all of the implications and many could be persuaded, exhorted or induced (only) by the benefits of the technology.

"It holds that an implementation of a new technology is acceptable if its excess of benefits over risks is sufficiently great so that those who enjoy this implementation could compensate losers." (*ibid.*)

Those theoretical remarks don't mean necessarily that the introduction of some in vehicle technologies involves some risky actions or that the balance is in the favor of the inconvenience or drawbacks. But two main reasons suggest to judge the problem in terms of Precaution principles:

- It is generally and usually accepted that any modification of a given system (social, environmental, etc.) could produce some imbalances in the system which are most difficult to be counterbalance and compensate, *a postiori*, rather than to be taken into consideration *a priori.*
- With the introduction of new technologies in the traffic domain, the driving conditions and the driving task will radically change and the adaptive capacities of human factors are yet to be known, especially in the situation of interactions between several types of driving tasks and several different systems, tools or services.

2.4.2. Acceptability Approach in other Domains

Then, briefly, even if both concepts belong to the same semantic family, in our context, the Acceptance must be considered as different from the concept of Acceptability. Furthermore, in the engineering sciences, acceptability can designate the positive rather than the negative implications, the a-priori agreement of the actors concerning a certain measure or action, or particular*ly, the potential a-posteriori agreement for a system as a whole.*

In fact, this approach by acceptability assessment is frequently used in other domains: nuclear power plants, social health, chemical hazards, political decisions.

Thus, the risk acceptability criteria, studies of "acceptable risk" or "risk adverse attitude" have been very active areas in the last few years in the Engineering and applied Sciences. The classic papers (Farmer, 1967; Starr, 1969) studied, for example, the problem of acceptability of new technologies respectively, in the context of reactor safety sitting. In this domain the term of "risk aversion," used relatively in the broad sense, reflects generally a conventional attitude towards risk. As in decision theory, a risk adverse decision maker finds uncertain consequences less preferable than a consequence with certainty, when the latter is the main value of the uncertain consequences.

Even if for Wu-Chien and Apostolakis (1981), the "boundary curve separating acceptable and unacceptable the region does not have to be a straight line on the logarithmic frequency-consequence space," the frequency of negative events should modify the Farmer's criteria of low frequency/high consequence. In this context, the impact function of a negative event is defined as the negative utility and characterizes globally one of the main criterion of risk acceptability.

This subject is currently being investigated more and more in several societal areas, with the exception of the in-vehicle devices implementation.

Based on one of the specificity of our time, concerning the relatively easy construction of new needs, and in order to attract external investments, often enough, the main concern of some industries or promoters is to show that the potential users are ready to receive, accept and pay for a proposed system. Even if in fact, a posteriori, it is proved that by a natural selection, the system is not accepted. In this case it is not the safety risk which is ignored but the economic and marketable risk.

2.5. Some Methodological Aspects: Assessment and Preventive Actions

In this context, many methods and approaches try to solve the problem of the consonance between the new technologies (in the transportation field) and the needs and acceptance/acceptability of the society: Validation studies, impact studies, demonstrators, surveys with users or other stakeholders, etc. But standardized methods agreed upon by all concerned actors don't yet exist.

In the assessment of driver acceptance of advanced transport telematics domain, for example, the literature shows that a really reliable instrument is far from emerging and even if the studies are aimed at the determination of level of user acceptance, the focus differs. Sometimes the systems are evaluated through pleasantness/usefulness attitudes (Kuiken and Groeger, 1993), the degree of use (Crosby *et al.*, 1993), the comfort and benefit (Beker *et al.*, 1995), the reliability, the relevance, the accuracy or intention to purchase, etc. (Schofer *et al.*, 1993; Crosby *et al.*, 1993; Barham *et al.*, 1993).

Concerning the safety implications of some social decisions or technologies implementations, the mains tools which seem to be the more adapted have been and are currently used: *surveys by questionnaires, checklists or guidelines techniques*.

Those methods suppose in fact the use of previous knowledge accumulated from different fields: Psycho-ergonomics, social representations, mechanisms of accidents or dangerous events, etc. In this context, a particular place concerns the ergonomic analyze of the "human activity" (in opposition with the "task"), where the main problem is "what the human beings really do" in specific situations.

The aim is to understand and to improve different human interventions or actions (e.g., avoidance of human errors). Abundant literature exists already in this field and driving activity or road accidents has been equally analyzed. (Leplat, 1985; Montmollin, 1986; Malaterre, 1987; Saad *et al.*,1992). Evidently, this kind of knowledge could very well be used in different preventive actions for a correct design of new tools or systems. It is important, for example, to note that the distinction between "active" errors and "latent" errors is particular significant in the domain of new complex systems or new technologies (Reason, 1993; Van Eslande and Alberton, 1997). If the "active" errors are linked to the direct control of the operator of a given system (through the interfaces characteristics), the "latent" errors are associated to the latent failure of the system was designed, conceived or decided without taking into account the "real activities" of the operators – but only their *supposed task* "who must be done" according to the designer imagination.

In the other domains supposing more potential safety implications, this kind of knowledge is also indirectly used through similar (but "quicker" and more concrete) methods. Erke *et al.* (1983) developed, for example, a checklist on occupational safety for use in military settings. Tuttle *et al.* already provided in 1974 a summary of checklists to help independent observers in detecting potentially hazards, and so-called "standardized" and "normalized" tools are periodically proposed by different actors. Some examples:

2.5.1. ITS Traffic Safety Supervised Methods

The assessments of already existing devices (or prototypes) are sometimes carried out with the help of specific safety checklists. One of them, for example, the Traffic-Safety Checklist, which takes into consideration the safety criteria was developed in the activities of the PROMETHEUS Safety Group. It was a contribution to the declared aim of PROMETHEUS, "to focus on the possible consequences for road safety of introducing new equipment in road traffic."

This approach towards a systemic and systematic a priori assessment of new technologies in road traffic seems to be one of the more adapted tools for increasing acceptability from the safety viewpoint. "It is a prediction tool developed to make the planner aware and to develop hypotheses concerning the influences of new equipment on road user behavior. It can be looked upon as a guidance procedure for manufacturers and experts" (Risser and Lehner, 1997). This method contributes to the avoidance of the development of systems which could induce some side effects and helps to improve some systems, initially *not too much reliable or useful.*

Based on this philosophy, in the field on PROMETHEUS, it was assessed, for example, by independent experts and heuristic/empirical methods, that some Common European Demonstrators (CED) are far from bringing the real safety benefits, or even worse, some CDEs could produce dangerous safety or side effects.

Other specific preliminary guidelines for human factor aspects of in vehicle crash avoidance warnings, for example, are also carried out. Major topics including level of warning, unique imminent crash warning signals, warning prioritization, compatibility with driver behaviors, warning message content, etc., are analyzed (Lerner *et al.*, 1996).

2.5.2. How are Assessed Intelligent Systems ?

It must be clearly noted, that the initial perception of the users concerning new ITS systems (Intelligent Transport Systems) to a great extent, will determine in good part the real acceptance of a new product launched on the market.

The technical information on the potential of a product in terms of safety should be available to the users during the whole development phase of the product. In this way, one collects, on the one hand, the reactions of the users after the demonstration of the technical advantages of the new product and, on the other hand, the reactions of the future customers after the launching of the new product on the market and its definitive acceptance by the market (Choffray and Lilian, 1980; Thirumalai, 1995; Zimmerman and Eliott, 1995).

Currently, a development strategy of a product should be preceded by a feasibility study, followed by field and operational tests, in order to reduce the risks and uncertainties of the development cycle. Even if those phases are not linked with the societal acceptability of the implementation of a given system, at least the acceptance by the users of the final product on the market is more or less ensured. "After evaluation successful strategies used by industry to market innovative and new products, a 'reactive' product development strategy, in combination with a 'proactive' and entrepreneurial concept strategy, appears to be best suited for the development of innovative ITS safety product" (Thirumalai, 1995: 2386).

In order to build prototypes, system concept development must be supported by a coherent strategy based on some key concepts, and the societal acceptability should be one of the most important concepts even if the marketability is in fact the main objective focused by the developers.

On the other hand, in the last years several extensive/deeper researches have been conducted which are relevant to ITS safety test and evaluation. Procedures or techniques for the evaluation of safety, ergonomical criteria, specific requirements to performance criteria, design restrictions, etc. are presently carried out through the world (Noy, 1997; Smiley, 1995; Morita, 1996; etc.). In fact, the ultimate aim of those researches is to help in the improvement of different systems in the developing phase, and to encourage ITS technologies which will have a positive influence on motor vehicle safety and finally on societal acceptability.

2.5.3. Harmonized Actions for ITS Evaluation – Proactive Activities

Other more consensual, structured and harmonized approaches initiated principally by the official organizations (e.g., CE) also try to orient the assessment of the new technologies in transportation through guidebooks, guidelines or focused studies. Some examples:

- CODE project provide a Guidebook for User Needs Analysis which identified some variables which take into consideration the basic ergonomic criteria in order to be satisfactory as regards the usefulness, accessibility, usability, comfort, satisfaction, safety, etc. (Robin-Prévallée *et al.*, 1996). This approach is supposed to be in anticipation of the future user acceptance and social acceptability.
- Concerning the same Telematics Applications, the CONVERGE project also provides an illustration of several categories of assessment (technical, socio-economic, user acceptance, cost-benefit analysis, etc.), accompanied by different techniques and methods (Zhang, 1996; Maltby, 1996).
- Similar concerted actions (workshops or meetings) as MEGATAQ or INUSE assessment assistance for developers are evidently very useful for the awareness of the potential problems raised by the ITS systems (Fleming and Mc Leod, 1997; Daly-Jones, Bevan and Thomas, 1997). But sometimes for the exhaustive concerns, those tools complicate the problem too much or they have a narrow vision.
- Given the great complexity of the problem, different public authorities become aware of the evolution of the new technologies and the potential impact to the societal level, especially from the safety viewpoint. While

if it is too early to propose new laws at the national or international levels, some regulation already exist.

As a guide and model for the national authorities, ECMT (1994) adopted for example a Statement of Principles concerning Ergonomics and Safety of in-vehicle information systems which begin to be implemented at the European level. Being aware of the potentially poor design of some systems which could adversely affect driver behavior and safety by distracting attention from the driving task, about the interference in the traffic between several systems, ECMT recommends a minimum criteria concerning, the responsibilities of drivers, designers, manufacturers, and suppliers, and suggests different ways for assessing systems before implementation.

- Even if in the domain of safety, authorities' actions are more reactive than proactive; i.e., the interventions are taken after comparative alternatives considering the benefits/cost/effectiveness, different organized activities start to win an important place. The governments and the stakeholders of ITS begin to be aware that the *a posteriori* intervention cycle is longer than the design cycle (Michon, 1993). In fact, governmental authorities have the responsibility for establishing vehicle safety policies, promulgating national policy initiatives or motor vehicle safety regulations on a scientific basis.

One of the particularly important activities at the international level concerns the actions and works developed in the framework of the International Harmonized Research Activities Working Group on ITS (IHRA), which tries to facilitate collaborative research in this areas (Noy, 1997).

Under the auspices of Enhanced Safety Vehicle (ESV), which is an international forum for the exchange of scientific and technological advances in vehicle safety, the IHRA group intends to harmonize research in ITS with the following aims, a) include crash avoidance interventions, b) define more clearly the role of government with respect to ITS safety, c) analyze ITS-driver interaction and the role of human factor, essentially in the unregulated area at the present time (*ibid.*).

The most critical factor since the most accidents involve human errors, is the human-machine interface, but the reliability and aptitude of the system is necessary even if not sufficient for successful implementation. "The ergonomics of the interface and human factors underlying driver-vehicle interactions are paramount to the realization of the full safety potential of ITS technologies" (*ibid.*).

If it is considered that "the scope of the research program in ITS is defined and limited by emphasis on three key elements: Government orientation, safety evaluation and driver-ITS interaction," this kind of activity could widely contribute to the inhibition of the technologies which are likely to have an adverse affect on safety and finally to the societal acceptability of ITS technologies.

2.5.4. Some Considerations on the Assessment and Guidance Actions

Two points are worth making concerning those approaches in the interface domains:

- The first is that what is being considered is not really systems but what is called set-ups, a system embedded in a context of use. This is significant; from a real user's viewpoint, there is usually very little point in evaluating a system in isolation. For example, many ISO 9000 series rules on quality assessment makes the same point, although from a rather different viewpoint. "The importance of each quality characteristic varies depending on the class of systems. Reliability is most important for a mission critical system, efficiency is most important for a time critical real time system, and usability is most important for an interactive end user software" (ISO91).
- The second point rises rather to the negative aspects; the users considered in constructing of such and such system type could be no real users but "virtual" users. But it is important to be aware of the dangers involved in deciding on behalf of some third party what it is he really wants or needs. In this field, other ISO rules provide guidance on human centered design activities, especially in the life cycle of computer-based interactive systems which could be a part of on board transport technologies. A collection of International Standards are given, for example, in the ISO/DIS 13407 (draft) which takes into consideration different ergonomically useful criteria but not necessarily adapted to the traffic technologies specifications.

This potential fragility is partially at least counterbalanced by the second type of evaluation, called in the committee's reports "technical evaluation by users," "users' acceptance," "users' satisfaction." In this context, an attempt is made to determine the user's real needs and to compare them with what can be offered by specific products in order to evaluate how satisfied the customer is likely to be with what is offered.

By analogy and according with King (1996), in practice on evaluation of the in-vehicle devices we find essentially three classes concerning who does evaluation or talk about it:

- Researchers or industries concerned with system elaboration: The researchers do not have all the time or the means to carry out any systematic investigation concerning what users might actually want. "The developers principally come into contact with users through their customer support services. In both cases, when a user is taken into account, it is an abstract, virtual user, whose needs are conformed to those the researcher or system developer thinks he would have."
- Customer associations, public authorities, public researchers: "Since what they are primarily interested in is the development of a core

technology, evaluation is seen as an assessment of a system's ability to perform a pre-determined task," thinking of an ultimate user, really users needs and societal benefits is too early and not necessarily relevant.

- Latent customer or investor of commercially purchasable systems: In this case, "the user is directly present, but concerned only with his own needs" (natural or artificial) and his predisposition to buy.

3. POTENTIAL IMPLICATIONS: SOME ILLUSTRATIONS

3.1. In-Vehicle Information Technologies – Driver/Information System Interaction

In recent years Information technologies has been growing rapidly in all sectors of society and the transport sector is affected too by these developments. The recently designed new applications dedicated to be located in the vehicles, and their potential implications attracted attention of several actors (researchers, decision makers, manufacturers) and numerous works are being carried out in this field. Indeed, a general survey of state of art shows that the new systems of information and assistance "may in fact give rise to a number of difficulties from the standpoint of their real effectiveness and safety"(ECMT, 1992).

It can be said that a thorough-going revolution is about to occur where driving tasks and facilities are concerned and the human factor must play a role as fundamental as that of technology itself.

In order to solve some human factor aspects, one of the first steps was to characterize the type of information currently transferred on-board the vehicle through the screens and displays which are really useful for the drivers. Only one example could illustrate this kind of approach. In this case additional tasks could occur for the drivers and the driving task could be affected. Organizing the presentation of information, select the really useful information taking into account the specificity of the driving task could be the reasonable compromise in order both to respect safety and take advantage of new type of information and services.

In this way, a proposal of a classification of on-board information (Petica, 1991) was integrated by ECMT (1994) as a basis of a Statements of Principles of Good Practice concerning Ergonomics and Safety of In-Vehicle Information systems.

3.1.1. Classification of On-Board Information Systems

A comprehensive analysis of the characteristics of the driving task and the projects for the development of information systems relating to the "intelligent" car shows that the information provided for drivers can be divided into two main categories:

a) Information systems related to vehicle use;
b) Information systems that is not related to vehicle use.

– The first category (a) may consist of the following classes:

Class 1. The class of *"intrinsic"* information, that is information obtained from the various functions of the vehicle itself. The physical medium of the interface consists of screens, warning lights and visual facilities in general. ·This class covers information relating solely to the condition of the vehicle and its particular characteristics. The information may be concerned directly with *safety:* tire pressure, condition of brake disks, position of handbrake, condition of ABS, direction, etc; or may *not be directly related to safety*: oil gauge, petrol gauge, water temperature gauge, etc.

Class 2. The class of exclusively *"extrinsic"* information which comes from the immediate or more distant environment. Information in this class has hitherto been exclusively the responsibility of the driver who *observed* it in the vicinity or *sough*t it by more or less appropriate means, the more conventional means relating to navigation, prior consultation of a map, and response to road direction signs and radio information. On-board driving aids and map-reading systems are designed to reduce uncertainty as regards navigation by providing information that was hitherto lacking. Existing systems proposed in most of the projects in the world use driving aids, displays and screens providing information such as: rear-view systems by microcamera, on-board map systems, (without guidance), real-time information on the traffic, auto-location systems, navigation, route guidance, infra-red image display for night driving, etc.

Class 3. Careful analysis of the already existing systems indicates the advent of a third type of on-board information, namely the *"relational"* or *interactive* type, more sophisticated in nature than that in Class 2. The interactive aspect concerns the condition of the vehicle, its position or speed in relation to environmental characteristics (condition of the road and traffic). The information is provided after automated "reasoning" so as to advise the driver on procedures to be adopted or decisions to be taken: all right to overtake, essential to brake, recommended route, etc. This information may take account of the psycho-physiological condition of the driver (fatigue, influence of alcohol) and the system may give advice or prohibit driving.

The information is in theory designed to improve the driver's capabilities and reduce errors, since his lack of reliability and his behavior depend on a great many variables. These information systems may tend to become actual systems of reasoning or, in the longer term, expert systems that can make driving automatic to some extent. They obviously use screens and keys for the purpose of dialogue and these can certainly give rise to problems in terms of use. For example, American manufacturers are concerned about the acceptance of users who prefer to make the decisions themselves and carry out their own maneuvers (Agnew, 1988) or elderly people who reject on principle anything that may significantly interfere with their customary behavior.

This class of information is illustrated by the driving and navigational aids systems and the intelligent co-pilot of the second generation now in the course of preparation whose main functions are: Monitoring and regulation of speed, tactical assistance in terms of allowing overtaking, operational aids by means of controlling the direction of the vehicle and interactive guidance systems.

- The second category (b) of information systems, not related to vehicle use, include systems which are designed to make the vehicle more user friendly or provide certain data of a specifically professional data or leisure-oriented nature: personal computer, radio-telephone and autoradio, on board television, etc. Even if this kind of systems could involve equally different social/safety aspects is not our intention to analyze it in this paper. However, the pernicious case of cellular phone will be raised later.

3.1.2. Comments and Inferences – How Should Be Improved Information Systems?

According to Petica some years ago (1991) and the previous observations, a number of considerations which could be valid even today, may be drawn in this field:

- Designed with a view to certain improvements, including greater safety, the systems themselves could raise safety problems where their use by drivers is concerned;
- The additional volume of information available to the driver in using these systems can be substantial and, if poorly organized, may have pernicious and perverse effects;
- The driver's loss of the power of decision and autonomy, since he will no longer be making his own "sensory" judgments, may be an additional source of conflict, both for himself and for other road users;
- In the initial stage the co-existence of conventional and "intelligent" vehicles, whereas the infrastructure will not be entirely "intelligent," will tend to make some users more vulnerable;
- Aside from a certain attraction to the novelty of the systems and natural curiosity, the potential user is not always able to improve his behavior and fundamentally change his approach to the use of the car;

- Possible faults in the technical systems may very well upset the driver and leave him powerless to handle the suddenly imposed cognitive constraints for which he is unprepared.

The crux of the problem relating to the introduction of on-board technologies as a source of information and assistance systems in general lies in fact in the choice of one of the two attitudes that might be adopted:

- To let things develop on their own, haphazardly in their diversity, knowing that some of the inconsistencies in terms of ergonomics and safety will gradually be rejected by the drivers themselves as they use the systems in what might be regarded as a process of natural selection. That attitude would solely entail financial and productive wastage (economic risk unacceptability);
- Take steps from the outside to harmonize the development of these systems, on the understanding that the risk of perverse and insidious effects endangering safety is by no means negligible (especially for "vulnerable" users) and that additional accidents may occur before the natural selection has been completed (safety risk unacceptability).

With respect to the problems set out above and in the light of the responsibility of public authorities to protect the society in general and the road user in particular, a set of proposals was suggested (*ibid.*).

Since this approach was initiated, several deeper ergonomic works was carried out in order to improve all of the information systems. One of the important studies commissioned by the U.K. Department of Transport and developed by ICE Ergonomics (1993) rendered very useful ergonomic data concerning: Information presentation, display design, control design and location, training and assessment of systems.

Those main activities were submitted to an expert council (Task Force) and finally materialized in a ECMT Declaration already mentioned *(ibid.)*.

This relatively small example however is pertinent and representative in the sense of the significance for the public authorities of the ergonomic implications and societal acceptability focused on safety.

3.2. Driver/Communication Systems Implications - Mobile Phone

The increasing need for communication has led to enhanced usage of new in-vehicle technologies not dedicated directly to the driving task and hence new sub-tasks for drivers.

A number of studies of the implications of this new driving task have already been carried out in different parts of the world, but the results and the measures taken have been fairly contradictory (Alm, 1990; Brookhuis, 1991; Nilsson, 1991; Petica, 1995; etc.). In this context, some more systemic approaches attempt to answer some questions about the risk of using car phones, based on current knowledge.

For example Petica (1995) shows that although the level of car phone ownership in France is relatively low in comparison to other countries, it is set to rise considerably in the years to come. One of the reasons is that "the mobile telecommunications sector is due to undergo a revolution which can be compared to that which aeroplanes and cars have experienced since the beginning of the century or which computers have experienced since 1950" (Note d'information, 1992).

3.2.1. Driving Task and Secondary Task: Negative Effects

These new conditions make it necessary to look at the problem of the use of in-car telephones in a different way as it at least seems probable that new road safety related problems will arise if there is no effective prevention.

From the psycho-ergonomic viewpoint, all of the experts agree that the interference of two simultaneous human activities creates performance problems when neither of these activities is completely automatic and when they both make use of the same sensory receptors, the same motor groups and the same psychological processes. The utilization of a car phone during driving is definitely a secondary task, it may cause disturbances because at a given moment it utilizes almost the same perceptive and cognitive mechanisms as are normally assigned to the principle task of driving.

On the other hand, it is essential not to ignore the beneficial effects of the communications in general and of a secondary activity when the primary task is monotonous or under-stimulates certain psychological functions. In these circumstances additional stimulation, of a reasonable level of difficulty, facilitates vigilance and the normal functioning of perceptual and cognitive structures.

According to the good part of the studies the most important parameters which seem to be definitely affected are as follows:

- The reaction time in an emergency situation and difficult driving conditions;
- A failure to respond to road conditions and response time;
- Variations in lane position and maintaining the same speed as the preceding vehicle;
- Checking in the rear view mirror (this becomes less frequent);
- The perceptual and mental load;
- The modification is more significant in the case of elderly persons.

The malfunction in the driving-communication-safety relationship increases with the inadequacy of the characteristics of the system:

- When the equipment is inappropriately positioned in the vehicle;
- When the equipment is ergonomically defective, even though aesthetically pleasing;

- When the sound quality of the conversation is poor;
- When some functions of the equipment are either useless or difficult for the user to implement because they are difficult to remember.

In fact the societal benefits and disbenefits must be evaluated both on the basis of the most pessimistic assumptions and research results, and on the basis of the number of such devices at national level. Until now, this number was relatively small, but this will soon cease to be the case, and for this reason, it will become important to consider additional risk factors.

Some factors can be regarded as definitely having a negative effect, and for other factors the risk has not yet been scientifically demonstrated or does not possess a high degree of statistical significance. This should also be considered in a wider frame of reference where the continuous need for communication, economic mechanisms, the practicality reasonable intervention, and more complex psycho-social phenomena (social desires and acceptance, public realization of certain limits of utilization) should be considered.

3.2.2. Latent and Active Solutions

As the result of a number of studies of the topic, in order to increase the social acceptability of the safety viewpoint, public authorities in several countries have investigated the possibility of regulating the utilization of car phones with regard to their potential safety implications.

An international survey for INRETS (Petica and Bluet, 1989) finds out the views of research institutes, public bodies and decision makers in the field of safety (approximately 100 organizations drawn from twenty-two countries). Sixty-six percent of those interviewed from major industrialized countries considered that the utilization of the telephone while driving was potentially dangerous and that this potential danger becomes real in the case of the basic model (a handset with built-in keyboard). Only 17 % felt that there is insufficient proof at the present time and 17 % hold no view. However, in most countries there are no specific regulations.

According to those questioned, solving this problem would involve the adoption and enforcement of a number of measures relating to regulatory and legal matters (22.2 % of views expressed), information and utilization advice (38.8 %), or to standardization and ergonomics (39 %).

Several countries already introduced specific regulations and it can be seen that different countries deal with the use of telephones in cars in different ways (Petica, 1995). The following generalizations can be made about the different strategies:

- Some specifications are already mentioned in legal texts. Some prohibitions are implicit, in view of the fact that any additional activity during driving is forbidden and punishable in the event of an accident;

- Strict regulations, which include details concerning the prohibition of car phone use while driving are not systematically applied, being either applied on an irregular basis or only in situations in which problems arise;
- Some recommendations do not have legal prohibitions but are directed at users to ensure that they are aware that the probability of a risk may increase considerably under some circumstances.

As a main conclusion of the mentioned study it is precised that the interference between the driving task and the telephone use could be considered, on the whole, that the probability of an accident increases significantly if telephoning takes place while driving. The scale of the risk is mainly related to the conditions on the road; it may be negligible or a two-to four-fold increase of risk. This estimate was in fact confirmed by later works and some statistical data. For example, recently in a large study (700 drivers) it was found that the risk of collision during phone use, is four times higher than in normal conditions and the danger is six times higher in the first five minutes of the communication. Consequently, about 6 % - 12 % of road accidents are due to telephone use (Redelmeter and Tibshirani, 1997).

In this context, could we conclude about at least a societal acceptability of using this device while driving if some important safety aspects are raised, without improving the phenomenon?

Even if the user acceptance of this kind of in-vehicle device is obvious, *the risk prevention in general and societal acceptability may be considerably improved and then road safety problems which result from new technologies in vehicles* may be significantly reduced if the public authorities adopt three types of measures:

- Measures which concentrate on increasing user awareness of risks;
- Measures which concentrate on the ergonomic safety of systems;
- Regulatory measures.

3.3. Driver/Monitoring Systems Needs and Acceptance – Driver Impairment Aids and Vehicle Control Systems

3.3.1. Objectives: What to Do? Why Assessment?

Another domain where the problem of user acceptance and societal acceptability could be illustrated is the field of the design and development of different systems in the European programs. One of the European projects, the SAVE project (System for effective Assessment of driver state and Vehicle control in Emergency situations), is a result of the hazardous situation arising from the number of accidents caused by driver impairment and the parallel development of new systems driving technology. As a result of the severe consequences of this kind of accidents – loss of alertness is the main cause of 34 % (Vallet, 1991) and 90 % of all traffic accidents could be attributed to human failure (Smiley and Brookhuis, 1987). The SAVE project intends to develop an integrated system which could solve, at least, a part of this problem. It is capable of detecting driver status, inform the driver/surrounding traffic, control/stop automatically the vehicle if the driver is seriously impaired, and call an emergency center for help.

The SAVE system operates thanks to various integrated systems (for further details, see Bekiaris, 1996). The most important sub-systems are the followings:

- *System monitoring driver alertness (Integrated Monitoring Unit);*
- *System which "takes the steering wheel" in case of emergency (Automatic Control Device);*
- *SAVE Warning System.*

In this field, the preliminary evaluation of users' needs, pre-acceptance and acceptability regarding driving aids, especially regarding the monitoring of drivers' alertness and automatic driving, is particularly important. It is a question of evaluating the subject's degree of approval of such and such an action in relation to the possibilities offered by a given system or product. Roughly speaking, there are two main methods of evaluating user needs and the potential acceptance. The first is basically fundamental research, such as the detailed analysis of accident data (Fontaine *et al.*, 1990; Malaterre, Fontaine and Van Eslande, 1991). The second type of needs analysis takes place during specific product development. Customer surveys are the best illustration of this (e.g., Barjonet, 1994).

On the other hand, the simple procedures based on the general feelings of users begin to be developed also (Van der Laan, Heino and De Waard, 1997). However, one can find a considerable number of criteria and results which are particularly relevant to the assessment of the general attitude of users towards specific systems (e.g., Rumar, 1987).

The first phase of user acceptance/acceptability was to clarify stakeholders' predisposition to the usefulness, the relevancy and the feasibility of such a system with respect to the SAVE system prototype concepts, even if the acceptance in the meaningful sense was not yet measured definitively. Primarily because this was the beginning of the SAVE project, and secondarily the product does not yet exist in terms of an operational tool.

Therefore, roughly speaking it could be considered that:

- On the one hand, the *a priori user Acceptance* was assessed, through the opinion of the ultimate users – standard drivers, elderly, disabled, professional/long distance, police drivers, etc.
- On the other hand, the *societal Acceptability* was tested through intermediate users such as the relevant Authorities (Driving Licence

Authorities, Medical Authorities, Police Authorities, International Legislation Authorities and Insurance companies).

Even if the distinction between the two concepts is difficult to establish, this strategy provided sufficient information in order to know in which directions the system must be improved and developed from both, the users' standpoint and societal viewpoint.

3.3.2. Methods: How Attempt to Predict and Improve Benefits

In order to cover the objectives of this kind of assessment, several convergent methods were used (Petica and Bekiaris, 1996):

- State of the art on various aspects concerning similar SAVE products or assessment methods;
- A users' needs questionnaire to make a survey in several (7) countries;
- Interviews, workshops and specialized questionnaires with various authorities representatives;
- Classical processing data for quantitative evaluation.

The survey by questionnaire was in fact the main tool for assessing ultimate users' opinions concerning different variables of driver impairment and the SAVE design, though several methods were used.

The criteria used were mainly emanating from the State of the Art survey. The main variables which have been used are the following: *1) Driver typology*: Essentially socio-economic indicators, driving practices, low alertness profile-identification of drivers likely to suffer from diminished alertness. *2) Opinions and needs: predispositions, tendencies*, attitudes towards the various concepts: Awareness and behavior, on-board new technologies, context of impairment, needs of assistance, etc. *3) Opinions and needs concerning the basic concepts* and functions of SAVE: Usefulness, acceptances, assessment of warning function, categories of drivers concerned, safety-responsibilities aspects, etc.

3.3.3. Ultimate Users' Feedback Pre-Acceptance

In order to suggest the global acceptance, some examples of synthetic results are given (for more details, see Petica and Bekiaris, 1996):

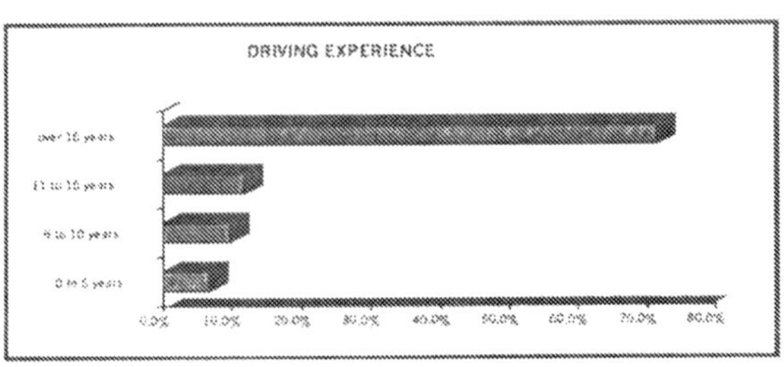

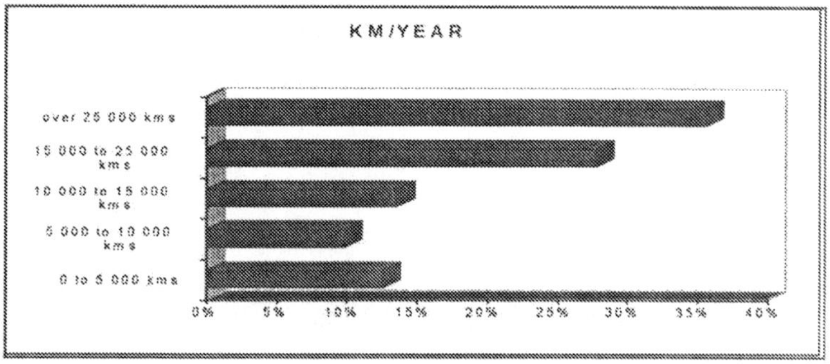

Figure 1. Classification of sample based on driving experience and km/year traveled

It would seem that the degree of trust we can place in the opinions expressed throughout the questionnaire is quite high if we consider the level of global road experience. A majority have more than 16 years experience (75 %) and drive more than 15000 km/year (a general average).

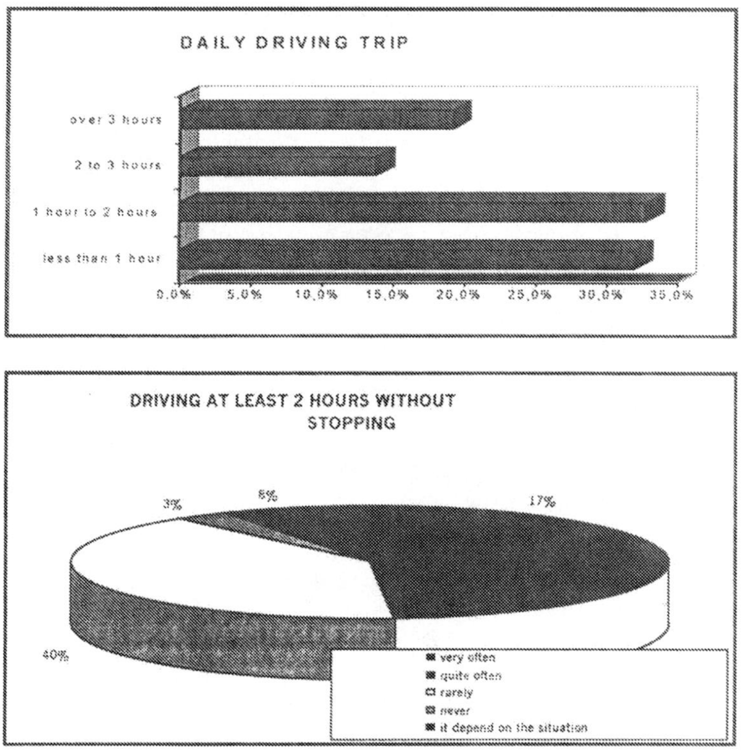

Figure 2. Classification of sample according to daily driving trips duration and behavior

The fact that 49 % of subjects drive often and fairly often (17 % and 32 %) more than 2 hours without stopping, may be interpreted as a global indicator of the social need for a system monitoring alertness, since we know what the effects are beyond this threshold, (first symptoms of fatigue).

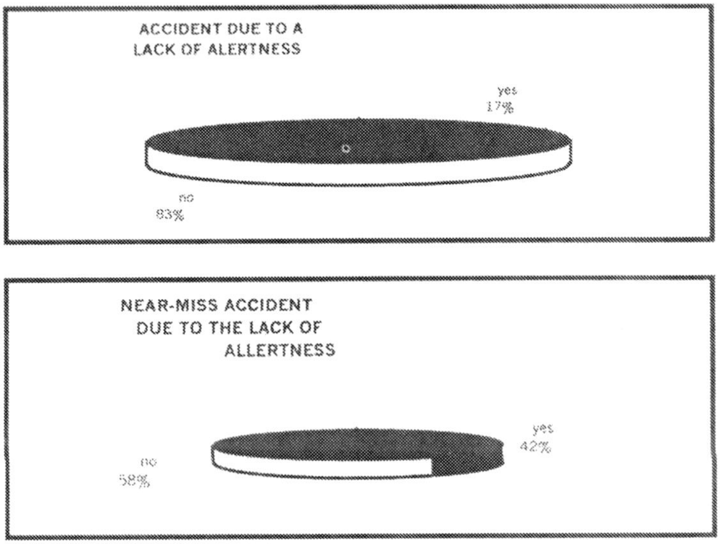

Figure 3. Rate of accidents and near-accidents (of sample)

Concerning problems of alertness, the replies received show that they are responsible for a major number of road incidents. Thus, as shown in Figure 3., 17 % of subjects state that they have had accidents due to diminished alertness and 43 % have been involved in near misses for the same reason, which seems highly symptomatic.

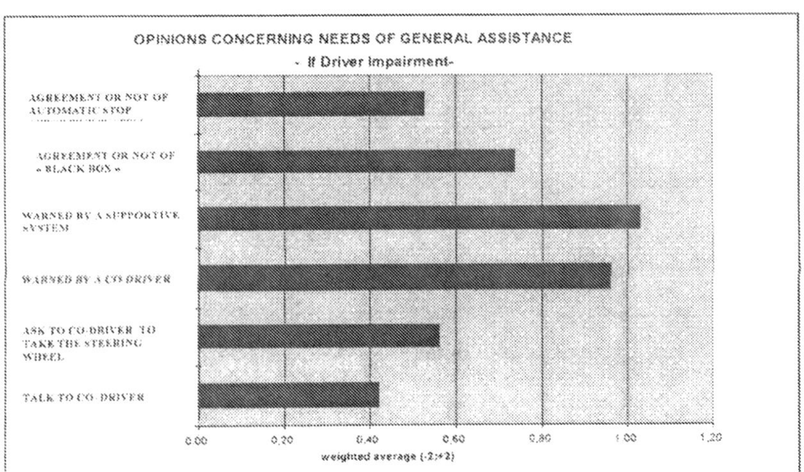

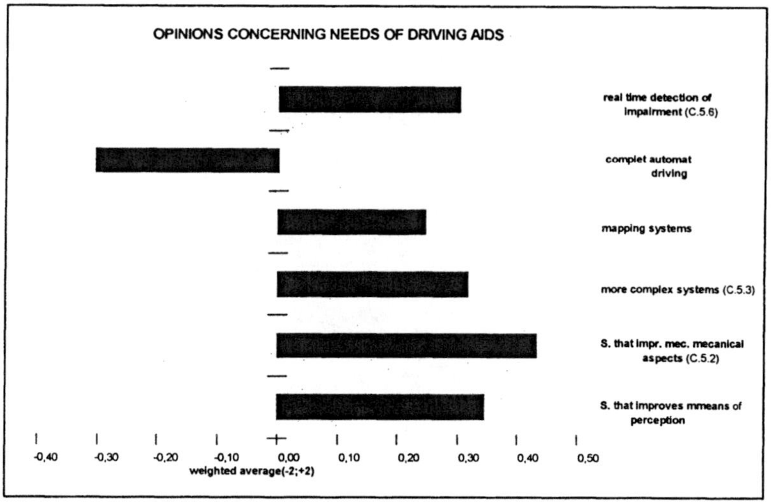

Figure 4. Opinions concerning needs of general assistance in case of driver impairment and needs of driving aids

Concerning the need for general assistance in this type of situation, the relation is again positive with preferences about the type of assistance evenly distributed. This implies that a human being and a technical system are equally trusted. Is this not excessive? The need for driving aids is slightly positive (between 0.25 and 0.45), with a definite rejection of automatic driving (-0.40). This must be taken into account during the development of SAVE.

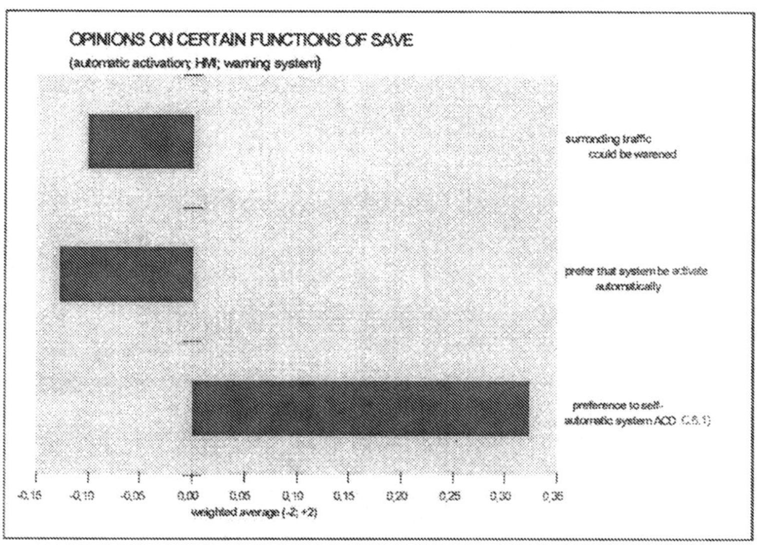

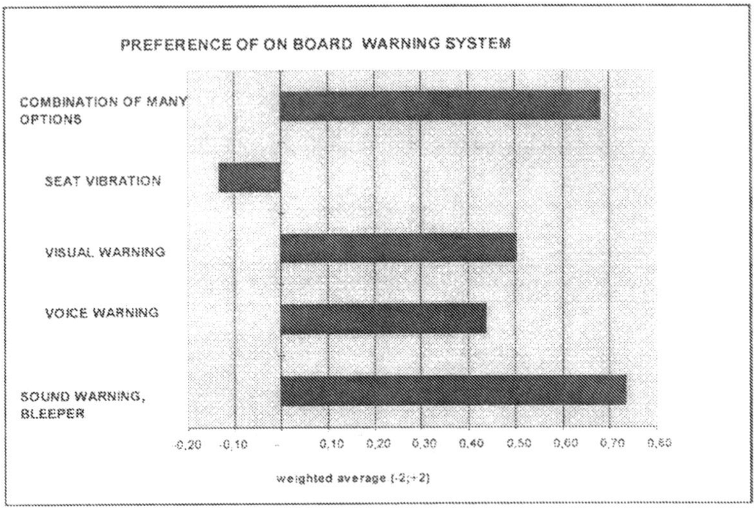

Figure 5. Sample opinion on proposed SAVE functions and regarding on-board warning systems

Figure 6. reveals the attitude of drivers in relation to the automatic functions offered by SAVE. Although a tendency to reject automatic driving is apparent in the replies to other questions, it would appear that if it is offered, it is preferable by far that the system be activated by the driver himself (0.30) and not automatically (-0.13). This has to be taken into account in the development of the system.

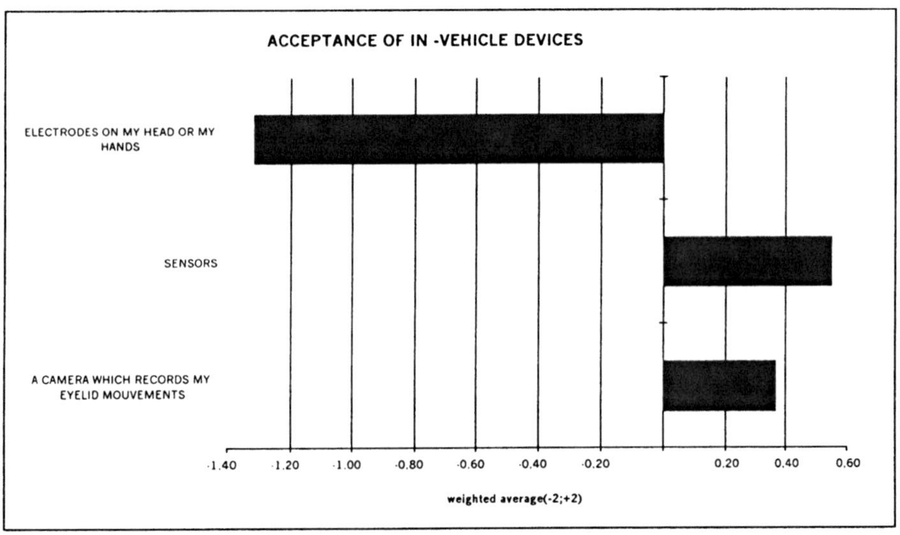

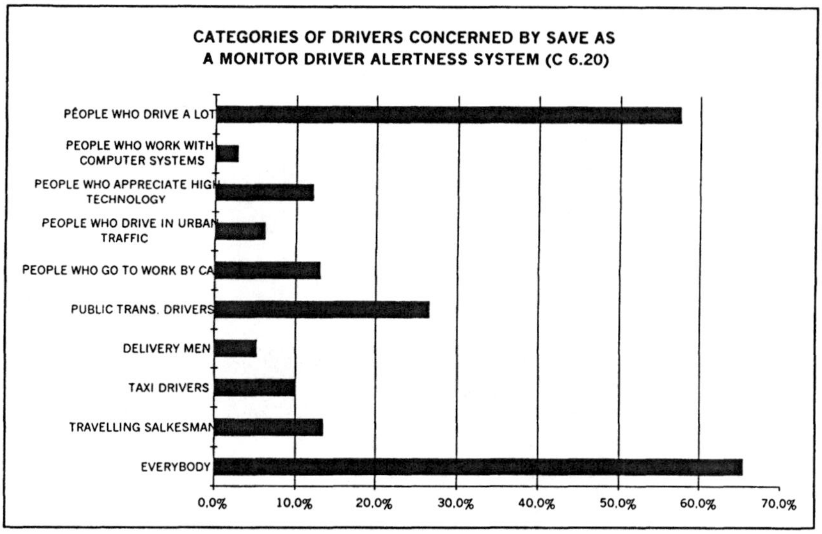

Figure 6. Users' acceptance of in-vehicle monitoring devices and categories of drivers concerned by SAVE

What is more, as might be expected, the devices which should certainly not be used to diagnose alertness are those comprising electrodes on the head or the hands (-1.35). It would be interesting to know whether this strong rejection is linked to the intrusive aspect of the apparatus, that is an invasion of privacy, or whether it is a question of aesthetics or supposed lack of comfort while driving. In contrast, sensors on the steering wheel or a camera seem acceptable, though only just (0.39; 0.59). As for the drivers most concerned by such a system, no category should be given preference. Everybody should be concerned (65 %) and especially people who drive a lot (58 %).

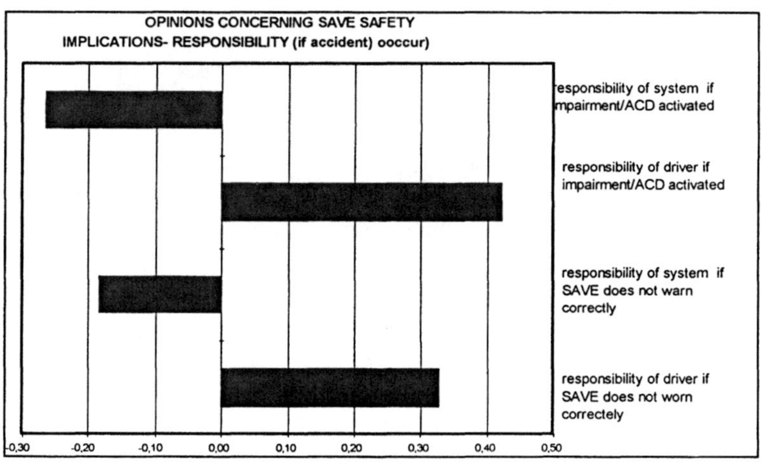

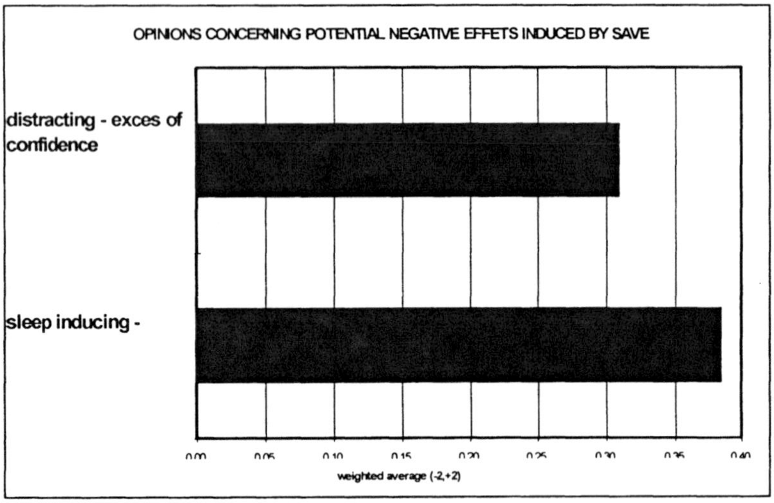

Figure 7. Users' opinion concerning SAVE safety implications/responsibility and potential negative effects

According with the first provisional conclusions of the survey, it seems that the users are attracted to this kind of device if some specific issues are integrated. The continuation of acceptance assessment on the basis of a restructured concept (as it was initially planed) is necessary. It is too early to conclude on a really acceptance and the next phases will confirm or not this tendency. In fact, the field of the new technologies suppose implicitly the generation and development of new needs where the real acceptance depends on the demonstration of the utility of a concrete product. But at the moment and for the public, this kind of device is not yet a product. It is only a concept and a sum of sub-concepts (sub -systems), involving different degrees of agreement.

3.3.4. Societal Acceptability Feedback: Through Authorities Representatives Opinion

If the different authorities or other stakeholders could be associated with the "barristers" or the representatives of the society, rather than the common users, some considerations must be raised:

- Despite the a priori social safety benefits of such a system (confirmed by ultimate users), the opinions of other intermediate users, key actors and different authorities in several European countries indicates more reservations for certain aspects (see Petica and Bekiaris *et al.*, 1996). Sometimes many important contradictions are noted between different stakeholders.

– Cultural or national differences between different actors or traffic specificity, the degree of integration/knowledge in this kind of projects, the specific type of "interest" in the field, the safety or commercial attitude, etc. could explain a very small systemic approach and consequently a very problematic degree of acceptability in this phase of development of new technologies. Briefly, and only to illustrate it, we give some issues of the workshops, interviews or specialized questionnaires performed in France:

- The difficulty in assessing acceptance is all the greater when the product is still virtual. It is extremely difficult to predict the reactions of the public to a product which does not even exist in the form of a prototype;

- It is important to distinguish between the acceptance of drivers in terms of a potential purchase and acceptance in terms of potential use in term of safety benefits;

- It seems that most stakeholders do not find the introduction of automatic driving acceptable. In SAVE, there are two distinct functions: *Driver assistance*, which would seem to be very acceptable, and *taking control* of the vehicle. The latter would seem to create a certain suspicion in users: Fear of false maneuvers, fear that the car would not be parked safely, fear of the loss of freedom of action, etc.;

- According to some experts, really users' acceptance will mainly depend on the cost/benefit ratio and essentially on its performance;

- On the technological level, the installation of black boxes in a great number of private vehicles seems rather illusory to certain experts;

- What effects (quantity, nature) the system will have on *surrounding traffic* has still to be defined. It is difficult to transpose the simulated effects on to real traffic. The reaction of non-equipped users to an automatic maneuver by an equipped car may be negative. Certain new types of disturbance may occur in traffic and fluidity may be affected;

- How the driver is alerted, where the information is displayed, has to be examined carefully, so as not to increase the mental load or affect his driving;

- In Europe, there is no legal definition of liability resulting from an accident where the presence of an on-board system could be one of the causes of the accident, but some elements seem to indicate that the driver would be responsible in case of accident;

-Several legal or sensitive problems have yet to arise; transfer of responsibility, freedom, interference with other users, limited reliability, false alarms, warning signal too early or too late, IMU not valid, limited efficiency, modification of the driving task, mental load, overconfidence, secondary effects, relationship between equipped and non equipped vehicles, conflicts among drivers, general effects (positive/negative) on safety, liability in case of undesirable events, failure of the system, accidents... –

modifications to regulations; limits, compatibility with other countries, protecting of the privacy in case of location of the vehicle, etc.

4. CONCLUSION

In conclusion it could be said that in-vehicle technologies applications may well contribute to an improvement of the traffic safety situations on the roads, if and only if the factors, the criteria and the principles already analyzed are jointly considered. Even if this process of continuous assessment could be long, difficult, slow and sometimes very expensive, it is the best way to reach the adequate balance of benefits/inconvenience. Therefore, it is suggested to:

- Take into consideration the real user needs, avoid the building of new needs by a continuous dialogue with the lay people;
- Take into consideration user acceptance, but don't forget as a systemic approach, the societal acceptability;
- Encourage any work or action oriented through ergonomic criteria toward social benefits as a whole, rather than for some categories of users or stakeholders;
- Assess the needs, the pre-acceptance degree, the general or particular opinions of the user every time new developments are envisaged;
- Consult and assess opinions of other stakeholders and non-direct end users;
- Simplify the use of the systems in order to avoid double tasks or workload increases;
- Continue offering collective, safety-relevant information for free;
- Avoid all the system implementations which could induce other side effects or unfortunate implications;
- Prevail user satisfaction and real benefits rather than economic interests or opportunities;
- Prove the usefulness of new systems for traffic safety;
- Consider any negative event observed on the road in the functioning of some technologies already introduced and take advantage of this return of experience by extension of new projects;
- Try to develop individual solutions instead of panaceas in road traffic;
- Extend only new technology applications that have proved to be efficient;
- Anticipate by a heuristic, inferential, experimental and criticism approach all the potential effects and even "rare events". Prevail rather "pessimistic scenarios" and excessive criticism than "optimistic scenarios" and unexpected inconvenient;

- Socialize, inform and warn the lay people about all knowledge concerning potential side behavior by using a new technology;
- Avoid abuse of telematics or other technologies or devices in collecting road pricing fees.

Moreover, the telematics helps or other in-vehicle devices are not in fact a panacea and will not be able to provide reliable assistance if the traffic continues to increase as heavily as it does now. In the future, in order to avoid in the future a very possible traffic collapse and to protect real traffic safety on the roads, it is vital evidently to improve the systems according with the aspects already analyzed, but at the same time increase the capacities of motorways, equally.

Once again, the best balance of benefits/inconveniences through a focus on:

- Safety traffic actions, as a main criterion;
- The general frame of Safety Culture;
- Jointly with Societal Precaution Principles;
- Through a Social and Safety Ergonomics approach,
  may induce and improve the *Societal Acceptability.*

5. REFERENCES

AGNEW, W. G. (1988), "Future Personal Ground Transportation", *Symposium of Southern California's Future,* Los Angeles.

ALLPORT, G. W. (1935), "Attitudes", in *A Handbook of Social Psychology,* vol.2., Worcester, Massachusetts: Clark University Press.

ALM, H., NILSSON, L. (1990), *Changes in Driver Behavior as a Function of Handsfree Mobile Telephones: A Simulator Study*, DRIVE Project V1O17, Report No. 47, 1990.

BARJONET, P. E. *et al.* (1994), *SARTRE, Social Attitudes to Road Traffic Risk in Europe,* Report on principal aspects, Paris: INRETS.

BARHAM, P. A. J. *et al.* (1993), *In-Vehicle Driving Tests-Research Protocols for EDDIT (part one), Elderly and Disabled Drivers Information Telematics*, Report V203/EDDIT/ Deliverable 16, Commission of the European Communities, Cranfield University, Bedfort, U.K.

BEKER, S. *et al.*, (1995), "User Acceptance and Willingness to Pay for Advanced Driver Support Systems", *TRAFFIC Technology Europe '95*, Berlin, 6-7 April 1995.

BEKIARIS, E., LYMBINAKI, M. (1996), "State of the Art", *SAVE internal report.*

BROOKHUIS, K. A. *et al.*, (1991), "The affects of mobile telephoning on driving performance", *Accident Analysis Prevention*, vol. 23, No. 4.

CROSBY, P. *et al.* (1993), "A Primer on Usability Testing for Developers of Travel Information Systems", *Transportation Research C2*, 143-157.

CHOFFRAY, J. M., LILIAN, G. L.(1980), *Awareness and Feasibility in Market Planning for New Industrial Products*, New York: Prentice Hall.

DALY-JONES, O., BEVAN, N., THOMAS, C. (1997), *Handbook of User-Centered Design,* INUSE project (IE 2016), Del. 6.2.1.

ECMT, Comittee of Deputies (1992), *Safety Aspects of On-Board Equipment*, group on transport, computers and telecommunications, CEMT/CS/TCT (92) 5, Restricted.

ECMT, Council of Ministers (1994), *Ergonomics and Safety of In-Vehicle Information Systems*, CEMT/CM (94) 20, Restricted.

ERKE, H., MINAS, G., ZIMOLOG, B. (1983), "Arbeitsbeanspruchung, Gefahrdungsinshatzung und Arbeitssicherheit an militarischen Arbietsplatzen", *Wehrpsychologische Untersuchungen,* Bd. 1.

FARMER, F. R. (1967) "Reactor Safety and Sitting: A Proposed Risk Criterion", *Nuclear Safety*, n°8.

FISCHHOFF, B. (1997), "Acceptable Risk – A Conceptual Proposal", *Risk Journal,* Vol. 8, n° 3.

FLEMING, A. M., MC LEOD, J. (1997), *A Structured Approach to Evaluating Telematics Applications*, MEGATAQ project (TE 2007), adapted from del. S4. 2.

FRANKLIN, P. (1997), RAPA, Risk Assessment and Policy Association, Law Center.

ICE Ergonomics Limited (1993), *The Design of In-Vehicle Information Systems,* Loughborough, internal report, draft.

ISO/DIS 13407, *Human Centered Design Processes for Interactive Systems*, Draft international standard, Voting begins on 1997/08/21, International Organization for Standardization.

ISO91 – ISO (1991), *Information Technology, Software Product Evaluation, Quality Characteristics and Guidelines for their Use*, Technical Report 9126, International Organization for Standardization.

LAAN, VAN DER, J., HEINO, A., DE WAARD, D. (1997), "A Simple Procedure for the Assessment of Acceptance of Advanced Transport Telematics", in *Transportation Research,* C, Vol. 5, n° 1, Elseveir Science Ltd.

KING, M. (1996), "Human Factors and User Acceptability ", in COLE, R. A. (ED.), *Survey of the State of the Art in Human Language Technology*, Oregon Graduate Institute of Science and Technology, USA.

KUIKEN, M. J., GROEGER, J. A. (1993), *Effects of Feedback on Driving Performance at Cross Roads and on Curves*, Report VK 93-12, Haren, University of Groningen, Traffic Research Centre, The Nederlands.

LEMAINE , J. M. (1972), "Initiation aux échelles d'attitudes", *Bulletin de Psychologie*, 25, 1971-1972.

LEPLAT, J. (1985), *Erreur humaine, fiabilité humaine dans le travail*, Paris: Armand Colin.

LERNER, N. D., KOTWAL, B. M., LYONS, R. D., GARDNER-BONNEAU, D. J. (1996), *Preliminary Human Factors Guidelines for Crash Avoidance Warning Devices*, NHTSA, U.S. Department of Transportation, DOT HS 808 342, Interim Report.

LORIGNY, J. (1996), "Une approche systémique et cognitive de la science en crise", *Revue internationale de Systémique,* vol. 10, n° 4.

Secrétariat général de la défense nationale (1992), *Note d'information sur l'évolution du radiotéléphone*, Direction STS, Sous-direction VST.

NOY, Y. I. (1997), "ITS Safety Test and Evaluation", *ITS Congress*, Berlin, 1997.

NOY, Y. I. (Ed.) (1997), *Ergonomics and Safety of Intelligent Driver Interfaces*, Lawrence Erlbaum Associates Inc.

MALTBY, D. (1996), *Checklist for Preparing a Draft Validation Plan*, CONVERGE project (TR 1101), Del. DV Q3.2, Telematic Application Programe, Transport Sector.

MALATERRE, G., FONTAINE, H., ESLANDE, VAN, P. (1991) *Analyse des besoins des conducteurs à partir de procès-verbaux d'accidents: Evaluation a priori des fonctions PROMETHEUS*, rapport INRETS n° 139, Paris: INRETS.

MALATERRE, G. (1987), *Les activités sous contrainte de temps: Le cas des maneuvres d'urgence en conduite automobile*. Thèse de Doctorat de 3ème cycle, Université Paris V.

MICHON, J. A. (Ed.) (1993), *Generic Intelligent Driver Support*, London: Taylor &Francis.

MONTMOLLIN, DE, G. (1984), "Le changement d'attitude", in MOSCOVICI, S. (Ed.) *Psychologie sociale*, Paris: PUF.

MONTMOLLIN, D, M., (1986), *L'intelligence de la tâche, Eléments d'ergonomie cognitive*, Paris: Peter Lang.

MORITA, K., MASHIKO, J. *et al.* (1996), "Legibility of Displayed Place Names by In-Vehicle Navigation Devices", *Journal of Light and Visual Environment*, Vol. 20, n°1.

NILSSON, L., ALM, H. (1991), *Effects of Mobile Telephone Use on Elderly Drivers Behavior – Including Comparisons to Young Drivers' Behavior*, DRIVE Projet V1O17, Report n° 53.

OCDE (1994), "Recherche en matière de routes te de transports routiers", *Amélioration de la sécurité routière grâce à la modification des attitudes*, Paris: OCDE.

PETICA, S., Bluet J. C. (1989), *Rapport sur l'enquête internationale concernant le radiotéléphone et la sécurité routière*, Rapport INRETS, DOC- 126224, SE 303.

PETICA, S. (1991), *Utilisation d'écrans à bord des véhicules*, Report INRETS, DOC-SE 329.

PETICA, S. (1995), "Safety Implications of In-Vehicle Devices-Risk Assessment of In-Car Radiotelephone Use", *Second world Congress in ITS*, Yokohama.

PETICA, S., BEKIARIS, A. *et al.* (1996), *Driver Needs and Public Acceptance of Emergency Control Aids*, SAVE project (TR 1047) Del. 3.1, Commission of the European Communities.

REASON, J. (1993), *L'erreur humaine*, Paris: PUF.

REDELMEIER, D. A., TIBSHIRANI, R. J. (1987), "Association between Cellular-Telephone Calls and Motor Vehicle Collisions", *The New England Journal of Medicine*, Vol. 336, n°7.

RISSER, R., LEHNER, U. (1997), "Evaluation of an ACC (Autonomous cruise Control) System with the Help of Behavior Observation", *ITS Congress*, Berlin.

ROBIN-PRÉVALLÉE, Y. *et al.* (1996), *Guidebook for User Needs Analysis*, CODE project (TR 1103), Telematic Application Program, Transport Sector.

ROTHENGATTER, J. A. *et al.* (1991), *Social Acceptance of Automatic Policing and Information Systems*, Report V I 033/DRIVE/Deliverable 7, Haren, The Netherlands: Trafic Research Center, University of Groningen.

RUMAR, K. (1987), *Social Impact, Acceptance*, PRO-GEN LP 71250.

SAAD, F., DELHOMME, P., ESLANDE, VAN, P. (1992), *Analyse des comportements en situation réelle de conduite: Le franchissement d'intersection*, Rapport de recherche INRETS, n° 158, Paris: INRETS.

SCHOFER, J. L. *et al.* (1993), "Behavioral Issues in the Design and Evaluation of Advanced Traveler Information Systems", *Transportation Research,* C2, 107-117.

SCOTT, W. A. (1968), "Attitude Measurement", in LINDZEY G., ARONSON, E. (Eds.), *Handbook of Social Psychology*, Reading, Mass: Adisson-Wesley.

SLOVIC, P., FISCHOFF, B., LICHTENSTEIN, S. (1982), "Why Study Risk Perception?", *Risk Analysis*, Vol. 2, n° 2.

SMILEY, A., BROOKHUIS, S. K. A. (1987), "Alcohol, Drugs, and Traffic Safety", in ROTHENGATTER , J. A., DE BRUIN, R. A. (Eds.), *Road Users and Traffic Safety,* Assen: Van Gorcum.

SMILEY, A. (1995), "Overview and Methodologies of the ITS Safety Evaluation Process", presented at the *Workshop on ITS Safety Evaluations*, ITS America, Reston.

STARR, C. (1969), "Social Benefit versus Technological Risk", *Science.*

TUTTLE, T. C., *et al.* (1974), *Psychological-Behavioral Strategies for Accident Control,* Report for the National Institute for Occupational Safety and Health, Contract n° HSM-99-73-61, Cincinnatti.

THIRUMALAI, K. (1995), "Strategies for Evaluating ITS safety Concepts for Marketable Products", *Second world Congress in ITS*, Yokohama, 2390.

THURSTONE, L. L. (1931), *The Measurement of Social Attitudes*, Chicago: University of Chicago Press.

WINKLER, R., NOWICKI, J. (1997), "Experience with the Implementation of On-Board Navigation Systems", *ITS Congress,* Berlin.

ZHANG, X. *et al.* (1996), *Guidebook for Assessment of Transport Telematics Applications*, CONVERGE project (TR 1101), Del. DV Q5.1, Telematic Application Program, Transport Sector.

ZIMMERMAN, C. A., ELLIOT, C. A. (1995), *ITS User Acceptance Research*, U.S.DOT, ITS America, Annual Conference Paper.

ESLANDE, VAN, P., ALBERTON, L. (1997), *Scénarios-types de production de "l'erreur humaine" dans l'accident de la route*, rapport de recherche INRETS, n°208, Paris: INRETS.

VALLET, M. (1991), "Les dispositifs de maintien de la vigilance des conducteurs de voiture", in VALLET, M. (ED.), *Le maintien de la vigilance dans les transports,* Caen: Paradigme.

WU-CHIEN, J. S., APOSTOLAKIS, G. (1981), "On Risk Aversion in Risk Acceptance Criteria", *Reliability Engineering,* n°2, 45-52.

PART FOUR
TRAFFIC PSYCHOLOGY IN PRACTICE: SOME SELECTED EXAMPLES

Chapter 16

DRIVER SELECTION

Psychological Assessments on Driver Aptitude as a Means of Driver Selection and Road Safety Measure in Germany

Iris Schanz-Lorenz
Psychologist, Stuttgart, Germany

1. INTRODUCTION

Have you ever been driving a car on a German "Autobahn" with cars overtaking you almost as fast as lightning? How dare you, not moving over to the right lane instantly in order to give way to one of those fast going 250 HP strong vehicles!

Have you ever tried to figure out the Italian system of road lanes? You think you are traveling on the only lane there is, while numerous cars on your left and right travel on imaginary lanes about 2 mm away from your outside mirrors. If you have survived the City of Rome, try to get on a circular road in Palermo, Sicilia!

French drivers seem to be more of the calm sort. They are able to travel slowly along a lonely country road, only speeding up when reaching a dangerous crossing.

You might suspect that everybody except yourself is guilty of reckless driving. According to statistics, less than 5 % of the driving population is registered due to driving offenses such as speeding, forcing one's way or driving while intoxicated. Moreover, it is only a small part of 3 % committing road offenses repeatedly (Jagow, 1995). Those drivers, already registered with a number of offenses are likely to relapse as a result of lasting maladjusted attitudes and habits. Persons registered with a score above 24 on a German road survey (0 points indicating zero registered traffic incidents) are likely to an almost 50 % recidivism rate (43.9 %) (Utzelman, 1985).

Discussing traffic psychological aspects, it is important to consider the large number of unregistered traffic offenses where non-appearing consequences contribute to a "wrong" way of learning, stabilizing negative habits and faulty behavior. Every risky car trip might consolidate negative behavior patterns and be judged "successful" by the individual if it has not resulted in an accident or denunciation. Recidivist road charges and an increase in traffic accident participation cannot exclusively be explained by the law of probability or situational factors. Road survey studies indicate distinct road accidents under almost identical driving conditions, infrastructure, rules and regulations, social, financial and economic background. There is good reason to presume that an increase in road accident participation is due to personality factors originating from individual preconditions and motivation structure. A reduced sense of accepting public rules and regulations might reflect a way to cover up an inappropriate sense of self-esteem, mistreating the rights of others. Aggressive tendencies can easily be acted out while participating in public traffic since the aggressor often stays anonymous. He trusts in the assumption that his own aggressive actions towards other road users will rarely result in sanctions such as license withdrawal.

The field of traffic psychology in Germany started out around the turn of the century until the end of World War II by selecting a positive sample of professional truck drivers and airplane pilots. With mass motorization after the end of World War II until the beginning of the 1970's, a negative selection of registered road offenders took place (Kroj, 1992). In cooperation with medical doctors, the professional psychologists working in the field of traffic safety started to establish psychodiagnostic criterions and methods in order to increase road safety measures (Kroj, 1992). A long way led to the integration of psychological and medical guidelines for assessments on driver aptitude (Müller, 1988). The Joint Advisory Council for Traffic Medicine at the Federal Ministry of Transport, Building and Housing and the Federal Ministry of Health agreed to the publication of these guidelines on April 21, 1999.

2. PUBLIC DEMANDS

It is the public's interest to see to a high standard of road safety measures in a specific country. Diagnosing individual misconduct and inappropriate behavior in traffic participation on the basis of a medical and psychological assessment on driver aptitude can be beneficial to both the individual and the public. The person who has committed road offenses repeatedly might get a better understanding of the underlying problematic psychodynamic behavior resulting in a criminal record. Ideally, ways of rehabilitation are pointed out to the candidate to reinstall the right or the aptitude to drive a vehicle after license revocation. Ways of rehabilitation can include individual psychotherapy, counseling, driver improvement measures (see Kroj and Dienes in this volume) or clinical treatment.

2.1. Police File

On the basis of the police file, the official administrative board of license affairs states doubts upon a person's aptitude to conduct a vehicle in public. The board of license affairs is in the position to grant or revoke driver documents according to the individual's conduct. In the case of misconduct, the administration board states serious doubts upon this person's driver aptitude. It is the traffic psychologist's task to transform these doubts into scientific hypothesis testing operations.

In cooperation with a medical expert, the traffic psychologist works towards clarification of these uncertainties by trying to give an answer to the questions raised by the administrative board. In posing the individual task, the board of license affairs considers past legal offenses. Fatal accidents are most frequently caused by drivers consuming illegal drugs or persons driving under the influence of alcohol.

2.2. Questioning

– Drunken driving

In the case of drunken driving the task for the traffic psychologist and her medical colleague will consist in discussing the following question posted by the administrative board:

"Is the person in question likely to commit a future drinking and driving offense or is he/she suffering impairment from past alcohol misuse which might influence driving abilities?"

Possibly, the candidate is still in possession of a valid driver document. The administrative board of license affairs must come to a decision upon the offender's right to drive a vehicle on public roads in the future after having failed in the past. Proving his/her aptitude and ability to take a responsible part in traffic participation is the task of the candidate himself. Both the traffic psychologist and her medical colleague end up with an expert opinion they put to disposal in a written report. The psychological assessment on driver aptitude provides information the board needs to come to a decision.

– Traffic violations

A person found guilty with a number of traffic violations lacking direct alcohol and/or drug consumption might be subject to the following official questioning the experts – traffic psychologist and physician – need to discuss:

"Is the person in question likely to break road safety rules and regulations in future?"

– Illegal drug consume

Drivers registered with an illegal drug offense might be confronted with the following:

"Is the person in question likely to drive a vehicle under the influence of drugs and/or is she/he suffering impairment from past drug abuse?"

The list of questions is to be extended for each individual impairment, double questioning is often realistic. It is important to realize that a pre-selection has already taken place. In the case of the alcohol drinking driver, traffic offenders with a blood alcohol concentration above 0,16 % BAC or recidivists with numerous drinking and driving offenses are the candidates for the psychological assessment. Going through a red light once is not sufficient to suspect unsocial, self-centered misconduct. A number of traffic violations need to occur before the administrative board can ask the offender to prove his/her aptitude or capacity by the support of a psychological assessment.

3. GUIDELINES FOR ASSESSMENTS ON DRIVER APTITUDE

Psychological assessments on driver aptitude can be asked for by the official board of license affairs under specific circumstances listed below (numbers and paragraphs in brackets referring to the German licensing law – Fahrerlaubnisverordnung FeV) (Argerion, Mc Cartley and Blacker, 1985; Moser, 1983). In general, it is the psychologist who draws responsible for the assessment. The medical personnel will add information, supporting psychological findings on the grounds of a medical check-up. When assessing distinct medical "cases," i.e., prognosing the driver aptitude for a person who suffered from an apoplectic fit, it is the medical colleague's task to discuss the officially induced questions. The traffic psychologist adds in return his psychological findings. In the following, it will be pointed out if the psychologist or physician signs responsible.

3.1. Assessments on Driver Aptitude *Can* Be Asked for if there are:

– Doubts towards somatic or psychological aptitude;
  Medical assessment (§ 11 Abs. 2 Satz 1 FeV);

Psychological assessment (§ 11 Abs. 3 Nr. 1 FeV) if other assessments are not sufficient;

- Doubts regarding visual sight;
  Medical assessment (§ 12 Abs. 8 Fev);
- Exception from the minimum age regulation;
  Psychological assessment (§ 11 Abs. 3 Nr. 2 FeV);
- Doubts based on weak results in the practical driving test (§ 18 Abs. 3 FeV);
  Psychological assessment (§ 11 Abs. 3 Nr. 3 FeV);
- Doubts based on a specific criminal record;
  Psychological assessment (§ 11 Abs. 3 Nr. 4 FeV);
- License application after several license withdrawals;
  Psychological assessment (§ 11 Abs. 3 Nr. 5 Buchstabe a FeV);
- License application after license withdrawal due to specific criminal offenses;
  Psychological assessment (§ 11 Abs. 3 Nr. 5 Buchstabe b FeV);
- Traffic offenses within 2-year probation period under certain conditions;
  Psychological assessment (§ 2a Abs. 4 Satz 1 StVG);
- Illegal possession of narcotic drugs;
  Medical assessment (§ 14 Abs. 1 Nr. 1 FeV);
- Infrequent consumption of cannabis;
  Psychological assessment (§14 Abs. 1 Satz 3 FeV);

3.2. Assessments on Driver Aptitude are *Usually* Asked for by the Board of License Affairs under the Following Circumstances:

- Reapplication after license withdrawal due to a high score within the German penalty system (i.e., a score of 18 is penalized with a 6-month license suspension);
  Psychological assessment (§ 4 Abs. 10 Satz 3 StVG);
  Exception (only Driving School Seminars): §4 Abs. 11 Satz 3 StVG;
- Traffic offenses committed by a person within the 2-year probation period after prior license withdrawal;
  Psychological assessment (§ 2a Abs. 5 Satz 5 StVG);

3.3. Assessments on Driver Aptitude will be Obligatory under the Following Premises:

- Exceptions from the minimum age regulation due to professional training;
  Psychological assessment (§ 10 Abs. 2 Satz 2 FeV);
- Alcoholic dependence;

Medical assessment (§ 13 Nr. 1 FeV);
- Alcoholic abuse;
  Psychological assessment (§13 Abs. 1 Nr. 2 Buchstabe a FeV);
- Repeated traffic violations under the influence of alcohol;
  Psychological assessment (§ 13 Abs. 1 Nr. 2 Buchstabe b FeV);
- Drinking and driving above a certain blood alcohol concentration (> 0,16 % BAC);
  Psychological assessment (§13 Abs. 1 Nr. 2 Buchstabe c FeV);
- Reapplication after license withdrawal due to reasons listed under 15-17;
  Psychological assessment (§13 Abs. 1 Nr. 2 Buchstabe d FeV);
- Clarification, if alcohol misuse subsided;
  Psychological assessment (§13 Abs. 1 Nr. 2 Buchstabe d FeV);
- Dependence on illegal drugs;
  Medical assessment (§14 Abs. 1 Nr. 1 FeV);
- Consumption of illegal drugs;
  Medical assessment (§14 Abs. 1 Nr. 2 FeV);
- Misuse of psychoactive prescription drugs;
  Medical assessment (§14 Abs. 1 Nr. 3 FeV);
- Regranting of application after prior license withdrawal due to reasons listed under 20-22 as well as 9 and 10;
  Psychological assessment (§14 Abs. 2 Nr. 1 FeV);
- Clarification if the person is still suffering drug dependency or consuming illegal drugs or consuming psychoactive substances;
  Psychological assessment (§14 Abs. 2 Nr. 2 FeV);

StVG: Straßenverkehrsgesetz
FeV: Fahrerlaubnisverordnung
Abs: Absatz
Nr.: Nummer

4. PROFESSIONAL QUALIFICATION OF THE TRAFFIC PSYCHOLOGIST

A 200 hours postgraduate training enables a solid professional qualification in the field of traffic psychology. The training curriculum includes basic scientific knowledge as well as specific practical reference. The postgraduate training is divided into seminars each containing 4-36 periods with a 45 minutes duration. About 2/3 of the total training period is dedicated to the acquirement of specific knowledge of the assessments on driver aptitude (legal procedure, methodology, guidelines, traffic medicine, quality management and other topics).

5. THE ASSESSMENT

5.1. Background Information – Analyzing Official Files

It is necessary to state the facts and figures which find consideration within the written report. Official files can include police records, previous assessments, penalties outside road traffic regulations, robbery, assault or willful destruction and others. Both the psychological and medical expert find the official files to their disposal before the actual assessment takes place. Previous documents might lead the traffic psychologist to assume that possibly behavior adaptation difficulties manifest themselves which cannot been seen in isolation. Instead, general attitudes towards social values and/or overemphasis of subjective needs can be detrimental for that individual's adjustment.

Empirical support stating the interaction of general criminal offenses and traffic violations is given by Argerion *et al.*(1985). Reason enough for the traffic psychologist to consider official records in his psychological preceding although personal data must be handled with care.

5.2. Third-Hand Information

The candidate is asked to prove his responsible behavior in traffic participation or diminish the doubts expressed by the administrative licensing board. Demonstrating his/her change in behavior can include the proof of unproblematic drinking patterns by means of normal blood measures. Even not necessarily this can indicate a change in alcohol consumption. For the former drug consumer a negative urine sample indicating a presently drug free life can help to demonstrate a change in lifestyle.

The offender is asked to reflect upon his personal development and psychological status. In order to support his efforts he could have arranged individual psychotherapy or attended group measures such as Driver Improvement (see Kroj and Dienes in this volume) to reduce the probability to relapse. In both cases, an informal attendance sheet with the signature of the psychotherapist or seminar leader can be helpful in order to evaluate the candidate's personal changes.

5.3. What Is Expected Of The Candidate?

Depending upon the official assessment task the traffic psychologist must consider a number of presuppositions and well established scientific findings the candidate must live up to. On the basis of the drinking and driving offense in combination with the measured blood alcohol concentration it becomes obvious that the person in question has developed a detrimental drinking pattern prior to the drinking and driving incidence. This problematic drinking pattern can result in another risk taking behavior. The risk to cause an accident with a blood alcohol concentration (BAC) of 0.08 % accelerates times four compared to sober driving (National Research Council, 1987) and raises dramatically with an increase in BAC: A measured blood alcohol concentration of 0.15 % may increase the accident risk up to 25 times (Müller, 1993). On the other hand, there is empirical evidence for marijuana consuming persons to be over represented among accident involved individuals (see, e.g., Daldrup *et al.*, 1987; Kauert, 1992; Brenner-Hartmann, 1995).

Depending on the measured BAC level after the drinking-driving incidence, research indicates an increased habituation to drinking large quantities of alcohol long before the offence takes place. These quantities are well beyond the social acceptable consumption level. Keeping control might progressively diminish. There is good indication that many drinking and driving situations took place before the actual registered offence (Stephan, 1986). Even first-time offenders with measured BAC levels above 0.16 % have developed a problematic drinking behavior and considerable physical alcohol toleration. Consequent effects in performance on drinking related tasks may even show without actual alcohol intake.

5.4. Prognosis

A positive prognosis on future traffic participation of the specific person must meet a number of criteria (Kroj, 1995). The candidate must be able to show a change in consumption patterns towards an unproblematic handling of the substance. Individual diagnosis can suggest a reduction in drinking amounts, finding alternatives in times of personal crisis or social gatherings. In the case of alcoholic addiction a total substance abstinence is indicated. With the infrequent user of cannabis, he must be able to show his ability to separate his cannabis consumption from driving a vehicle.

Depending upon the candidates status the experts suggestion to the administrative board of license affairs includes the revocation of the right to drive a vehicle in public traffic or having a drivers license reinstalled after withdrawal. It is exclusively the job of the license administration or the justice to grant or revoke driving permissions.

Two or three weeks after the date of the assessment, when all the information and laboratory results are readily available to the traffic psychologist he sets up a written report of findings, adding the results of the physician. It is up to the candidate himself to produce the results at the official board of license affairs. Often, candidates with a negative prognosis at present prefer to keep the results to themselves or wait for necessary changes in behavior to stabilize long enough before turning in the assessment at the administrative board.

5.5. Hypotheses Testing Procedure

For the professional traffic psychologist there are a number of steps to consider and work through before coming to a final conclusion and suggestion. The psychologist needs to transform the doubts and questions raised by the administrative board into psychological relevant working hypotheses (Nickel, Utzelmann and Weigelt, 1990).

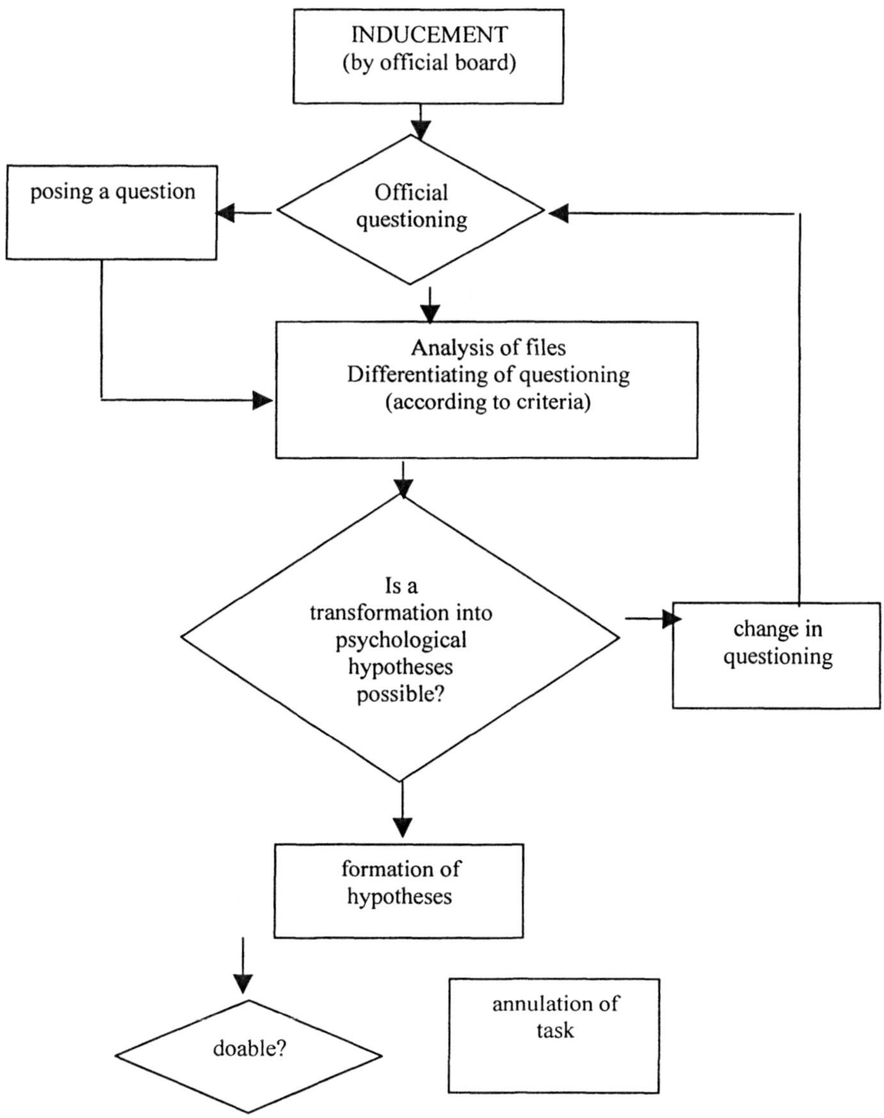

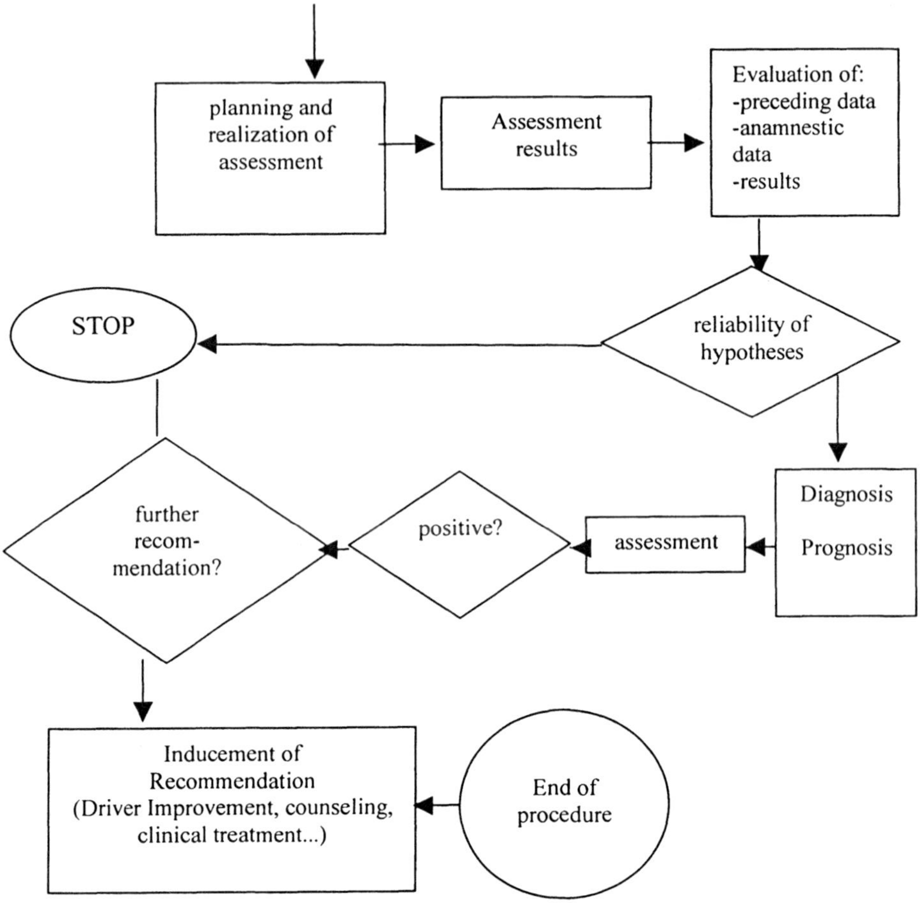

Source: Nickel, W.-R., Utzelmann, H.-D., Weigelt, K.-G., Schematisch Darstellung des Ablaufs- und Entscheidungsprozesses bei verkehrspsychologischen Fragestellungen, Abb. 2, S.42, 1990.

One by one the traffic psychologist proceeds down the scheme. Working with a DWI-driver (driving while intoxicated) the first hypothesis must consider the drug toleration of the candidate. Does the candidate realize his drug habituation? Is he realistic about his past drug abuse?

5.6. Negative Outcome

An almost 40 % negative prognosis of the psychological assessments on driver aptitude might be the result of unrealistic views of the past drug consummation habits. The candidates often report the consumption of having "a beer now and then" without getting a realistic viewpoint of the actual intake or even misuse in the past. In creating an honest and friendly assessment atmosphere the traffic psychologist can react to non-obvious information. In order to come to a valid result the psychologist must be able to understand the drug toleration on the basis of the candidate's statement. By getting an objective point of view the client might be able to realize his past or present drug problem which could enable him to change the detrimental behavior.

5.7. Illegal Drug Consumption

In the case of an illegal drug consumer, psychological hypothesis testing needs to include the discussion if the candidate is willing or able to sustain a drug free lifestyle. Even when producing a negative urine sample on the date of the assessment a total drug abstinence is not guaranteed. The psychologist has to be able to get a clear picture of the drug consumer's past and present habits: Types of drugs consumed, intensity of drug abuse and socio-psychological background information. All this information will add to the hypothesis testing procedure.

A combination of different drugs consumed are common. The dangerous effect of combining alcohol intake and THC (haschisch or dagga) is especially detrimental to driving related tasks with an increase in accident involvement (Daldrup *et al.*, 1987).

Trying to estimate the recidivism rate of drug abuse, it is important to consider the non-restraining effects of alcohol intake. Illegal drugs often are consumed by persons already known as heavy alcohol drinkers (Hurrelmenn and Hesse, 1991). The probability to consume different illegal drugs after habituation of cannabis intake accelerates dramatically (Geschwinde, 1991). Social surveys provide an indication that only 1/5 of the habituated haschisch consumers are exclusively THC users (Kleiner *et al.*,1992). A sample of persons involved in road accidents showed up to 25 % cannabis intake prior to driving (Kauert *et al.*, 1992). An increase of road accident involvement is based on a variety of effects the consumers may not be able to influence after consumption. Drug concentration and/or absorption may influence consumer effects. Different effects may occur even when dosage or situational factors are kept at a constant level. Personal factors, pre-experiences, attitudes, personal motivation or multiple consumption of other drugs can influence the drug experience. Unlike the decomposition of the blood alcohol concentration THC reduction takes place in irregular intervals. Unpredictable flash-back phenomena are well described (Täschner *et al.*, 1994). A number of complex mixtures of pharmacological, medical, psycho-functional and personal motivation may influence the future drug intake.

Comparing the effects of alcohol consumption and illegal drug intake prior to a driving task is confusing (Ladewig, 1987). The only valid indicator for a positive prognosis is given by a total drug free lifestyle.

5.7.1. Heroin Consumption and Traffic Participation

The consumption of heroin in combination with traffic participation is detrimental on the grounds of an acute intoxication resulting in an euphoric and sedative nature. Apathic behavior and unrealistic performance perceptivity usually follow heroin consumption. Information processing is slowed down, attention and concentration skills are affected. With a progression in addiction rules and regulations are neglected. Convulsion, vomiting, restlessness and anxiety may accompany yielding effects. The need to provide more drugs often result in car rides although the person lacks self-control and adequate attention in driving.

5.7.2. Cocaine Consumption and Traffic Participation

The consumption of cocaine and amphetamines does have comparable effects in traffic behavior: The subjective feeling of increased performance can lead to a risk taking behavior without realizing personal limitations. Cocaine is rarely used exclusively. Instead, a polytoxic drug consuming pattern is common. With the decreasing effects and the "coming down" phenomena, paranoia, clinical depression and even suicidal actions may appear. Responsible behavior in traffic participation subsides.

Catamnestic studies after clinical treatment indicate a high risk in relapse. A cross European study estimates a 23-43 % success rate resulting in a drug free life after therapy (Brenner-Hartmann, 1995). According to the psychological guidelines on drivers' aptitude assessments, the experts must be able to state if the person in question is able to abandon illegal drug consumption.

5.8. Diagnostic Interview

At the beginning of the psychological interview it is necessary to explain the inclination and objectives to the candidate. He must understand the doubts expressed by the administrative board. It is essential to discuss the underlying assumptions discussed above. Finally, the client must understand the conditions allowing the experts to come to a positive prognosis.

A friendly and open-minded atmosphere can allow the candidate to feel as comfortable as possible when talking about unpleasant parts of his life. After "getting the hard facts," the candidate must be given enough time to reflect upon the faulty behavior, the problematic background as well as possible changes after the incident(s) and future intentions. A 60-90 minute individual diagnostic interview will allow the professional traffic psychologist to work through the hypothesis testing scheme he/she keeps in mind.

It is the psychologist's task later on to see how well details and pieces of information fit into an overall picture. In reaching a final conclusion the psychologist considers the official file, police records, psychometric tests and written questionnaires. This data adds to the information gathered in the psychological diagnostic interview.

To avoid misunderstanding and to obtain reliable data it is vital to cross-question and provide feed-back on the conclusions the expert draws. This will allow the candidate to react instantly.

5.9. Medical Assessment

A medical assessment is carried out by a physician according to the questions posted by the administrative licensing board. In the case of previous drug abuse the drawing of blood samples or urine testing can become necessary. Sight and coordination tasks help to test the candidate's driving aptitude. Anamnestic data as well as previous drug consumption patterns are considered by the physician before summing up the findings.

The medical guidelines for assessments on driver aptitude describe criteria on fundamental assessment indications, persons with motor system disabilities, hypotonia, coronary heart disease, cardiac output deficiency due to congenital or acquired heart defects or other causes, consequences of injuries to and diseases of the spinal cord, parkinsonism and other extrapyramidal diseases including cerebellar syndromes, diseases of the neuromuscular peripheral tissue, states following injuries to and operations on the brain, congenital brain damage and brain damage acquired during early infancy, organic psychoses (organo-psychological disorders), chronically cerebro-organic psycho-syndromes (dementia and organic personality changes), senile dementia and personality changes due to pathological ageing processes, affective psychoses, schizophrenic psychoses and kidney diseases (Lang, 1975; Hermann, 1978).

After ending the medical and the psychological assessment, both the traffic psychologist and the medical expert share their results before defining the diagnosis and stating a prognosis.

5.10. Psychometric Testing

Psychometric tests may add information if previous alcohol and/or illegal drug abuse has led to a deterioration of driving related tasks. The individual is asked to perform single and/or group testing. Standardized and normed psychometric tests showing least possible fluctuation when retesting are common practice. Test performance must be closely related to driving related demands. Results reached by the candidate give an idea of the individual's performance abilities in comparison to a normative sample under similar conditions. The results reached by the candidate should, however, not be overstressed: Individual fluctuations together with the test situation may end in an under average score.

In the case of unsatisfactory results, indicating rather weak psycho-functional aptitude, the traffic psychologist may ask for a real situation driving task. An engineering instructor will accompany the candidate and deliver a written report on the performed driving skills. This information can add to the already existing data.

5.11. Questionnaire Methods

Compared to psychological interview methods questionnaires offer a good way of standardization and group comparison. Critics add, the missing individuality and situational factors as well as underlying strategies impede interpretation. The use of questionnaires may be helpful to reassure other psychological and medical findings. Covering questions on individual alcohol consume patterns, illegal drug consummation or biographical anamnesis, the questionnaire adds to the overall information gathered within the diagnostic interview. Collecting all data will lead to a medical evaluation on the one hand and a more detailed psychological evaluation on the other.

A 15-20 pages written assessment will be mailed to the candidate in a few weeks time. A positive prognosis may well end in the administrative licensing board regranting a driver license after road violence or drug abuse in the past. A negative prognosis may lead to further recommendations. The candidate may be advised to seek psychological counseling or a clinical treatment before applying for another assessment on driver aptitude – hopefully being more successful next time.

6. DISCUSSION

The psychological and medical assessments on driver aptitude as a means of driver selection are regarded an appropriate proceeding in the attempt to keep a high standard of traffic safety in Germany. At the same time the differential diagnostic assessment is subject to criticism. Ethical objections towards diagnosis (Müller, 1984) and specific aspects in driving selection diagnosis have been mentioned. The introduction of a uniform problem-solving proceeding within a high standard of quality management system across the European countries is required. However, a vast expenditure of the procedure does not seem realistic. The differential diagnosis applied by the experts in driver aptitude assessments is based on the assumption of the diversity and non-homogeneity of the people we are dealing with. DWI-drivers (driving while intoxicated) cannot be treated as a homogeneous group (Perrine, 1990). The individual must be understood on the grounds of his or her biographic background. The assessment on driver aptitude as lacking individuality has been a critical point in the past. In the case of the alcohol drinking driver different demands in attitude and behavior alteration might become necessary. Distinct problematic drug consuming patterns call for a differential way of behavior changes.

The psychological and medical assessments are not limited to the selection of drivers with a negative traffic prognosis. Instead, ways of restoring the right to drive a vehicle in public can be suggested. The legal framework within a specific country determines official questioning and inducement of the assessment. The experts' opinion on the probability of recidivism is called for. Their task is to translate official doubts into appropriate psychological and medical operations allowing either the validation or falsification of the underlying hypothesis.

The official administrative licensing board is free to take into account the results of the assessment. The character of the psychological and medical assessment is only a suggestive one. The final decision process is imposed on the administrative board.

The candidate may benefit from its outcome in getting his/her license reinstalled by distracting official doubts. In the case of a negative prognosis, the candidate might pick up the experts' recommendation and bring about vital changes in his or her life. Alternatives to the past detrimental behavior can be initiated thus helping the person on his/her way.

7. PERSPECTIVES

While the European unification process continues, an adjustment of driver selection after delinquent traffic participation is necessary. Taking up the juristic aspects in European law provides the chance of mutual acceptance and transparency. The implementation of the assessment on driver aptitude can be beneficial to traffic safety measures on European roads. Regional differences, traditions, mentalities or alterations in legal issues can be taken into account. There is a vast potential of experiences already accumulated, ready to recall and to be extended for the needs of the future traffic safety requirements.

8. REFERENCES

ARGERION, M., MC CARTLEY, O., BLACKER, E. (1985), "Criminality among Individuals Arraigned for Drinking and Driving in Massachusetts", *Journal of Studies on Alcohol*, 46, 525-530.

BRENNER-HARTMANN, J. (1995), "Fahreignungsbegutachtung bei Drogenmißbrauch. Ist eine zum Alkohol analoge Bewertung möglich?", in RISSER, R. (Ed.), *35. BDP-Kongress für Verkehrspsychologie*, Bonn, Deutscher Psychologen Verlag.

Bundesanstalt für Strassenwesen (1999), Berichte, Begutachtungs-Leitlinien zur Kraftfahreignung, Mensch und Sicherheit, Heft M 115.

Bundesanstalt für Strassenwesen (2000), Berichte, Begutachtungs-Leitlinien zur Kraftfahreignung, Mensch und Sicherheit, Heft M 115.

DALDRUP, T. *et al.* (1987), "Cannabis und Alkohol im Straßenverkehr", *Blutalkohol*, 24, 144-156.

Fahrerlaubnisverordnung (FeV), Beschlussfassung vom 19.06.1998, Vogel Verlag, München.

FEUERLEIN, W. (1989), *Alkoholismus – Mißbrauch und Abhängigkeit*, 4, überarbeitete Auflage, Stuttgart, Thieme.

GESCHWINDE, T., *Rauschdrogen. Marktformen und Wirkungsweisen*, Berlin: Springer.

HERMANN, (1978), in NICKEL, W. R., UTZELMANN, H. D., WEIGELT, K. G. (1990), S. 39.

HURRELMANN, K., HESSE, S. (1991), "Drogenkonsum als problematische Form der Lebensbewältigung im Jugendalter", *Sucht*, 37, 240-253.

JAGOW, J. (1995), "Grundlagen und Instrumente des Mehrfachtäter-Punktesystems", *Neue Zeitschrift für Verkehrsrecht*, 1,11-14.

KAUERT, G. *et al.* (1992), "THC – Blutspiegel und Fahrtauglichkeit. Unveröffentlicher Vortrag, gehalten auf der 71", *Jahrestagung der Deutschen Gesellschaft für Rechtsmedizin,* in Berlin.

KLEINER, D. *et al.* (1992), "Erfahrungen mit Cannabiskonsumenten – Ergebnisse einer Umfrage bei Kliniken und Drogenberatungsstellen", *Sucht*, 38, 7-17.

KROJ, G. (1992), *Perspektiven der Unfallforschung und Sicherheitsarbeit*, Verlag H. Vogel, München, S. 92.

KROJ, G. (Ed.) (1995), *Psychologisches Gutachten Kraftfahreignung*, Bonn: Deutscher Psychologen Verlag.

KUNKEL, E. (1991), "Die Eignungsuntersuchungen bei den medizinisch-psychologischen Untersuchungsstellen", *Zeitschrift für Schadensrecht*, 12, 325-330.

LADEWIG, D. (1987), "Katamnesen bei Opiatabhängigkeit", in KLEINER, D. (Ed.), *Langzeitverläufe bei Suchtkranken*, Berlin: Springer.

LANG, (1975), in NICKEL, W. R., UTZELMANN, H. D., WEIGELT, K. G. (1990), S. 39.

MOSER, L. (1983), "Kriminalität und Verkehrssicherheit", *Blutalkohol*, 20, 465-469.

MÜLLER, A. (), in *Blutalkohol,* 30/93, S. 66.

MÜLLER, A. (1984), in NICKEL, W. R., UTZELMANN, H. D., WEIGELT, K. G. (1990), S. 39.

NICKEL, W. R., UTZELMANN, H. D., WEIGELT, K. G. (1990), "Schematische Darstellung des Ablaufs- und Entscheidungsprozesses bei verkehrspsychologischen Fragestellungen", in NICKEL, W. R., UTZELMANN, H. D., WEIGELT, K. G., Abb. 2, S. 42.

PERRINE, M. W. (1990), "The Continuum of Drinking Drivers in America", in NICKEL, W. R. (Ed.), Driving Behavior and Traffic Environment, Köln.

STEPHAN, E. (1986), "Die Legalitätsbewährung von nachgeschulten Alkoholersttätern in den ersten zwei Jahren unter Berücksichtigung der BAK", *Werte. Zeitschrift für Verkehrssicherheit*, 32, 2-8.

TÄSCHNER, K. L., *et al.* (1994), "Folgen des Cannabiskonsums – Ergebnisse einer Expertenbefragung zu Straßenverkehrsdelikten und "Echopsychosen" (Flash-back)", *Versicherungsmedizin*, 46, 2-4.

UTZELMANN, H. D., HAAS, R. (1985), *Evaluation der Kurse für mehrfach auffällige Kraftfahrer,* H. 53, Schriftenreihe Unfall- und Sicherheitsforschung Straßenverkehr, Bergisch-Gladbach.

Chapter 17

THE TRAFFIC PSYCHOLOGICAL JOB IN THE GERMAN-LANGUAGE AREA

Professor Walter Schneider
Birgit Bukasa
KfV, Intitute for Traffic Safety Research, Austria

1. INTRODUCTION

For 50 years now, the application of psychological methods on the solution of risk problems in motorized traffic can be noted within the German-language area, which has led to the development of full-time employment for psychologists. This report aims to make some statements on their professional profile under the demands of state and society and to demonstrate a European prospect.

2. THE TRAFFIC PSYCHOLOGICAL EXPERT IN THE FEDERAL REPUBLIC OF GERMANY

In § 2 (driving license) n° 8, the German road traffic law ("Strbenverkehrsgesetz-StVG") determines:

"If facts become known that give reason to doubt the aptitude or ability of the applicant, the authority on driving licenses may instruct the applicant to produce a certificate or reference by a medical specialist or public health officer, a certificate by an officially recognized appraisal office for driving aptitude or by an officially recognized expert or examiner for motorized traffic within a reasonable period of time."

The experts or examiners mentioned within this law are usually engineers by their basic qualifications, as stipulated in the "motor vehicle expert law." Only exceptionally can the professional field of examiner also be opened to qualified psychologists. This has happened in the past in few isolated cases. From the point of view of traffic psychology, the exclusive basic qualification of engineers can only be seen as a historically determined assessment without factually supported justification, considering the qualifications of traffic psychologists for this work.

Only the afterwards settled "driving license decree" (Fahrelaubnisverordnung-FeV), defines as a condition for the acceptance of officially recognized appraisal offices for driving licenses, that they include qualified psychologists in their working team (enclosure 13 to § 67, paragraph 2 FeV): The acceptation by the relevant state authority requires among others that "the necessary personnel equipment with a sufficient number of physicians, psychologists and officially recognized experts and examiners for motorized traffic is guaranteed..."

"The qualification as psychologist requires: A diploma in psychology, at least two years practical working experience (usually within the clinical psychology, industrial psychology), at least one year experience in the appraisal of driving aptitude."

These conditions of independent work as an expert within the officially recognized authority imply that "one year experience in the appraisal" under guidance, technically speaking "under supervision," has to be done.

Before going into detail on the criteria of suitability which have to be applied to traffic psychologists in an officially recognized examination office according to § 13 FeV, the systematic connection with the law (StVG) and the decrees (FeV) require a closer look at other traffic psychological fields of work, which are mentioned there and thus legally protected.

2.1. Traffic Psychological Consultation According to § 72 Fev

The FeV also determines a third professional field for traffic psychologists in § 71.

2.2. Courses and Seminars for the Elimination of Aptitude Deficiencies

Next to these settings in laws and decrees on the employment of qualified psychologists in the Federal Republic of Germany, those of the Federal Republic of Austria shall be regarded, which came into legal force earlier than in Germany.

2.3. Legal Development of Traffic Psychology in Austria

In Austria, traffic psychological examination offices, as well as traffic psychologists employed there, have to meet legal qualification demands.

With the driving license-health decree ("Führerscheingesetz-Gesundheitsverordnung FSG-CV"), which came into effect on November 1st, 1997, the conditions for traffic psychologists and traffic psychological examination offices within the field of driving aptitude diagnostics were determined sufficiently. The decree aims to guarantee appraisals, which are as homogeneous and qualitatively high as possible. The alterations to the former legal position are to be seen before the background of the EUY-instruction 91/439/EWG made by the board, which obliges the member states to standardize their regulations on the minimum requirements for health suitability of motor vehicle drivers.

Initially, a short survey of the relevant laws:

3. TRAFFIC PSYCHOLOGICAL EXAMINATION OFFICES

A traffic psychological examination office, which can be either an institution or an association of self-employed psychologists, must contain at least six traffic psychologists and be able to carry out traffic psychological examinations in more than one federal state at a time (§ 19, paragraph 3 and 4 FSG-GV).

The examination office is required to document their standards of examination realization and administration, decision criteria, realization of education and extended vocational training as well as the nationwide coordination in a manual. In the case of self-employed psychologists, all examination offices are required to apply the same testing methods and to evaluate similarly (§ 19, paragraph 3 and 4 FSG-GV).

4. EDUCATION AND EXTENDED VOCATIONAL TRAINING OF TRAFFIC PSYCHOLOGISTS

The following professional conditions of content must be fulfilled as pre-conditions (§ 20, paragraph 1 and 2 FSG-GV):

- Title "psychologist" according to the psychology law;
- At least 160 hours of traffic psychology theory as well as 1600 hours of practical training within a traffic psychological examination office, including:

  Realization of at least 100 exploration discussions in the presence of a traffic psychologist,

  Providing at least 150 psychological evidences.

Alternatively to this theoretical and practical training, academic research work within the field of traffic psychology for at least three years and the providing of altogether 150 traffic psychological evidences are also acknowledged.

Until the expiration of the provisional regulations (November 1st, 2001), clinical and health psychologists which have been self-employed for at least 5 years, can do a shortened theoretical and practical training in traffic psychology. The extent is determined by a commission of experts individually (§ 24, paragraph 2 FSG-GV).

In order to maintain their qualifications, traffic psychologists are committed every year to:

- Give proof of at least 8 hours extended vocational training in the field of traffic psychology;
- To discuss at least one case in full detail under supervision;
- To undergo collective supervision within the traffic psychological examination office in which they are employed (§ 20, paragraph 4 FSG-GV).

Traffic psychologists which have independently been giving traffic psychological evidences for at least four years within a traffic psychological examination office, are entitled to give practical training (§ 20, paragraph 3 FSG-GV).

5. AUTHORIZATION AND QUALITY ASSURANCE

The authorization as traffic psychological examination office is given by the minister for science and traffic. In cases of suspicion of mismanagement or inadequacy, as well as for verification of personnel, factual and professional conditions, he receives advice by a commission of experts, which is composed of one representative each from the ministry for science and traffic, the ministry for employment, health and welfare/psychological field, the professional association of Austrian psychologists/division traffic psychology, the universities/traffic psychological field, the public health officers entrusted with the examination of driving licenses and the Austrian medical association (§ 21, paragraph 1 FSG-GV). The adequateness of testing methods applied is evaluated only by three psychological representatives of the commission of experts (§ 21, paragraph 3 FSG-GV).

6. CONSEQUENCES

The new legal regulations, established in cooperation with the division traffic psychology of the Austrian professional association of psychologists, represent an up-valuation and a better embodiment of traffic psychology:

Within the field of theoretical training, the division traffic psychology is especially challenged to make contributions of professional content, and even more in connection with the setting of a necessary binding curriculum division should also provide a sufficient offer of extended vocational training. With the extensive practical training, the examination offices face noticeable additional work.

The professional ability of the commission of experts is relevant for the efficiency of the new regulations. Thus, a status which is merely honorary, consultative is less favorable, the integration of traffic psychologists is positive.

7. PROSPECTS

Similar decrees in other working fields of traffic psychology, especially a decree on additional vocational training for the FSG according to § 4, paragraph 9, are to be expected. From the point of view of the division, a maximum coordination in the field of education and extended vocational training for traffic psychologists is essential.

This explanation of the situation of traffic psychological experts in Austria already shows, how high the demand for qualification for the functions which are legally reserved to traffic psychologists is. The following two passages will give an account of qualification planning in Germany:

8. QUALIFICATION AS TRAFFIC PSYCHOLOGIST IN GERMANY

As can doubtlessly be deducted from the cited German regulations for the exertion of traffic psychological consultative and rehabilitation work in 1. to 3., a university degree in psychology is not sufficient for the qualification as a traffic psychologist, even if traffic psychology is part of the education at the respective university.

According to the demands set by the decrees in Germany and Austria, a systematic further qualification followed by an assessment in the sense of successfully done extended vocational training is required.

While the Austrian legal conditions include a qualification confirmation by an office of state, the question on the personal certification remains open in Germany.

§ 72 FeV lays down a private-law institution for the "proof of experience in traffic psychology" (see above): the division traffic psychology in the professional association of German psychologists ("Berufsverband Deutscher Psychologinnen und Psychologen-BDP"). Until now, they have not observed any certification assignments. As they are also no legal entity – only the BDP, the total professional association is – a legal question remains to be answered: If the certificate required is given in the name of the BDP, the signature authorizations of the BDP are also valid. The chair or the appointed secretary of the BDP would then have to give out further signature authorizations to the board of the division traffic psychology and enter this into the association register or the division traffic psychology would itself have to become a legal entity, in order to provide the legally relevant certificates as demanded in the decrees.

Apart from the professional conditions, they will have to establish legally relevant authorizations.

In contents, the proof is original for the division of traffic psychology according to letters a) and b), the proof according to letter c) contains as possible proof of qualification the certification as clinical psychologist, which can only be produced as special reference in a certificate of the division traffic psychology.

In n° 3, the wording of the decree in § 73 "accreditation and certification" points out the possible certification by an office accredited according to EN 45013.

This office can carry out personnel certification for traffic psychologists, if it is thus accredited. At this time, though, there is no certification office accredited in the Federal Republic of Germany. The further development is not yet clarified.

9. THE TRAFFIC PSYCHOLOGICAL EXPERT IN AN APPRAISAL OFFICE FOR DRIVING APTITUDE (§ 67, PARAGRAPH 2)

The acknowledgement as examination office is bound – similar to the Austrian regulations – to personnel qualification and the sufficient filling. Enclosure 13 to § 67, paragraph 2 states in the matter of actual acknowledgement of the examination offices.

The demands on qualification of psychologists working therein was already cited and explained in section 1. It remains to be added that from the point of view of the board of the division of traffic psychology, the one year education "under supervision" should be used for additional education, which meets the demands of both other professional fields, for which a working team of the division has also provided a curriculum draft, which is to be integrated into the teaching in cooperation with the German society for psychology ("Deutsche Gesellschaft für Psychologie – DGPs –: The representatives of the DGPs in the board of the confederacy (of DGPs and BDP) have suggested "given the legal amendments within Europe which can be expected..." "and the thereby following impulses for the professional field of traffic psychologists, to consider the organization of a course for extended vocational training for traffic psychology. After thorough discussion of all factors it was decided to await the process of opinion – forming within the BDP on this question and to re-discuss the subject thereafter." (Cited from Report Psychologie 1998, p. 47).

The curriculum draft is currently being discussed in the relevant committees. It remains to be decided, who might function as a responsible body for the course of extended vocational training and what content of teaching shall become a definite part of the curriculum, which shall thus be not cited here in detail. It is certain, though, that all demands made by decrees and laws will have to be covered by the content of teaching.

This also applies to the demands on examination and appraisal, put down in enclosure 14 to § 11 FeV, paragraph 5, § 76, paragraph 3, which shall be related in the next passage.

10. APPRAISAL GUIDELINES IN GERMANY

Apart from the maxims in enclosure 14 the procedure within the examination and appraisal is settled in detail by guidelines for the examination of driving aptitude. These are a joint work of physicians and psychologists that have worked together in a parity committee and will submit the result of their work to the ministers for traffic and health in the spring of 1998.

The coming into effect of the appraisal guidelines, together with the FeV cited here in detail, is expected for the January 1st, 1999.

11. TRAFFIC PSYCHOTHERAPY AND TRAFFIC THERAPISTS

In the German traffic regulations (StVG) and the afterwards settled licensing arrangement for individuals in motorized traffic (FeV), measures of psychotherapeutic or medical treatment of individuals for the restoration of aptitude for the driving of motor vehicles are not included, as they are not part of their area of responsibility.

The fact that an interest in the integration of psychological psychotherapists into the grouping of traffic psychologists exists, is affirmed by a majority of those qualified psychologists who are members of the division of traffic psychology in the BDP, but not yet consensus in the entire association.

The psychological psychotherapy has been defined by a proper psychotherapist law since 1998. The certification as clinical psychologist as specified in § 72 FeV under N° 4, letter c, will have to be lead to a license in conformity with the law. This will be a long process. The professional psychologic testgat given by the BDP thus temporarily remains the only alternative: Clinical psychologist the adequate proof of qualification of § 72, n° 4, letter c.

Whether all qualifications for the exertion of traffic therapy are thus named, can however be doubted because of the simple fact that the traffic psychological advice (§ 72 FeV), which is almost certainly less demanding than the traffic psychological rehabilitation, contains qualification characteristics which surpass the qualification of a "clinical psychologist BDP."

The restoration of aptitude for the driving of motor vehicles after loss of this aptitude due to mental deficiencies or behavioral disorders that can only be consolidated with external help, is to be assigned to the curative field and thus bound to legal admission, according to the general understanding of the conditions for exertion of this profession. This admission is unobtainable even for a certified traffic psychologist. From the point of view of traffic psychology, only the most suitable combinations and partial qualifications will have to be discussed and decided upon.

The rehabilitation of people involved in accidents can serve as an example of a rather promising application of traffic psychological psychotherapy, for example, tram-drivers who are not able to conduct vehicles safely due to a psychological shock.

There are bound to be many specializations for traffic psychotherapy, each which will require additional qualification.

This might then give cause to make use of these qualification profiles in other European countries.

12. PROSPECTS FOR EUROPEAN DIMENSIONS

Due to the legal integration of this profession, the German language area provides approaches which can be used on a European level for the development of traffic psychological professions in Europe.

For the employment as the expert for questions or occasions concerning administrative or criminal law, the possibility of certification by an office accredited according to EN 45013, which is at this point only suggested in the German traffic law, seems more promising than the direct licensing by the government (as applied in Austria), because of the fact that the United Kingdom, which is bound to play a deciding role, does not have "an arrangement which can be compared to the German system of public appointment and swearing-in" (Oberlander, 1997).

In accordance with the historical legal background in the English-language area, self regulation is preferred:

“Self regulation in Great Britain means that the professional associations function at the same time as professional organizations. These institutions name members that offer their work as experts in a respective partial field. That is the “Royal Institute of Chartered Surveyors” can name experts for the field of construction respectively property examination. Members of the institute are required to have passed respective exams.

Apart from the professional and special associations, university institutes, foundations and associations of different kinds can possibly assist in the search for experts. Here, we need to point out registers, which name experts – more precisely, expert witnesses. The “Institute of Arbitrators” is rather unusual, it primarily serves the tending of the arbitration system. They also offer educational course for the qualification of experts.

The “Academy of Experts” was found in 1987, aiming to set professional standards for the expert system and to promote the professional ethics. Here, the “British” principle of consumer protection becomes evident: The circle of qualified experts is narrowed down by membership in professional associations on the basis of qualification proof and the guarantee of minimum standards in professional and ethical respect.

Also in Germany, within the professional field of experts, certification as proof of qualification gains acceptance in many working areas for experts, e.g. also in the field of traffic for the accident reconstruction expert, for whom in 1998 the expert system institute has published a syllabus and learning schedule: “Professional demand profile for traffic accident experts” (“Fachliches Anforderungsprofil für certification by their society of certification”).

The certification of traffic psychologists by an office accredited according to EN 45013 is one of the most interesting options for a future organization of the job profile of traffic psychology.

To avoid a splintering of traffic psychological working areas through a majority of access conditions, the acquiring of a traffic psychological certificate by a traffic psychologist with the necessary license for a curative profession seems possible, from the point of view of consumers probable, and – for the preservation of a hopefully not realizable market survey – from the point of view of suppliers surely also desirable.

Conclusion

TRAFFIC PSYCHOLOGY FOR THE 2000'S: PROFESSION AND SCIENCE

Heikki Summala
Department of Psychology, Traffic Research Unit, University of Helsinki

1. INTRODUCTION

Traffic psychology is an applied field of psychology that focuses on the human in transportation systems and so doing actually covers a multitude of areas in psychology from perception to personality. Through its social aims it can also be characterized as multidisciplinary having close cooperation with other disciplines. This is of course nowadays the case with many fields of science, as boundaries are crossed and a multidisciplinary approach is necessary for solving the problems of society and mankind.

The recent growth of traffic psychology is clearly tied to its aims. Decreasing traffic accidents, congestion and pollution are the aims, in which traffic psychology and traffic psychologists have been striving and will have to strive to have a direct effect. Therefore, on one hand we need academic basic research to describe, explain and predict human behavior in traffic while, on the other hand, we need applications that aim at changing that behavior. These efforts should cover all levels of transport behavior. Thus traffic psychology must not merely be confined to how drivers or pedestrians act in traffic but what is essential is how they make trip decisions and even how they choose their traffic mode and vehicle.

As a science, traffic psychology must aim for a good theory which can lead to practical decisions. Developments in theory formation, research and education in traffic psychology is also needed to guarantee that traffic psychologists as a profession can prove to be useful, to give right advice, be it about driver education or driver selections, car, road environment, traffic control devices or the design of traffic policies. The area and field of application of traffic psychology has to be understood as being very wide, and traffic psychology has a duty towards all questions arising from man's action in traffic.

At present the profession of a traffic psychologist is most extensively practiced in German speaking countries in the areas of driver selection and improvement (Kroj and Dienes, Lorenz, and Schneider and Bukasa in this book). Plenty of other opportunities are still left. However, as a profession traffic psychologists thrive only if they can give right answers to current topics. Obviously they will have to surpass the knowledge of an enlightened engineer, traffic instructor or doctor in questions about safety of traffic environment and people in it. What makes this aim more difficult to achieve today is among others a special feature in traffic behavior, especially when it comes to driving a car, also known to complicate decision making in traffic politics. Almost all the people have a vast experience in driving (or cycling and walking in traffic) and this experience shapes their attitudes and makes them to feel themselves as experts. Traffic psychology is not alone. For example, in work and organizational psychology psychologists in fact compete with many other professions such as engineers or economists, and new management models, best sellers and educational programs come about from very diverse starting points. It is not surprising therefore that the profession of psychologists attempts to secure their status with psychologist laws.

It is also essential however that traffic psychology science and education should develop in a way to let psychologists on their part, together with other professionals, have their influence in solving multidisciplinary problems in the area of transportation.

2. THE ESSENTIAL CHALLENGES OF TRAFFIC PSYCHOLOGY

It is very clear that traffic psychologists have more than enough aims for the 21st century. Approximately 45.000 people per year die in traffic accidents in EU countries alone and decisions and measures to be taken presuppose a lot of traffic psychological expertise.

One of the permanent questions of a theoretical and practical importance is the lifespan of the car driver and the changes within. The development and decline of driver skills is directly connected to the safety problems of young and elderly drivers, the first question being what the skills are with which a car is driven safely in traffic, how they develop and how they decline.

3. SKILLS AND MOTIVES

Skills and their development have for a long time been described as a hierarchical model, which as applied to traffic instruction, starts with learning to handle the control devices and moves on to controlling traffic situations (see Huguenin and Rumar, in this book). Even though there has been an increasing amount of research on the perceptional skills of the driver during the last few decades (see Cavallo and Cohen, in this book), we still do not know how skills exactly develop and how the development of skills or strategies could be speeded up in young drivers. A classic example is how drivers along with experience learn lane-keeping with peripheral vision, thus releasing resources for observation of more distant objects, which also enables anticipation of traffic situations (Mourant and Rockwell, 1972). We do not, however, as yet know what exactly the mechanisms and strategies are that drivers learn during their first 30.000-50.000 kilometers, or 5 to 7 years of driving that also means leveling off the learning curve indicated by accident involvement (eg. Riemersma, 1987; Summala, 1987; Summala, Nieminen and Punto, 1996). Depicting such mechanisms makes it possible to plan new teaching methods into the driving school curriculum.

Work on such mechanisms is also needed for the development of driver behavior models. One influential tradition comes from the concept of safety margin and David Lee's influential article (1976) presenting a model about distance keeping in terms of the concept of time to contact . Along with the time-to-lane-crossing measure (Godthelp, Milgram and Blaauw, 1984) these kinds of time based submodels are already applied in designing intelligent driver support systems as well as in more general theories on driver behavior (Summala, 1996a, 1997a).

On the other side, motivational factors have acquired an important weight in traffic psychology research since what could be called a fundamental principle in traffic psychology was formulated, namely that crucial to safety is what the driver actually does in any given traffic situation rather than his skills or traffic conditions as such (Näätänen and Summala, 1974). Motivational determinants can be partly included in the process models noted above, indicating different thresholds of action, but not exhaustively. Therefore, the connections between personality, life-style factors, driving style, and deviant driving need further efforts (see also Åberg and West, this book; Hatakka, 1998; Lajunen and Summala, 1995; Parker, Reason *et al.*, 1995).

One of great challenges still left for the 21st century continues to be young male drivers who are skillful drivers but have a very high fatality rate just after licensing. The great questions are what to do with them, what are our means to influence youngsters' driving and risks in driving education, what are the means to influence deviant driving habits by interventions such as driver improvement.

4. TESTING OF DRIVING SKILLS AND ABILITIES

One of the big challenges in traffic psychology is testing driving ability and measuring and predicting of what can be called sufficient skills or sufficient ability.

Motorized countries have traditionally relied on a driving test in order to receive a driver's license and it seems to be a rather general opinion that the test would also be a valid indicator in screening risky drivers and estimating the driving ability of elderly and unhealthy drivers (eg. Risser, 1985; West *et al.*, 1993; Gully, Whitney and Vanosdall, 1995; Lundberg, Johansson *et al.*, 1997). On the other hand traffic medicine has long defined diseases that exclude receiving a driver's license and also defined the effects of different drugs on driving and given instructions to doctors. There is also a long tradition in traffic psychology aimed at developing laboratory tests that could predict driving ability and that could be used in making decisions about driver's licenses.

This has, however, turned out to be a rather difficult task for many reasons. Observations merely between different tests and accidents have not led very far due to the fact that the accident criterium is rather difficult due to random variance among others. Even though early studies (eg. Häkkinen, 1958) could predict accidents from the psychomotor tests fairly well, the best results were for professional drivers who are tied to their time schedules and not able to compensate or change behavior after an accident and, furthermore, a larger average number of accidents gives better chances to use accident criterium than among private car drivers.

New interest in the development of laboratory tests has arisen with increasing focus on elderly and unhealthy drivers.

The ageing of the driving population in the western countries has caused an unforeseen research activity, aimed, on one hand, at changing the traffic environment and the car to be better suited to ageing drivers and, on the other hand, at seeking methods to be able to predict driving ability and, finally, at developing educational methods to support elderly drivers in maintaining safe mobility. This is, undoubtedly, one of the big practical challenges to traffic psychology. What is especially needed now are extensive studies combining laboratory testing, well designed on-road driving tests that also quantify performance, knowledge of the driving practice and exposure to accidents in different conditions and, last, knowledge of incidents and accidents. In a well-designed study, accident output should be prospectively predicted.

Here we bump again into the multidisciplinary nature of traffic psychology. The fairly lean results of studies up to recent times seem to suggest that expertise and various methods from many professions should be applied in assessing driving ability. Regular medical examinations made by general practitioners have not proved cost-effective (The Finnish Association for Traffic Medicine, 1993; Hakamies-Blomqvist *et al.*, 1996), and not even reporting of excluding diseases to licensing authorities by doctors is well accepted until yet. Search for sensitive and specific cognitive tests for predicting accidents makes some progress but is far from complete (eg. Janke and Eberhardt, 1998) and there is further need for research to develop and to test the validity of the driving test (eg. Dobbs *et al.*, 1998). A follow-up system of driving ability should obviously use all these components, as a matter of fact, with the principle that as many nets should be cast in water as possible (Summala, 1996b). This would mean that the police, when monitoring traffic, as well as the doctor, in his reception, would send possible cases of decreased driving ability into follow-up examinations. Part of these cases can clearly be directly reported to the driver's license authorities and part can be guided to follow-up examinations.

5. INTEGRATION OF TRAFFIC PROFESSIONALS

The important question here is what the testing system itself is like. This question should really bring about co-operation of many professions combining the knowhow of the traffic psychologist, the traffic instructor and the traffic doctor as well as the ophthalmologist or optometrist. Organizationwise, this could mean, for example, development towards a "traffic department store" which, rather than being an licensing agency, would resemble the present private medical center with its doctors, psychologists and therapists, etc. Thus at least three groups of professionals would work under the same organization: traffic instructors, who, on one hand, give basic training to the beginning drivers and, on the other hand, test and train the elderly drivers; traffic psychologists, who take part in the testing of the young and the elderly, and in driver improvement activities; and the doctors specialized in traffic. Getting the human-centered expertise of the field of traffic under the same roof would greatly intensify the co-operation of different professions and would be very important for the future of traffic psychology.

This kind of a knowhow-center could very well extend the scope of activities even more. Selection and periodical tests of professional drivers would be one of its activities, and it might even cover questions of traffic design, which very often are local. Even though one important aim of traffic psychology within traffic science (see later) is to produce and to develop design guidelines for a safe road environment, it is necessary to take care of defects in the existing environment, in collaboration with traffic engineers. The knowhow-center could prove to be an important factor of safety in its own region and neighborhood. This means expanding towards traffic engineering and more generally towards traffic science. If we consider the versatile basic education of the traffic psychologist, his/her expertise should be utilized as extensively as possible.

6. TRAFFIC PSYCHOLOGY AS A PART OF TRAFFIC SCIENCE

The actual question about the driving ability of the ageing and unhealthy drivers clearly includes a contact surface to neuropsychology as well as to neurology, gerontology and other fields of medicine. Another important contact surface for traffic psychology are the technical sciences, traffic engineering and traffic science as their top. One of the aims of academic research towards a more profound knowledge of man in traffic and modeling of human behavior in traffic can be seen as a part of traffic science. As noted above, for example, it is not easy to predict accident involvement and it is indeed very difficult to estimate especially the effects of the changes in the traffic system to the accident output.

An essential aim and tool would be a computational simulation model (eg. Pursula, 1998). These kinds of simulation models, especially applied to capacity problems and traffic management, do already exist, but their submodels on driver behavior are still fairly superficial. Such simulation models cover, for example, a motorway, a lower class two-lane road with junctions, or city traffic, and can be extended to a larger area incorporating trip generation and route selection processes.

One task of traffic psychology is to produce better and better models on how drivers perform basic activities, how they keep distance to the lead car and how they react to deceleration of the lead-car or to a car coming from a secondary road, etc. What is also essential is how people divide and direct attention, as well as the influence caused by in-car devices; at the moment the use of a mobile phone is one factor affecting the driver's actions extensively but industry is introducing new equipment at increasing rate that may require much more visual attention than the classic cell phone. The next step for these models is to include population variance in them, which is. how people of different ages and driving experience act. Developing the simulation models further will give us improved estimates on how, for example, the ageing of population will change traffic in different surroundings or how different driver support systems, and the penetration of them in the vehicle fleet, will affect traffic behavior and accidents. The important output of this kind of computational simulation are accidents which can be estimated in different scenarios.

It is obvious that this kind of an aim will require an exceeding amount of theoretical and empirical research in the field of traffic psychology, but it is bound to raise this field of science as science beside others.

7. TRAFFIC PSYCHOLOGY AND COGNITIVE ERGONOMICS

Psychologists have long been active in optimizing driver interfaces whether they be outside the car such as traffic sign systems or inside it, like instrument panels. The concept of intelligent transport system (ITS) meant a major boost in research on driver behavior that also focused on the development of driver models to be used as a reference model for intelligent support systems (Nilsson, Harms and Peters, this book; Summala, 1997b). Several books and a massive number of research reports were published and models of the general architecture of the intelligent vehicle and highway system were presented in the 1990's (eg. Michon, 1993; Parkes and Franzén, 1993; Peacock and Karwowski, 1993; Noy, 1997).

This kind of work continues both in the car industry and at road test sites funded by governments, and will provide much work for competent psychologists both in design and research. Still another area where traffic psychology is much needed is the development of design standards that covers both in-car interface and roadway design (see eg. Theeuwes, this book).

8. TRAFFIC PSYCHOLOGY AND SUSTAINABLE DEVELOPMENT

Traffic psychology must not, in the next millennium, be restricted to how people cope in the present traffic or how the traffic environment should be changed in order for people to cope with it and be well attended to. A more important aim for mankind is reasonable mobility that would save the environment. This will raise transport behavior in general to be an important research subject - what factors regulate transport behavior and trip decisions along with moving in traffic. Let it be noted that getting a car influences later transport behavior extensively. The car is an exquisitely convenient vehicle, but the aim should lie in the sensible use of the car, and traffic psychology should be in the forefront in realizing this aim.

The meaning of a car and a driver's license in a young person's life, especially a young man's, has, for this whole century been great, and along with the increased prosperity and motorization the driver's license has practically become the property of everybody, men as well as women. Typically the driver's license will be acquired as soon as it is possible, even though a car would not be needed for a practical purpose where public transport is well arranged. The peak accident rate right after acquiring the driver's license is partly due to lack of skills and partly for motivational factors especially specific to young males that are not easy to affect. However, in the last decade there has been an especially favorable development in the figure depicting the deaths of young males in Sweden, which would seem to be due to the fact that especially young men do not acquire driver's license as soon as it is possible (Krantz 1999, Summala, 1998). This is an example of a very delightful development which should lead to the car being used only when it is really necessary. When young people form a family and get children, a car will be especially useful, and it is also the phase in life when they are already socialized and have passed the phase when risks are taken and limits are tested. An important task for traffic psychology would be promotion of this development to have an effect on the decisions of acquiring a driver's license and a car.

Europe can boast of things being rather well in this respect as most EU countries have a good network of public transport and the car is necessarily needed only in the countryside. There is, however, a lot of pressure to lower the licensing age or at least lowering the practicing age. France and Sweden have made it possible for 16 year old youngsters to practice driving with their family members, and Finland is also pondering about adopting a similar model. Licensing age would still be 18 as in the EU countries, thus independent driving will be possible at the age of 18. This is argued with safety factors - youngsters would get plenty of supervised practice with their parents. Because of self-selection, however, conclusive results on the whole safety benefits of this model are still lacking although youngsters entering to such a program show clearly lower accident rates as compared to those not selecting it (Gregersen *et al.*, 2000). Selection of such programs by youngsters and parents, due to demographic, social, and economic factors as well as safety-consciousness, known in traffic psychology for a longer time (Näätänen and Summala, 1976, p. 80-85), results in group differences even without any program effects.

9. TRAVEL HABBIT FORMATION

One important aspect is also easily forgotten in these kinds of plans. When a youngster is allowed to drive from the age of 16 with family members, it may greatly affect his/hers (and the whole family's) future trip behavior, and the proportion of trips done with the car may increase. This would be countereffective for the principle of sustainable development in transportation. Trip behavior being strongly determined by habit (Forward, 1998; Gärling *et al.*, 1998; Verplanken *et al.*, 1997), habit formation will undoubtedly be among important research subjects for the 21st century.

10. MOBILITY WITHOUT A PRIVATE CAR

There has been much discussion indeed lately on mobility and on the fact that the focus should not be on safety alone but the mobility of population should be also taken into account. Especially for the ageing population mobility is held to be an important factor for the quality and activeness of life.

A certain dichotomy can be seen between two continents considering mobility and the car. In the U.S., with its mainly private transport system, the car and driver's license are really important for mobility, and also the earlier age of licensing is understandable. In Europe the basic structure of society is different and public transport usually fairly well arranged. Mobility and safety should here have a different relationship to the private car than in the States. Youngsters going to school do not typically need the car nor a lift as public transport, walking and cycling are sufficient for school trips. Where an own car or public transport is no more suitable, mobility can be assisted by partly supported summoned minibuses and taxes, as is the case in Finland where a specific law even provides 18 subsidized leisure trips per month for the handicapped people.

We can, however, say that part of public transport is all the time standing at the edge of a cliff: a private car is for an able individual such a practical vehicle that individual decision making easily turns towards the use of a car instead of public transportation. Additionally, public transport and private cars provide plenty of overlapping capacity in the transport system. For example, during the public transport strike in the Helsinki metropolitan area in 1998 approximately one third of trips subject to strike were done as passengers in private cars, which only meant an increase of 0.25 occupants in private car passenger figures. An increase in the occupant figure from 1.3 (driver + 0.3 passengers), which is the present mean in the Helsinki area, to 2.0 would be enough to cover the whole of public transport. This would not, however, be according to the ideals of sustainable development nor would people be willing to give up good public transport (Tuovinen, 1999).

Traffic psychology will, during the next millennium, have to participate strongly in planning the future of society and in guiding the trip behavior of people according to the principles of sustainable development (see also Risser). This is even more important in motorizing countries with a strong tendency to continue the development gone through in the highly motorized countries.

11. IMPORTANT STEPS FOR TRAFFIC PSYCHOLOGY AS SCIENCE AND PROFESSION

Finally, two important issues should be noted. One criterium for the status of a scientific discipline is publication policy, which is the practice that new results are published in international (peer reviewed) journals, to guarantee proper accumulation of knowledge and to avoid inventing the wheel again, among other things. The quantity and quality of international publishing also makes this field more attractive for young students choosing their special area: a vital condition for the prosperity of traffic psychology in the future is successful recruitment of competent students.

Reflecting the growth of human-centered research on traffic behavior, new journals are coming in this area. Along with traditional journals with broad coverage that also publish extensively in traffic safety and behavior, such as *Accident Analysis and Prevention*, *Ergonomics*, and *Human Factors*, there are new journals more specifically focused in traffic psychology. The first journal specifically devoted to traffic psychology started recently (*Transportation Research F: Traffic Psychology and Behaviour*), and another human factors oriented journal devoted to transport (*Transportation Human Factors*) has just taken off.

Secondly, as important for the science as well as profession, programs in traffic psychology should be increasingly provided at universities that give fairly broad basic knowledge in this area along with connections to the neighbor sciences.

12. REFERENCES

DOBBS, A. R., HELLER, R. B., SCHOPFLOCHER, D. (1998), "A Comparative Approach to Identify Unsafe Older Drivers", *Accident Analysis and Prevention,* 30:363-370.

Finnish Association for Traffic Medicine (1993), *Ajokortti ja terveys*, (Driving Licence and Health), Helsinki: Finnish Association for Traffic Medecine.

FORWARD, S. (1998), *Behavioural Factors Affecting Modal Choice*, Swedish National Road and Transport Research Institute.

GÄRLING, T., GILLHOLM, R., GÄRLING, A. (1998), "Reintroducing Attitude Theory in Travel Behavior Research", *Transportation,* 25:129-146.

GODTHELP, H., MILGRAM, P., BLAAUW, G. J. (1998), "The Development of a Time-Related Measure to Describe Driving Strategy", *Human Factors*, 26:257-268.

GREGERSEN, N. P., BERG, H. Y., ENGSTRÖM, I., NOLÉN, S., NYBERG, A., RIMMÖ, P. A. (2000), "Sixteen Years Age Limit for Learner Drivers in Sweden – Evaluation of Safety Effects", *Accident Analysis and Prevention*, 32:25-35.

GULLY, S. M., WHITNEY, D. J., VANOSDALL, F. E. (1995), "Prediction of Police Officers' Traffic Accident Involvement Using Behavioral Observations", *Accident Analysis and Prevention*, 27:355-362.

HAKAMIES-BLOMQVIST, L., JOHANSSON, K., LUNDBERG, C. (1996), "Medical Screening of Older Drivers as a Traffic Safety Measure – A Comparative Finnish-Swedish Evaluation Study", *Journal of the American Geriatrics Society*, 44:650-653.

HÄKKINEN, S. (1958), "Traffic Accidents and Driver Characteristics", *Scientific Researches* n° 13, Finland's Institute of Technology.

HATAKKA, M. (1998), "Novice Drivers' Risk and Self-Evaluations", *Annales Universitatis Turkuensis B 228*, University of Turku.

JANKE, M. K., EBERHARDT, J. W. (1998), "Assessing Medically Impaired Older Drivers in a Licencing Agency Setting", *Accident Analysis and Prevention*, 30:347-361.

KRANTZ, L. G. (1999), "Rörlighetens mångfald och förändring", (Diversity and Change in Mobility. Peoples' Daily Travel in Sweden 1978 and 1996), *Meddelanden från Göteborgs Universitets Geographiska Institutioner Serie B 95*, Gothenburg: School of Economics and Commercial Law in Gothenburg University.

LAJUNEN, T., SUMMALA, H. (1995), "Driving Experience, Personality, and Skill and Safety-Motive Dimensions in Drivers' Self-Assessments", *Personality and Individual Differences*, 19: 307-318.

LEE, D. N. (1976), "A Theory of Visual Control of Braking Based on Information about Time-to-Collision", *Perception*, 5:437-459.

LUNDBERG, C., JOHANSSON, K., *et al.* (1997), "Dementia and Driving: An Attempt at Consensus", *Alzheimer Disease and Associated Disorders*, 11:28-37.

MICHON, J. (Ed.) (1993), *Generic Intelligent Driver Support*, London: Taylor & Francis.

MOURANT, R. R., ROCKWELL, T. H. (1972), "Strategies of Visual Search by Novice and Experienced Drivers", *Human Factors*, 14:325-335.

NÄÄTÄNEN, R., SUMMALA, H. (1974) "A Model for the Role of Motivational Factors in Drivers' Decision-Making", *Accident Analysis and Prevention*, 6:243-261.

NÄÄTÄNEN, R., SUMMALA, H. (1976), *Road-User Behavior and Traffic Accidents*, Amsterdam & New York: North-Holland/American Elsevier.

NOY, Y. I. (Ed.) (1997), *Ergonomics and Safety of Intelligent Driver Interfaces*, Mahwah, New Jersey: Lawrence Erlbaum.

PARKER, D., REASON, J. T., MANSTEAD, A. S. R., STRADLING, S. G. (1995), "Driving Errors, Driving Violations and Accident Involvement", *Ergonomics*, 38:1036-1048.

PARKES, A. M., FRANZÉN, S. (Eds.) (1993), *Driving Future Vehicles*, London: Taylor & Francis.

PEACOCK, B., KARWOWSKI, W. (Eds.) (1993), *Automotive Ergonomics*, London: Taylor & Francis.

PURSULA, M. (1999), "Simulation of Traffic Systems – An Overview", *Journal of Geographic Information and Decision Analysis*, 3:1-8.

RIEMERSMA, J. B. J. (1987), *Visual Cues in Straight Road Driving*, Soest: Drukkerij Neo Print.

RISSER, R. (1985), "Behavior in Traffic Conflict Situations", *Accident Analysis and Prevention*, 17:179-197.

SUMMALA, H. (1987), "Young Driver Accidents: Risk Taking or Failure of Sskills?", *Alcohol, Drugs and Driving*, 3:79-91.

SUMMALA, H. (1996a), "Accident Risk and Driver Behaviour", *Safety Science*, 22:103-117.

SUMMALA, H. (1996b), "Tarvitsemmeko autonkuljettajien ikäkausitarkastuksia?", (Do we Need Medical Checks for Elderly Drivers?), *Suomen Lääkärilehti* (Finnish Medical Journal), 51: 2032-2037.

SUMMALA, H. (1997a), "Hierarchical Model of Behavioral Adaptation and Traffic Accidents", in ROTHENGATTER, T., CARBONELL, E. (Eds.) *Traffic and Transport Psychology,* Amsterdam: Elsevier.

SUMMALA, H. (1997b), "Ergonomics of Road Transport", *IATSS Research*, 21(2):49-57.

SUMMALA, H. (1998), "Behavior and Risk Typology: Disaggregation of Accident Statistics and Behavior", *Transportation, Traffic Safety and Health IV*, Tokyo, Japan, October 21-22.

SUMMALA, H., NIEMINEN, T., PUNTO, M. (1996), "Maintaining Lane Position with Peripheral Vision during In-Vehicle Tasks", *Human Factors*, 38:442-451.

TUOVINEN, E. (1999), *Pääkaupunkiseudun joukkoliikennelakon vaikutukset matkakäyttäytymiseen*, (The Effects of the Public Transport Strike on Travel Behavior in the Capital Area), Thesis, Traffic Psychology, University of Helsinki.

VERPLANKEN, B., AARTS, H., KNIPPENBERG, VAN, A., KNIPPENBERG, VAN, C. (1994), "Attitude versus General Habit: Antecedents of Travel Mode Choice", *Journal of Applied Social Psychology*, 24: 285-300.

WEST, R., FRENCH, D., KEMP, R., ELANDER, J. (1993), "Direct Observation of Driving, Self Reports of Driver Behaviour, and Accident Involvement", *Ergonomics*, 36:557-567.

Index

Printed in the United Kingdom
by Lightning Source UK Ltd.
125992UK00001B/34/A

9 780792 374794